Mechanical Engineering Series

The Mechanical Engineering Series presents advanced level treatment of topics on the cutting edge of mechanical engineering. Designed for use by students, researchers and practicing engineers, the series presents modern developments in mechanical engineering and its innovative applications in applied mechanics, bioengineering, dynamic systems and control, energy, energy conversion and energy systems, fluid mechanics and fluid machinery, heat and mass transfer, manufacturing science and technology, mechanical design, mechanics of materials, micro- and nano-science technology, thermal physics, tribology, and vibration and acoustics. The series features graduate-level texts, professional books, and research monographs in key engineering science concentrations.

More information about this series at http://www.springer.com/series/1161

Yogesh Jaluria

Advanced Materials Processing and Manufacturing

Springer

Yogesh Jaluria
Department of Mechanical & Aerospace Engineering
Rutgers, The State University of New Jersey
Piscataway, NJ, USA

ISSN 0941-5122 ISSN 2192-063X (electronic)
Mechanical Engineering Series
ISBN 978-3-030-08340-3 ISBN 978-3-319-76983-7 (eBook)
https://doi.org/10.1007/978-3-319-76983-7

Printed on acid-free paper

This Springer imprint is published by the registered company Springer International Publishing AG part of Springer Nature.
The registered company address is: Gewerbestrasse 11, 6330 Cham, Switzerland

Preface

Manufacturing has been a critical ingredient in the growth and development of human civilization through the ages. Raw materials that occur naturally, such as wood, or those that are obtained through mining and other means, like copper, iron, and silver, have been employed to produce useful items. Starting with simple tools and utensils, manufacturing techniques have been expanded extensively to produce a wide range of materials and products. Materials processing, which is basically a subset of the overall manufacturing system, focuses on the processes undergone by a given material to obtain desired shapes, properties, and characteristics. Classical processes such as forging, forming, casting, and cutting have been modified over the years to meet the challenges posed by new and emerging applications and materials.

This book is focused on the area of advanced materials processing and manufacturing. Though the main interest lies in advanced materials and applications, the relevant processes are often based on and can be linked with the classical methods. Therefore, the book presents traditional manufacturing methods to examine the underlying basic principles and then moves on to a range of new materials processing techniques needed to fabricate items that are of particular interest in the modern world, such as optical fibers, specialized polymers and alloys, and microelectronic components. Since the area of manufacturing and materials processing is a vast one, only some important processes are considered in detail and the basic mechanisms that apply to a much wider range of processes are discussed. Of considerable importance is the link between the processes to which a given material or component is subjected and the characteristics of the final product. This aspect is presented in detail. In recent years, there has been a growing emphasis on the design and optimization of systems because of growing worldwide competition and the development of new processes and techniques. This aspect, which is crucial to the development of the relevant manufacturing system to achieve the desired process, is discussed in some detail. Several important practical issues arise in any manufacturing process and these are outlined.

This book is written at the first-year graduate level. It can also be used as reference book for senior undergraduate thermal science courses, such as those on heat transfer and thermodynamics, and for elective courses in manufacturing. It can be used for reference by practicing engineers as well. The book is largely written for mechanical engineers. However, the material is suitable for courses and as reference in other engineering disciplines, such as chemical, aerospace, industrial, and materials engineering.

The book is directed at advanced materials processing and manufacturing, employing examples from diverse manufacturing techniques. The modeling of the process and the system is an important aspect in the study of any manufacturing process. The mathematical modeling of the system leads to governing equations that are generally solved numerically because of their complexity. The results obtained from numerical simulation, as well as those from experimentation, are presented to indicate the dependence of the characteristics of the final product on the transport processes to which the material is subjected. Experiments are also used to study the properties and characteristics of the product. All these results provide the inputs that are needed for designing the appropriate manufacturing system, as well as for optimization.

The first part of the book is devoted to the basic concepts in manufacturing and materials processing. It brings out the importance of this field, the advancements seen in recent years, the governing mechanisms, new and advanced techniques, and analytical, numerical and experimental methods that may be used to obtain solutions for typical systems and processes. These fundamental aspects bind different manufacturing processes together and provide the basis for modeling, simulation, prediction, and design of the relevant systems. The second part of the book focuses on a range of important manufacturing processes. These include processes like casting and crystal growing that are based on phase change, continuous processing, polymer processing, thin film deposition, optical fiber drawing, and several other traditional and new and emerging processes. The important considerations in these processes, the dominant mechanisms, and typical results from simulation or experimentation are outlined. The last part focuses on the practical aspects of system design, considering important issues such as feasibility, sensitivity, uncertainty, design, and optimization. Examples are taken from the manufacturing processes discussed in the second part to present the main ingredients in the design and optimization of manufacturing systems. The improvement of the design to obtain the best possible solution under the given constraints is an important consideration and is discussed.

Therefore, the book offers a systematic approach to a study of advanced materials processing and manufacturing, proceeding from the basic aspects through traditional and advanced manufacturing methods to system design and optimization, using up-to-date examples to present the current information and techniques available in this important field. Examples range from simple systems to large, complex practical systems. The main thrust of this book is to develop and discuss the basic considerations that arise in materials processing and manufacturing, as well as the appropriate methodology. Readers are encouraged to use their own backgrounds, imaginations, and available literature for considering other processes and systems.

The material included in this book has been used in elective courses for graduate and senior undergraduate courses at Rutgers University and at the Indian Institute of Technology, Kanpur, India. The topics and examples have thus been tested in a classroom environment. Several research projects directed at material processing have been used for much of the material presented here. Industrial interactions on manufacturing have also formed the basis for practical aspects discussed in the book.

Many colleagues, friends, and students have contributed in significant ways to the development of this material and to my understanding of the manufacturing process. Of particular help have been the many discussions I have had with Professors V. Sernas, C.E. Polymeropoulos, and N.A. Langrana of Rutgers University on the study and design of manufacturing systems. My interactions with students and industry over many years on various manufacturing processes and on practical design problems have also provided important insights into these processes.

This book is dedicated to my late parents, who were always a source of encouragement, support, and inspiration. Certainly, the greatest contribution to this work has been the unrestricted, endless support and encouragement of my wife, Anuradha. The patience and understanding of our children, Pratik, Aseem, and Ankur; their spouses, Leslie, Karishma, and Russell; and our grandsons, Vyan, Nalin, Zev, and Jai, have been of immense value. Without their love and support, it would not have been possible for me to find the time and drive to complete this book.

Piscataway, NJ, USA Yogesh Jaluria

Contents

Part I Basic Considerations

1 Introduction . . . 3
- 1.1 Materials Processing and Manufacturing . . . 3
- 1.2 Different Materials and Products . . . 6
- 1.3 Underlying Transport Phenomena . . . 8
- 1.4 Important Manufacturing and Materials Processing Techniques . . . 9
  - 1.4.1 Traditional Manufacturing Processes . . . 10
  - 1.4.2 Recent and Advanced Materials Processing Methods . . . 13
- 1.5 Practical Aspects . . . 16
- 1.6 Basic Thrust and Scope of the Book . . . 16
- References . . . 17

2 Mathematical Modeling of Manufacturing Processes . . . 19
- 2.1 Basic Aspects . . . 19
- 2.2 Governing Equations . . . 22
- 2.3 Common Challenges in Mathematical Modeling . . . 23
  - 2.3.1 Variable Properties . . . 23
  - 2.3.2 Viscous Dissipation . . . 26
  - 2.3.3 Buoyancy Effects . . . 27
- 2.4 Additional Transport Mechanisms and Considerations . . . 30
  - 2.4.1 Non-Newtonian Fluids . . . 30
  - 2.4.2 Surface Tension Effects . . . 31
  - 2.4.3 Particulates and Powdery Materials . . . 32
  - 2.4.4 Phase Change . . . 33
  - 2.4.5 Chemically Reactive Materials . . . 34
  - 2.4.6 Complicated Geometry . . . 36
  - 2.4.7 Combined Mechanisms . . . 37

2.4.8 Complex Boundary Conditions 37
2.5 Other Challenges 38
References 38

3 Numerical Modeling and Experimentation 41
3.1 Analytical Solution 41
3.2 Numerical Modeling 44
3.2.1 Accurate Imposition of Appropriate Boundary Conditions 45
3.2.2 Numerical Modeling of Complicated Regions 47
3.2.3 Modeling of Combined Mechanisms and Complex Transport Phenomena 49
3.2.4 Multiscale Modeling 52
3.3 Numerical Solution 53
3.4 Verification and Validation of the Model 55
3.5 Modeling of Manufacturing Systems 56
3.6 Numerical Simulation 57
3.7 Experimentation 59
3.8 Link Between Transport Processes and Product Characteristics 63
References 64

Part II Different Manufacturing Processes

4 Manufacturing Processes with Phase Change 69
4.1 Analysis 70
4.1.1 Interface Tracking 70
4.1.2 Enthalpy Method 73
4.2 Simple Conduction Models 74
4.3 One-Dimensional Solidification 79
4.4 Results for Multidimensional Problems 82
4.4.1 Interface Tracking: Transformed Grid Method 82
4.4.2 Enthalpy Formulation 83
4.4.3 Conjugate Transport 85
4.4.4 Experimental Studies 87
4.5 Continuous Casting 88
4.6 Alloys 93
4.7 Crystal Growing 99
4.8 Other Processes and Materials 101
4.9 Summary 101
References 103

5 Continuous Materials Processing 107
5.1 Basic Process 108
5.2 Heat Transfer Within the Moving Material 110
5.2.1 Analysis 111
5.2.2 Analytical and Numerical Results 113

5.3 Convective Transport Due to a Moving Surface 117
5.4 Conjugate Transport 120
5.4.1 Boundary-Layer Formulation 120
5.4.2 Solution of the Full Equations (Elliptic Formulation) 123
5.5 Buoyancy and Forced Flow Effects 126
5.6 Transient Effects 133
5.7 Experimental Work 136
5.8 Additional Aspects 137
5.9 Summary 141
References 141

6 Polymer Processing 145
6.1 Material Properties 146
6.2 Single Screw Extrusion 149
6.2.1 Two-Dimensional Model 150
6.2.2 Fully Developed Flow 153
6.2.3 Developing Flow 155
6.2.4 Three-Dimensional Transport 157
6.2.5 Axial Formulation 160
6.2.6 Tapered Screw 160
6.3 Residence Time 163
6.4 Mixing 165
6.5 Experimentation 165
6.6 Twin Screw Extrusion 169
6.7 Combined Heat and Mass Transfer and Chemical Reactions 172
6.7.1 Moisture Transport 173
6.7.2 Chemical Reactions 175
6.8 Die Flows 180
6.8.1 Coupling of Extruder with Die 180
6.8.2 Transport in Complex Dies 181
6.9 Additional Aspects in Extrusion 185
6.10 Injection Molding 186
6.11 Thermoforming 189
6.12 Summary 192
References 193

7 Thin Film Deposition: Micro-/Nanoscale Fabrication 197
7.1 Introduction 197
7.2 Basic Aspects 201
7.3 Characteristic Results 203
7.4 Metal-Organic Chemical Vapor Deposition (MOCVD) 209
7.4.1 Basic Process 211
7.4.2 Gallium Nitride 211
7.4.3 Chemistry for GaN Deposition 212

7.5 Validation . . . 214
7.6 Flow and Temperature Distributions . . . 217
7.7 GaN Deposition . . . 224
7.8 Transient Effects . . . 227
7.9 Characteristics of Deposited Thin Films . . . 232
7.10 Summary . . . 234
References . . . 235

8 Manufacture of Optical Fibers: Drawing and Coating Processes . . . 239
8.1 Optical Fiber Systems . . . 239
8.2 Optical Fiber Manufacturing . . . 242
8.3 Preform Fabrication . . . 243
8.4 Drawing of Optical Fibers . . . 244
8.4.1 Basic Equations . . . 246
8.4.2 Characteristic Results . . . 248
8.4.3 Multilayer Fibers and Dopants . . . 252
8.4.4 Hollow Fibers . . . 259
8.4.5 Furnace Wall Temperature Distribution . . . 264
8.5 Process Feasibility . . . 265
8.6 Thermally Induced Defects . . . 268
8.7 Fiber Cooling . . . 272
8.8 Fiber Coating . . . 274
8.9 Multiscale Modeling . . . 282
8.10 Summary . . . 283
References . . . 284

9 Other Manufacturing Processes . . . 287
9.1 Heat Treatment . . . 288
9.1.1 Basic Aspects . . . 289
9.1.2 A Batch Annealing Furnace . . . 292
9.1.3 Surface Heat Treatment . . . 297
9.1.4 Case Hardening . . . 300
9.2 Joining and Bonding Processes . . . 300
9.2.1 Soldering . . . 301
9.2.2 Welding . . . 307
9.3 Fabrication of Microchannels and Microdevices . . . 313
9.4 Additional Manufacturing Processes . . . 317
9.5 Summary . . . 320
References . . . 320

Part III System Considerations

10 Simulation, Design, and Optimization of Manufacturing Systems . . . 325
10.1 Modeling and Simulation . . . 327
10.2 Inverse Problem and Process Feasibility . . . 330

10.2.1 Inverse Problem 330
10.2.2 Feasible Domain 331
10.3 Design Optimization 331
10.4 Uncertainties 339
10.5 Additional Design Aspects and Strategies 342
10.5.1 Knowledge-Based Design 342
10.5.2 Concurrent Experimentation and Simulation 346
10.6 Summary 347
References 347

Index 349

Part I
Basic Considerations

The first step in discussing manufacturing and materials processing is the presentation of an overview of the basic processes and systems that are of interest, indicating the main aspects that need to be considered and listing some of the major materials processing techniques. Both traditional and recent methods for manufacturing are of interest. Though the materials considered are often quite different, the basic approach for modeling, simulation, and experimentation are often quite similar, as are the design and optimization of the relevant manufacturing system. Therefore, the first part of the book focuses on the mathematical modeling of the processes and discusses the nature of the governing equations and boundary conditions, multiscale modeling, important underlying mechanisms, challenges that often arise, and different approaches that may be employed to study the process and obtain quantitative results. This is followed by a detailed discussion of numerical modeling and experimentation, considering different methods for obtaining accurate and realistic results to understand the basic characteristics of the process. The effects of important parameters and operating conditions are thus determined. These can then be used to predict the outputs from a given system or process and also form the basis for system design and optimization.

Chapter 1
Introduction

1.1 Materials Processing and Manufacturing

Manufacturing, which refers to the mass production or fabrication of useful components and products starting with available raw materials, is an engineering field that has spearheaded the progress and economic growth of humanity over many centuries. Materials processing is another term, which has often been used as synonymous to manufacturing. However, materials processing involves transforming raw materials into usable forms and is an important constituent of the overall manufacturing system, which has several additional components needed for mass production. In most cases, the material undergoes several processes before the final, desired product is obtained. Materials processing has been around for ages, with operations like casting and forging being used for thousands of years. In recent years, a wide range of new materials, such as ceramics, composites, advanced polymers, specialized alloys, semiconductors, optical materials, and nanomaterials, have been developed, along with new techniques for materials processing. New and emerging applications in energy, environment control, communications, transportation, biotechnology, and other engineering fields demand high product quality, specific characteristics, and consistency, along with reduced costs. This poses challenging constraints and requirements on the materials processing and manufacturing systems. Several books, such as those by Schey (1999), DeGarmo et al. (2007), Groover (2012), and Kalpakjian and Schmid (2013, 2016), discuss different manufacturing processes, relevant concerns, and existing methods.

The changes in the material as it undergoes processing are largely determined by the basic transport mechanisms that govern the temperature, pressure, concentration, and other physical and chemical conditions within the material as functions of time and location. The transport phenomena thus govern the characteristics of the product as well as the rate at which the process occurs. Therefore, the two main aspects of materials processing, product quality and rate of production, are strongly influenced by the transport phenomena, which are mainly determined by fluid flow, heat and

Y. Jaluria, *Advanced Materials Processing and Manufacturing*, Mechanical Engineering Series, https://doi.org/10.1007/978-3-319-76983-7_1

mass transfer, and thermodynamics. The transport mechanisms are also important in the study, prediction, operation, control, design, and optimization of the process and the system (see, for instance, Ghosh and Mallik (1986), Jaluria (2008), and Dieter and Schmidt (2012)).

The importance of transport phenomena is particularly evident in new and emerging areas of materials processing. Materials such as polymers, alloys, ceramics, composites, semiconductors, and optical materials need energy input for fabrication since changes in the local conditions, such as temperature and pressure, are needed for the desired transformations in the material. Chemical conversion, for instance, is strongly dependent on the local temperature, pressure, and shear. Plastics and glass need to be heated to particular temperatures for forming or shaping. Similarly, micro-/nanoscale devices, thin films, additive manufacturing, electronic circuitry, and coating are largely based on thermal transport to achieve the desired characteristics. There are many other processes, such as cutting, grinding, milling, and cold forming, in which energy removal is necessary in order to maintain the appropriate temperature level to avoid damage. Thus, it is seen that the transport phenomena are an important aspect in most manufacturing processes and are particularly crucial in advanced materials processing that involves most of the new and emerging materials and techniques (Shin 2013). Though this book is directed at manufacturing and materials processing, the focus is on the underlying transport phenomena and the basic aspects that govern product characteristics and productivity. Different manufacturing processes are considered, but details on existing operations, techniques, specific applications, and materials are given only if relevant to the basic considerations. Some references that may be mentioned regarding the transport phenomena involved with materials processing include Poulikakos (1996), Kou (1996), Poirier and Geiger (1998), and Jaluria (2013).

The structural changes in the material, which determine its behavior, characteristics, and properties, largely occur at micrometer or nanometer length scales, where physical and chemical effects lead to transformations in the material. The operating conditions, such as energy input, temperature, pressure and mass inflow, are imposed on the system at macroscopic length scales. The system design parameters are also generally at larger macro- or engineering length scales. Thus, materials processing involves different length scales, one at the smaller scales where structural changes occur and larger scales where the operating conditions are applied. Along with different length scales, different time scales are also correspondingly involved. This implies that the governing transport phenomena at micro-/nanoscale must be investigated if the existing processes are to be improved or optimized to obtain desired product quality and production rate. This information is particularly valuable in developing new processing techniques. Additionally, the modeling and analysis of these micro-/nanoscale processes must be linked with the transport phenomena at larger length and time scales. Thus, multiscale modeling is necessary in most practical processes (see, for instance, Li (1985), Jaluria (2009, 2016), and Chiu et al. (2011)).

Experimental data are critical in the study of manufacturing processes. These are needed for model validation, for physical insight, and for providing inputs to the modeling of complex problems, for which the boundary conditions, coupled mechanisms, and other aspects may not be available or easily amenable to analysis. In addition, experimentation is needed to obtain the relevant material properties that are crucial to both experimental and numerical results.

In many cases, heat and mass transfer is associated with fluid flow, which is important in several materials processing operations such as casting, soldering, welding, crystal growing, extrusion, chemical vapor deposition, spray coating, injection molding, wire and fiber drawing, and fabrication of composites. The buoyancy-driven flows that arise in the molten material in crystal growing, for instance, can significantly affect the quality and purity of the crystal. Consequently, the quality and characteristics of the semiconductor devices fabricated from the crystal are affected. The transport processes that arise at the solid-liquid interface also determine the rate of production. Similarly, the fluid flow generated in the liquid melt in mold and continuous casting strongly influences the shape, stability, location, and behavior of the solid-liquid interface, which then impacts on the microstructure of the casting obtained.

The flow of molten metal in welding and soldering is critical to the geometry, location, and dimensions of the joint formed. This flow is largely driven by surface tension and gravitational and viscous forces, and the bond obtained solidifies due to heat loss. In optical fiber drawing, the neck-down of glass from a much larger diameter of a preformed cylinder to the fiber diameter during the draw process is governed mainly by the viscous flow of glass, which is softened by increase in temperature. In thin film fabrication by chemical vapor deposition, the flow, the heat and mass transfer, and the associated chemical reactions are critical in determining the deposition rate, film uniformity, and its quality and characteristics. The heat transfer in furnaces and ovens that are used for heat treatment of materials such as steel, aluminum, and glass determines the thermal cycle undergone by the material. This temperature variation, in turn, determines the characteristics and properties of the product. The formation and flow of droplets in sprays are important in rapid fabrication based on deposition. These processes and several others are discussed in greater detail later. However, it is seen that the nature, magnitude, and behavior of the transport processes are of crucial significance in the resulting product quality, rate of production, and system performance.

The link between different processing techniques and the basic transport mechanisms has generally not been established to the level needed for significant improvement of existing operations and development of new ones. In most cases, the effort has focused on specific manufacturing systems and the associated practical issues, procedures, and circumstances. It is thus important to investigate the basic transport phenomena as applied to manufacturing, in order to understand the main underlying mechanisms and expand the applicability of various techniques and the range of materials and products obtained. Another aspect that has not received significant attention in the literature is the dependence of product quality, process control, system behavior, and optimization on the transport processes. It is also necessary to establish the present status in transport phenomena related to materials processing and to determine the future work needed in this area to meet critical demands in manufacturing. This involves linking practical engineering systems with the basic transport phenomena and thus have a strong impact on an area that is of particular importance today.

This book is directed at the important issues that arise in manufacturing, focusing on the transport phenomena, particularly heat and mass transfer and fluid flow, that are involved in materials processing and linking these with the product characteristics and with the system. A fairly wide range of processes is considered in order to identify the basic mechanisms that arise and study their effects on the processed material.

1.2 Different Materials and Products

A critical element in any discussion or study of manufacturing is the material undergoing processing to obtain the desired product. Intermediate steps may involve processing of raw or available materials to obtain desired material characteristics for further processing. Over the last few decades, considerable work has been done on materials and an extensive range of new materials have been developed to meet the growing needs in many different new, emerging, and challenging application areas. The materials that are of interest in engineering products and systems may be classified in terms of the following broad categories:

Metals and alloys
Polymers
Composite materials
Ceramics
Semiconductor materials
Fluids

Figure 1.1 gives a schematic of different materials used in engineering processes and systems, along with a few important examples. These materials are subjected to various processing techniques, such as forming, machining, casting, and heat treatment, in order to convert the raw materials to their useful forms.

Metals have been used extensively in a wide range of applications because of their strength, toughness, high electrical and thermal conductivity, cost, availability, and ease in processing. Thus, copper has been used for pipes and electrical connections due to its malleability and high electrical conductivity, aluminum for transportation systems due to its low weight, and gold in electronic circuitry due to its corrosion resistance. Alloys have expanded the range of applications due to modifications in the properties of various metals by variations in composition and are now among the most widely used materials in practical systems. Steel, in its different compositions, is among the most widely used material in many different applications, ranging from construction to automobiles. Solder, which is often an alloy of tin and lead, with silver being added in some cases and tin-silver-copper alloys being used for lead-free solders, is used for making electrical connections. Aluminum alloys are used in airplane parts and so on. Polymers include plastics, rubber, fibers, coatings, and other similar materials. They have found extensive use due to low cost,

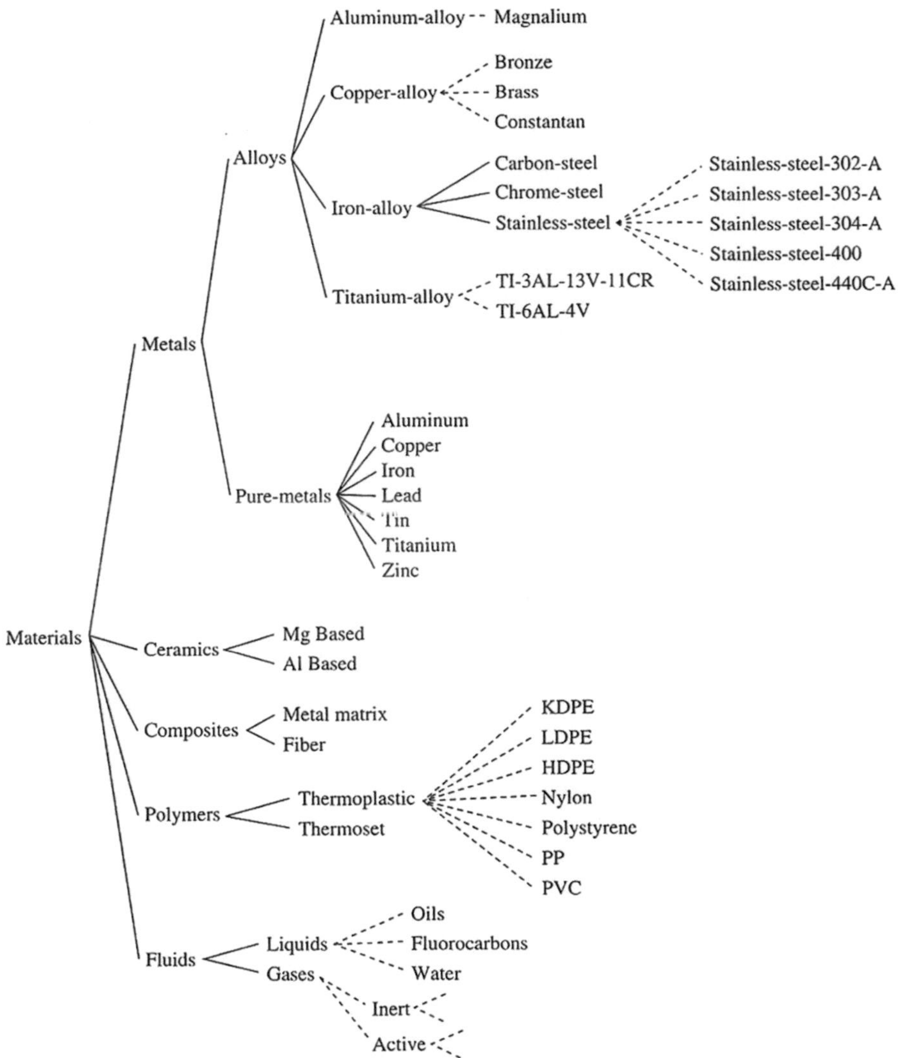

Fig. 1.1 Different types of materials commonly used in engineering systems and processes

low weight, high thermal and electrical resistance, easy fabrication, and versatility. Food materials, which are generally biopolymers, are also included in this category.

Ceramics are generally formed by the fusion of powders such as aluminum oxide, beryllium oxide, and silicon carbide under high temperature and pressure. Glass and other optical materials are also included under ceramics and are valuable due to transparency. Optical fibers have found extensive use in telecommunications, data transmission, and diagnostics. Ceramics are widely used due to their high temperature resistance, low electrical conductivity, low weight, relatively low cost, and corrosion resistance.

Composite materials are engineered materials formed by embedding reinforcing agents like glass fiber, carbon, ceramic, and metal in a binder, which usually consist of different polymers, as the base material. These materials have replaced metals and alloys in many applications due to high strength-to-weight ratio, fatigue resistance, and a substantially broad range of properties obtained by the selection of reinforcing elements and binders. They can often be custom-made for a desired set of properties. Currently, composite materials are used in car body moldings, pressure vessels, rotor blades, printed circuit boards, and a host of other applications.

Semiconductor materials, which include elements like silicon, germanium, and carbon, as graphene, diamond, carbon nanotubes, and other basic structures, as well as compounds like gallium nitride, gallium arsenide, and silicon carbide, have obviously made the revolution in computers, processors, and electronic hardware possible through their use in electronic devices like transistors, microelectronics, and integrated circuits. Fluids, such as coolants, lubricants, and inert gases are of interest during processing as well as in some manufacturing systems. There are various other materials such as wood, concrete, cement, and stone that may be of interest in some applications. New materials are continuously being developed to meet the demands for specific properties and characteristics and to improve the materials used in existing systems. A more detailed discussion of engineering materials and their properties is given in most books on materials, processing, and design (see, for instance, Dieter (2000), Jaluria (2008), and Kalpakjian and Schmid (2016)).

1.3 Underlying Transport Phenomena

Many of the processing techniques needed to convert raw materials into useful forms or into finished products involve transformations, which generally occur at the micro- or nanometer length scale, and input of energy, mass flow, shear, pressure, force, and other physical quantities, generally applied at the boundaries that involve engineering or macro-scales. Among the most important basic processes are those that involve transport phenomena, which include:

Energy transfer
Mass transfer
Momentum transport

The resulting heat and mass transfer can increase or decrease the temperature and pressure within the material, leading to changes in the material properties and characteristics. The material changes are brought about by chemical reactions, structural changes, conversion, and other transformations occurring at the smaller length scales. Thermodynamics is an area which is intimately connected with such transformations. Similarly, material flow affects the transport within the material as well as at the boundaries. Fluid flow in materials processing is often closely coupled with the heat and mass transfer processes.

The importance of the transport phenomena is seen in processes such as casting, which involves melting of a given material by the input of thermal energy followed by solidification as the energy is removed to achieve a desired shape or configuration, and thermoforming, where a given item made of, say, plastics is heated to a temperature at which it can be bent or formed and then cooled to obtain the final shape. Such transport processes are of particular interest in new and emerging materials and in advanced materials processing techniques. Materials such as glass and polymers need the thermal input for processing. Manufacturing techniques like 3D printing and laser welding require consideration of energy transport and temperature levels attained. Therefore, it is critically important to understand the underlying transport phenomena in most of these important manufacturing techniques and develop tools to analyze, model, and experimentally study the process to improve existing processes and develop new ones to meet challenges arising due to new materials and manufacturing techniques.

Thus, a study of materials processing or manufacturing requires a consideration of the energy, mass, and momentum transport at the boundaries and within the material. These changes are then linked with variations in material structure and characteristics. It is, therefore, necessary to couple the conditions imposed at the boundaries at the macro or engineering scale with the structural changes occurring at the micro- or nanoscale. This multiscale problem has to be modeled and simulated to obtain the inputs needed for design, prediction, control, and optimization of the overall process and manufacturing system. Experiments are needed for validation of the model as well as for providing inputs that are not easily obtainable from modeling.

1.4 Important Manufacturing and Materials Processing Techniques

Several manufacturing processes have been briefly mentioned earlier to indicate the importance of transport phenomena and the need for detailed study of the underlying transport mechanisms that lead to changes in the material. The field of materials processing is quite extensive, and only a few processes have been mentioned so far. Some of these will be considered again and in detail to present important aspects and techniques, along with characteristic results. Other processes will also be discussed and the related transport phenomena presented. In these processes, the basic thermal issues will be presented and related to those arising in other materials processing systems. However, before specific manufacturing processes are discussed, let us consider some of the common processes used today. Both traditional processes that have been used for centuries and more recent processes are given. References such as Ghosh and Mallik (1986), Schey (1999), Kalpakjian and Schmid (2016), Degarmo et al. (2007), and Poirier and Geiger (1998) discuss various manufacturing processes and existing systems.

1.4.1 Traditional Manufacturing Processes

A list of common traditional manufacturing or materials processing techniques in which transport phenomena, particularly fluid flow and heat and mass transfer, are of particular importance is given below, along with a few well-known methods as examples.

1. Heat treatment: annealing, hardening, and surface treatment
2. Casting: mold casting and continuous casting
3. Forming: hot rolling, wire drawing, metal forming, extrusion, and forging
4. Joining: welding, soldering, brazing, explosive bonding, and epoxy bonding
5. Glass processing: glass blowing, forming, and annealing
6. Chemical processing: cooking, curing, and baking
7. Coating: spray coating and polymer coating
8. Gas flame processing: heating, cutting, and welding
9. Machining: cutting, grinding, and drilling

Energy transport, particularly heat transfer, is critically important in all the foregoing manufacturing processes, including machining where heat removal to maintain the temperature level is the main concern. Figure 1.2 shows sketches of many of these processes. These include mold casting, heat treatment, soldering, continuous casting, extrusion, and hot rolling. Mold casting, shown in Fig. 1.2a, involves pouring molten material into an enclosed region of given shape and dimensions, known as a mold. The material solidifies as energy is removed from the liquid by heat transfer at the boundaries. A liquid-solid interface moves with time as solidification progresses. The time-dependent shape and location of the interface are determined by the flow and heat transfer processes in the mold, until the entire fluid solidifies. The mold has to be designed keeping in mind density change on solidification, additional material needed for further processing, and other practical issues. The heat transfer processes, which affect temperature levels and transport rates, can thus be used to control the rate of solidification, the overall solidification time, and the characteristics and quality of the casting. Casting has been one of the most extensively used manufacturing processes over the years, and considerable work has been done for a wide range of materials.

Figure 1.2b shows the desired temperature variation for annealing, which is a process used to relieve internal stresses in a material, restore ductility, and reduce hardness. It is used extensively for metals, glass, ceramics, and other materials that have internal stresses built up due to cutting, forming, or other operations. Basically, the material is heated up to the annealing temperature, which is generally the recrystallization temperature at which the distorted grain structure of the material is replaced by a strain-free one. This temperature is a function of the material, and, at this point, the microstructure of the material becomes relaxed and is able to realign itself. The material is held for a finite time, known as soaking time, to allow the transformations to occur and then cooled very gradually in air or in a furnace. Acceptable deviation from the desired temperature variation with time is also shown. The soldering process in Fig. 1.2c shows the flow of a preformed solder, supplied in the shape of a doughnut that is heated and melted. Due to surface tension forces, along with gravitational and viscous forces, it flows and forms a bond, as shown, between the pin or terminal and the plated-through hole in a circuit board. Surface tension effects are particularly significant due to the small clearance between the terminal and the hole. Once the bond is formed, it cools down by heat removal until it solidifies and forms a solder joint. This process and other soldering techniques have been of particular interest in electronic systems. Welding has similar considerations with respect to melting, flow, and solidification. Continuous casting

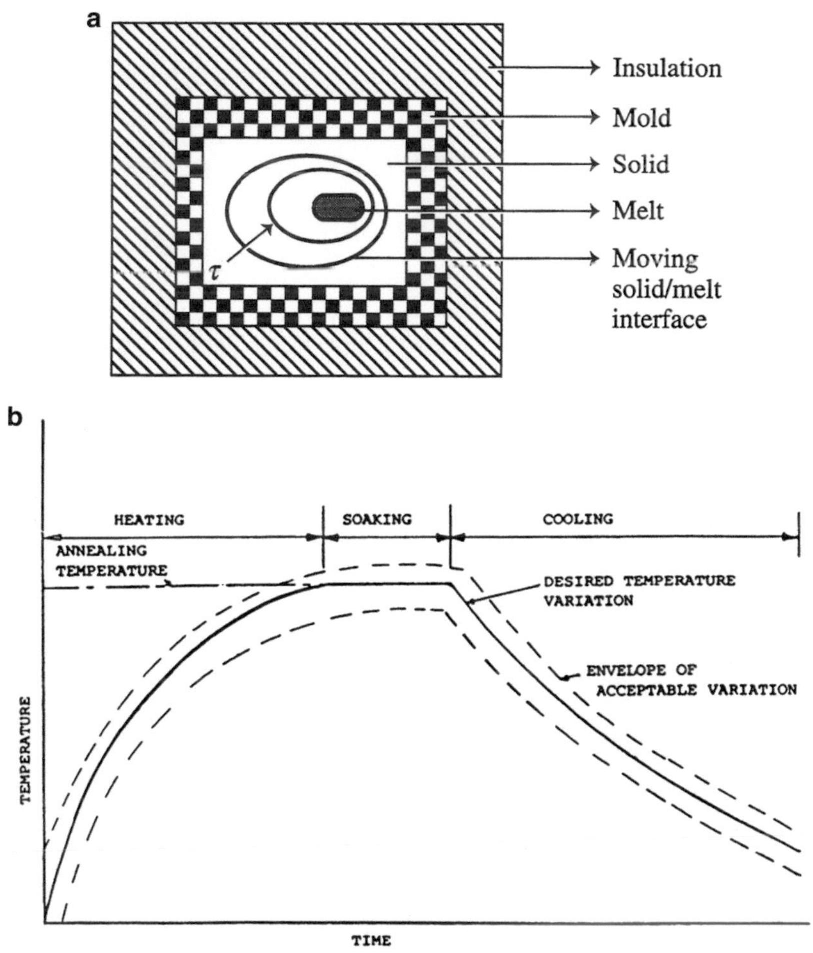

Fig. 1.2 Sketches of a few traditional manufacturing processes in which the underlying transport phenomena are significant. (**a**) Mold casting, (**b**) heat treatment, (**c**) solder joint formation, (**d**) continuous casting, (**e**) extrusion and hot rolling processes for a heated material

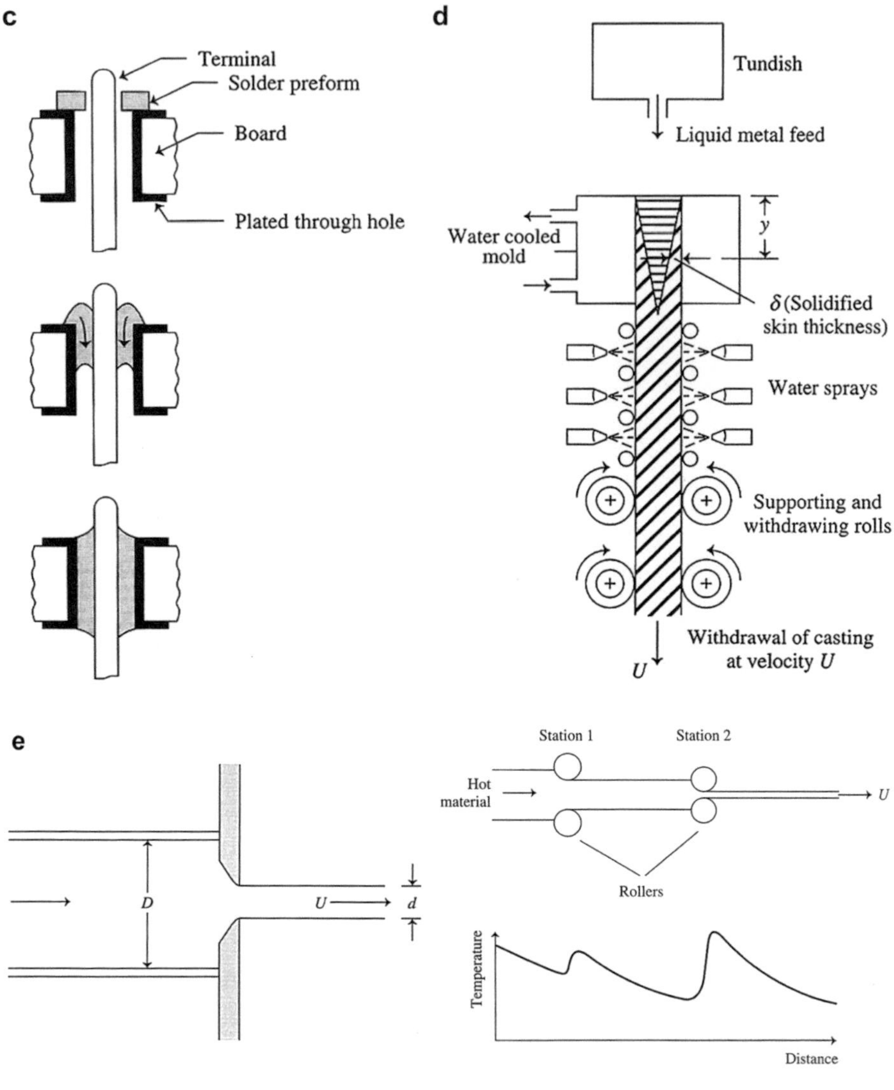

Fig. 1.2 (continued)

is a process which involves solidification of molten material over an essentially stationery interface while the material is continuously fed, as shown in Fig. 1.2d. It is widely used in the manufacturing of metal and alloy sheets and rods.

Similarly, the extrusion and hot rolling processes shown in Fig. 1.2e involve a continuously moving material as it undergoes thermal processing due to heating or cooling. In extrusion, the material is generally heated and forced through a die to obtain a particular shape and dimension. Hot rolling applies at temperatures above the material recrystallization temperature and cold rolling at temperatures below this level. The moving material is reduced in dimension as it passes through rolls at appropriate temperature levels. Generally, local heating is needed for hot rolling and cooling for cold rolling. This is a fairly common process used in the manufacture of sheets and rods, used, for instance, in automobile and cable industries. The references given earlier may be consulted for further information on these systems. Additional details are also provided later on in the book as the basic aspects and underlying transport phenomena are considered.

1.4.2 Recent and Advanced Materials Processing Methods

Many other manufacturing and materials processing techniques have been developed in recent years because of advances in new materials and the emergence of many applications areas that have brought in additional challenges. Some of the common advanced materials processing techniques are listed below, along with a few examples.

1. Polymer processing: extrusion, injection molding, and thermoforming
2. Reactive processing: food processing and rubber manufacture
3. Powder processing: powder deposition, sintering, sputtering, and ceramics
4. Fabrication of semiconductor devices: crystal growing and epitaxy layer deposition
5. Optical materials: optical fibers, silica glass, and optical devices
6. Chemical vapor deposition: GaN, TiN, GaAs, SiC, and selective surfaces
7. Laser processing: laser heating, cutting, and welding
8. Alloy casting: specialized alloys and micro-/macro-segregation
9. Rapid prototyping: 3D printing, near net shape, stereo lithography, and additive manufacturing
10. Coating: thermal sprays and polymer coating
11. Composite materials processing: metal and ceramic composites and reinforced plastics
12. Fabrication of micro- and nanoscale materials and devices
13. Others: ultrasonic and electrochemical machining, fluid jet cutting, and micro-gravity processing

Figure 1.3 shows the schematic of some of these processes. These include optical fiber drawing, Czochralski crystal growing, chemical vapor deposition for thin film fabrication, and twin-screw extrusion for polymers and reactive materials like food. In optical fiber drawing, shown in Fig. 1.3a, a specially fabricated cylindrical silica glass preform is heated and drawn into a fiber. The fiber is cooled and coated, generally with a polymer, before being rolled into a spool. In Czochralski crystal growing, sketched in Fig. 1.3b, molten material such as silicon is allowed to solidify across an interface as a seed crystal is withdrawn, resulting in a pure crystal from which electronic devices are fabricated. This process has been one of the major steps

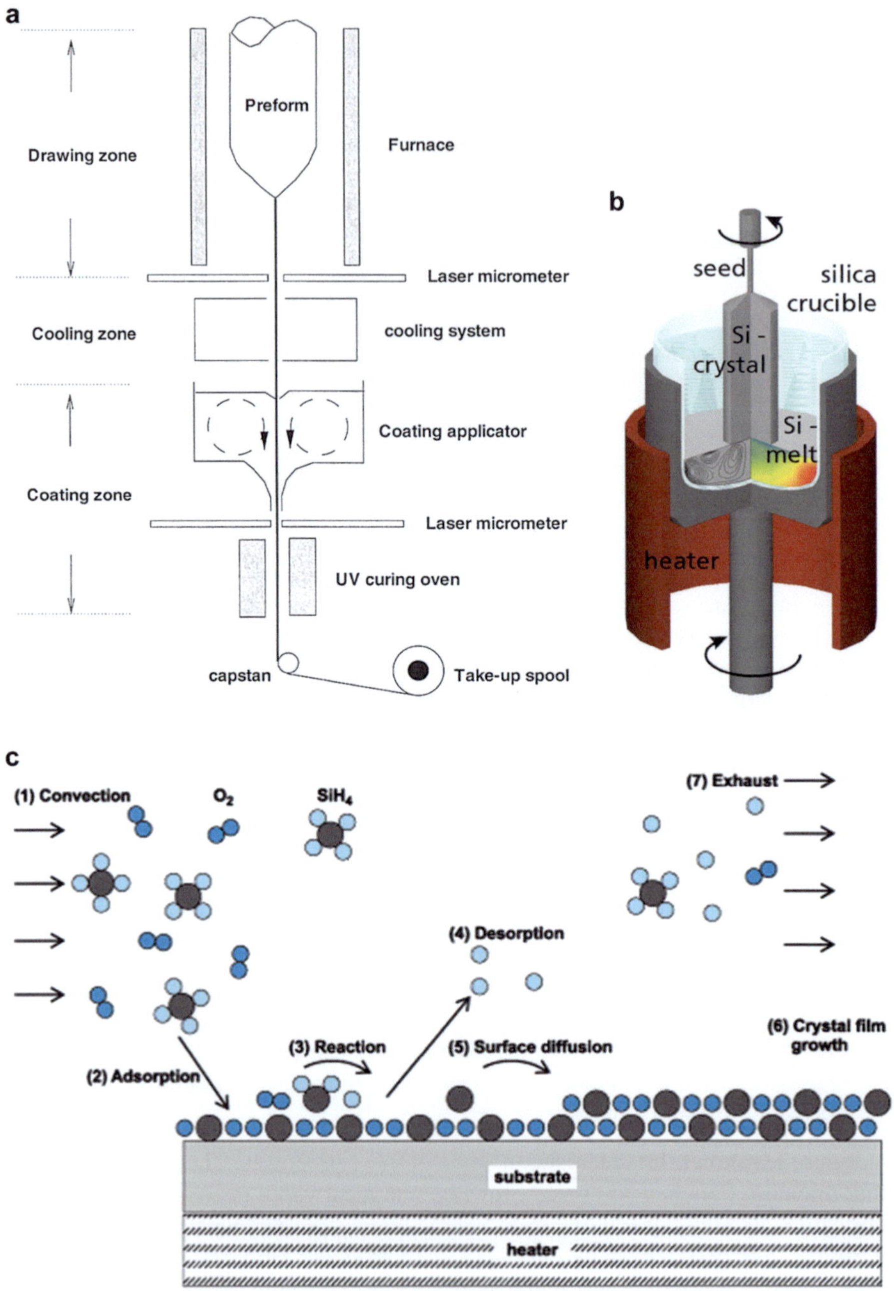

Fig. 1.3 Sketches of more recent materials processing systems. (**a**) Optical fiber drawing, (**b**) Czochralski crystal growing, (**c**) chemical vapor deposition, (**d**) twin-screw extrusion

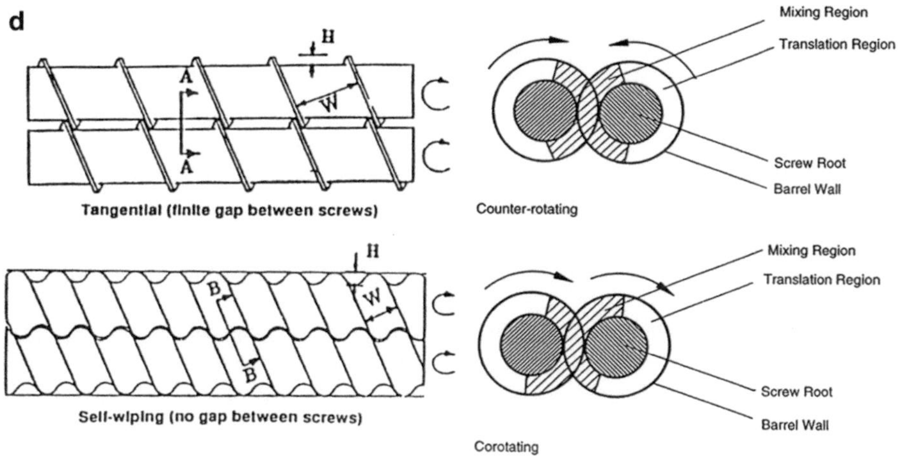

Fig. 1.3 (continued)

in the overall fabrication of electronic chips. Chemical vapor deposition (CVD), shown in Fig. 1.3c, is used for obtaining high-quality thin films, from which semiconductor devices are fabricated, or for surface coating. Reactive gases are fed into the reactor with carrier gases and the resulting reactions in the gas phase as well as at the surface result in a deposited film that grows with time and yields the desired thin film. Polymer screw extrusion is a process in which materials such as plastics are melted or softened and forced through a suitable die to obtain specific shapes and dimensions. A single-screw extruder consists of a screw rotating in a cylindrical barrel, with an inlet at one end and a die at the other. The pressure rises due to the screw motion and constriction provided by a die at the exit, a higher pressure being the result of a narrower die. Extrusion is an extensively used manufacturing process for polymers and food. It is also used in many cases to feed the molten or softened polymer under pressure into an injection molding system, which is another important manufacturing system and is discussed later. Twin-screw extrusion, shown in Fig. 1.3d, has become important in the last few decades mainly because of superior mixing characteristics that lead to more uniform products. This is particularly attractive in the extrusion of reactive materials, such as food. Different configurations of twin-screw extruders have been developed for different applications. A particularly important one for food extrusion is the corotating self-wiping profile in which one screw wipes the surface of the other screw, thus reducing the chances of food being entrapped and thus getting spoilt over time.

In all the manufacturing processes mentioned here, the quality and characteristics of the final product, as well as the rate of fabrication, are strongly dependent on the underlying transport phenomena. Therefore, detailed analysis, modeling, simulation, and experimentation are needed to obtain the transport rates as functions of location, time, temperature, flow, pressure, and other variables in order to determine the design parameters and the operating conditions that would lead to the desired product quality and characteristics, as well as high production rates. The choice of operating conditions for a desired output is an inverse problem in which the outputs are known and inputs are to be determined. This is generally a fairly complicated problem without a unique solution. This aspect is discussed in greater detail later in the book.

1.5 Practical Aspects

Several important basic and practical issues arise in manufacturing and materials processing that need a detailed consideration. Some of the relevant basic issues have been mentioned earlier and will be discussed in greater detail later. Important practical issues that are of particular interest in manufacturing are listed below:

Product quality and characteristics
Rate of production or fabrication
Process feasibility
System efficiency
Manufacturing costs
Inverse problem for choosing appropriate conditions
Design of appropriate manufacturing system
System optimization and control
Optimal operating conditions
Uncertainties in design and operating conditions
New products and methods

The transport mechanisms that govern materials processing form the basis for evaluating and achieving these practical aspects. For example, product characteristics and quality are affected by the temperature distributions, the production rates are determined by the energy transfer rates, and feasibility may be determined by material flow and constraints on temperature, pressure, and stress. Efficiency and environmental effect are determined by energy supply and heat losses. Costs are obviously related to the rate of production, besides operating and capital costs. Uncertainties affect the final design and operating conditions. Also, a better quantitative understanding of the underlying transport processes could lead to new products and improved techniques.

1.6 Basic Thrust and Scope of the Book

This book is focused on materials processing and manufacturing, particularly on advanced manufacturing related to new and emerging materials processing techniques. Since the subject matter is vast and involves a wide range of manufacturing techniques, the book largely considers processes in which the underlying transport phenomena are of particular significance. Detailed discussions of existing manufacturing systems and current practice are not given, unless relevant to the consideration of the basic processes involved.

Therefore, the book is mainly concerned with the basic transport mechanisms that govern materials processing. By considering a fairly wide range of materials and processes, the main challenges in mathematical modeling, numerical simulation, and experimentation are brought out. The basic transport phenomena are linked with the practical aspects such as product quality and productivity. The dependence of material characteristics, quality, and behavior on physical and chemical conditions to which it is subjected is considered to determine the material transformation as it goes through the manufacturing process. Of particular interest are the prediction, control, design, and optimization of the process in order to achieve the desired results. Examples are taken from several important processes to illustrate basic approaches and typical concerns and results. It is important to improve existing processes and develop new ones in order to meet challenges arising from new and emerging application areas, new materials, environmental concerns, and energy conservation.

Starting with a detailed study of the relevant transport phenomena, the book considers a range of different and important manufacturing processes. It presents approaches to address various challenges that arise in different processes. It also presents results for typical conditions to indicate the current status and the modeling and experimental work that is needed. Such results form the basis for addressing the practical aspects of interest in given applications. The major emphasis is on developing general methods that may be adopted for different manufacturing processes and the common thread that runs through most of these without getting into excessive details of the system. The basic approaches and results are then linked with the practical issues mentioned earlier, particularly choice of operating conditions through the solution of inverse problems, system design, and optimization.

A fundamental understanding of transport phenomena is assumed, though a brief review of the governing principles and equations is given for completeness and for facilitating further discussions on modeling and experimentation. The book is written at an advanced senior or early graduate level and may be used as a text or reference book in various courses. It is also suitable for engineers involved in manufacturing and materials processing. The references given in the book may be used to obtain more detailed information on specific manufacturing techniques and systems.

References

W. K. S. Chiu, C. P. Grigoropoulos, B. Q. Li (eds.), Special issue on advanced thermal processing. J. Heat. Transf. **133**, 030301 (2011)

E.P. DeGarmo, J.T. Black, R.A. Kohser, *Materials and Processes in Manufacturing*, 10th edn. (Wiley, New York, 2007)

G.E. Dieter, *Engineering Design: A Materials and Processing Approach*, 3rd edn. (McGraw-Hill, New York, 2000)

G.E. Dieter, L. Schmidt, *Engineering Design*, 5th edn. (McGraw-Hill, New York, 2012)

A. Ghosh, A.K. Mallik, *Manufacturing Science* (Ellis Horwood, Chichester, 1986)

M.P. Groover, *Fundamentals of Modern Manufacturing: Materials, Processes and Systems*, 5th edn. (Wiley, New York, 2012)

Y. Jaluria, *Design and Optimization of Thermal Systems*, 2nd edn. (CRC Press, Boca Raton, 2008)

Y. Jaluria, Microscale transport phenomena in materials processing. J. Heat Transf. **131**, 033111-1–03311117 (2009)

Y. Jaluria, Thermal issues in materials processing. J. Heat Transf. **135**, 061701-1–06170114 (2013)

Y. Jaluria, Heat and mass transfer in materials processing and manufacturing. Adv. Heat Tran. **48**, 1–94 (2016)

S. Kalpakjian, S. Schmid, *Manufacturing Engineering and Technology*, 7th edn. (Pearson, Upper Saddle River, NJ, 2013)

S. Kalpakjian, S. Schmid, *Manufacturing Processes for Engineering Materials*, 6th edn. (Pearson, Upper Saddle River, NJ, 2016)

S. Kou, *Transport Phenomena and Materials Processing* (Wiley, New York, 1996)

T. Li (ed.), *Optical Fiber Communications, Vol. 1: Fiber Fabrication* (Academic Press, New York, 1985)

D.R. Poirier, G.H. Geiger, *Transport Phenomena in Materials Processing* (Wiley, New York, 1998)

D. Poulikakos (ed.), *Transport Phenomena in Materials Processing, Advances in Heat Transfer*, vol 18 (Academic Press, San Diego, 1996)

J.A. Schey, *Introduction to Manufacturing Processes*, 3rd edn. (McGraw-Hill, New York, 1999)

Y.C. Shin, Special issue: Thermally assisted manufacturing. J. Manuf. Sci. Eng. **136**(5), 060201-1 (2013)

Chapter 2
Mathematical Modeling of Manufacturing Processes

The importance of heat and mass transfer and of the associated fluid flow and material transformations, including chemical reactions, in a wide range of materials processing techniques has been discussed in the preceding chapter. In many circumstances, such as melting and solidification that are encountered in casting and crystal growing, the heat transfer to and from the material is at the very core of the process, since it determines the rate of phase change. In addition, the temperature distribution and the buoyancy-driven flows that arise in the molten material due to temperature and concentration differences affect the characteristics of the solid-liquid interface and the microstructure of the product. In processes like food extrusion, hot rolling, thermal spray coating, and soldering, the thermal transport again determines the rate of fabrication and the characteristics of the final product. Thus, it is necessary to develop mathematical models for these processes in order to understand the basic mechanisms and thereby lay the foundation for analysis, numerical simulation, and experimentation. Computational models are developed on the basis of the mathematical models. Numerical solution techniques are generally needed since complexities that arise in common materials processing systems make it very difficult to obtain analytical results, which are applicable largely for very idealized and simplified systems. Mathematical models also guide the design of relevant experiments and the appropriate data to be obtained. They help in generalizing the experimental and numerical results, ultimately leading to greater insight into the basic processes involved and the framework to use the results for improving existing processes and for developing new ones.

2.1 Basic Aspects

A wide variety of complexities and considerations arise in materials processing that set these processes apart from many other basic and applied areas of fluid mechanics, thermodynamics, and heat and mass transfer. Table 2.1 lists some of the important

Y. Jaluria, *Advanced Materials Processing and Manufacturing*, Mechanical Engineering Series, https://doi.org/10.1007/978-3-319-76983-7_2

Table 2.1 Important considerations in transport phenomena associated with materials processing

1. Coupling of transport with material characteristics
Different materials, properties, behavior, material structure
2. Variable material properties
Strong variation with temperature, pressure, shear, concentration, other variables
3. Complex geometries
Complicated domains, multiple regions
4. Complicated boundary conditions
Conjugate conditions, combined modes
5. Interaction between different mechanisms
Surface tension, heat and mass transfer, chemical reactions, phase change
6. Multiscale modeling
Microstructure conversion, mechanisms operating at different length and time scales
7. Complex flows
Non-Newtonian flows, free surface flows, powder and particle transport
8. Inverse problems
Non-unique multiple solutions, iterative solution
9. Different energy sources
Laser, chemical, electrical, gas, fluid jet, heat
10. System optimization and control
Link between transport processes and system design and optimization

considerations in the transport processes associated with materials processing. All these considerations make the mathematical and numerical modeling of the process and the associated system for materials processing quite involved and challenging. Special procedures and techniques are generally needed to satisfactorily simulate the relevant boundary conditions and material property variations.

Modeling is one of the most crucial elements in the study, design, and optimization of thermal materials processing systems. Since practical processes are generally very complicated, simplifications, idealizations, and approximations are needed to make the problem solvable. This process of simplifying a given problem so that it may be represented in terms of a system of equations for analysis is termed *mathematical modeling*. Similarly, a simplified physical arrangement or model may be developed for experimentation, which may also be carried out on actual systems, whenever possible. Once a model is obtained, it is subjected to a variety of operating conditions and design variations. If the model is a good representation of the actual process under consideration, the outputs obtained from the model characterize the behavior of the actual system. This information is used in the design of the system or process, as well as in obtaining and comparing alternative designs by predicting the performance of each design, ultimately leading to an optimal design.

A systematic procedure may be followed to obtain a mathematical model for a given process (Jaluria 2008). Some of the main considerations employed in developing a suitable mathematical model are given below and applied in various applications later.

1. *Steady* versus *time-dependent.* Materials processing is generally a time-dependent problem. However, depending on the response of the various components and the variation of the boundary conditions, the problem may be treated as steady, quasi-steady, periodic, or transient.
2. *Spatial dimensions needed.* It is always desirable to reduce the number of spatial dimensions needed to characterize the behavior of a particular component. Depending on a particular dimension being much larger than the others, a one- or two-dimensional model may be satisfactory instead of a full three-dimensional one. Similarly, symmetry about an axis could be used to obtain an axisymmetric model.
3. *Lumped mass approximation.* This approximation implies no spatial dependence, i.e., uniformity across the component. It can be invoked, for instance, if the Biot number, $Bi = hL/k$, where h is the convective heat transfer coefficient, L a characteristic dimension, and k the thermal conductivity, is much smaller than 1.0. Frequently, small metallic pieces undergoing thermal processing can be treated as lumped, considerably simplifying the analysis.
4. *Simplification of boundary conditions.* Common approximations are uniform temperature and heat flux, smooth surface, negligible curvature effect, and negligible surface roughness. Step changes in boundary conditions are often used, rather than the actual variation over short times.
5. *Material properties*. In many cases, a given material property may be assumed to be constant if the change over the given ranges of temperature, pressure, and other variables is small compared to the average value of the property. Simple functional dependence on the independent variables can also be developed on the basis of a curve fit of property data and used in the analysis.
6. *Neglecting smaller effects.* Radiation may be neglected if the temperatures are relatively low, volume changes due to heating may be neglected in liquids and solids, friction may be neglected if the frictional effects are small compared to other forces, and so on. Rough estimates are made to justify such simplifications.
7. *Idealizations.* Analysis may be carried out for ideal systems, such as systems with no losses, no leakages, no inefficiency, or zero friction, and then the effects may be added, if significant, afterward.
8. *Conservation laws.* The models must satisfy the conservation laws, and these, along with simplifications such as those indicated above, yield the governing equations for the process. The general formulation is discussed later in this chapter and the formulation for specific manufacturing processes later in the book.
9. *Additional simplifications.* The governing equations may be further simplified by using scale analysis to determine the dominant effects and neglecting smaller ones. For instance, at low speeds and high viscosity, the viscous effects dominate, and the momentum changes may be neglected. For short times, the material may be treated as semi-infinite. For turbulent flow in a channel, the velocity may be

assumed to be uniform to focus on the heat transfer. Several such simplifications will be discussed later while considering different materials processing systems.

Numerical models are based on the mathematical model and allow one to obtain, using a computer, quantitative results on the system behavior for different operating conditions and design parameters. Only very simple cases can usually be solved by analytical procedures, and numerical techniques are needed for most practical systems. Numerical modeling refers to the restructuring and discretization of the governing equations in order to solve them on a computer.

It must be mentioned that it is necessary to thoroughly understand the concerns, intricacies, and basic considerations that characterize different materials processing techniques in order to make a significant impact on the field. The dependence of the properties of the final product on the transport processes must be properly understood and characterized so that analysis or experimentation can be used to design processes to achieve desired product characteristics and production rates. This is the only way that research and detailed study of the underlying processes can stay at the cutting edge of technology in materials processing and significantly affect the future developments in this field.

We will first consider the basic aspects of modeling transport processes that govern materials processing, followed by a consideration of some of the additional complexities that arise in specific processes. Some of the important challenges encountered in mathematical modeling are briefly discussed, leaving further details and solution methods to be presented along with various manufacturing techniques later in the book.

2.2 Governing Equations

The basic equations that may be used to model the flow and thermal transport in materials processing are based on the conservation of mass and energy and the force-momentum balance that give rise to the well-known equations for fluid flow and heat transfer. These may be written for a general three-dimensional process with variable properties, as (Burmeister 1993; Oosthuizen and Naylor 1998; Jaluria 2013)

$$\frac{\partial \rho}{\partial t} + \nabla.(\rho \bar{V}) = 0 \tag{2.1}$$

$$\rho\left(\frac{\partial \bar{V}}{\partial t} + \bar{V}.\nabla \bar{V}\right) = \bar{F} - \nabla p + \nabla.\left[\mu\left(\nabla \bar{V} + \nabla \bar{V}^{T}\right)\right] - \frac{2}{3}\nabla\left(\mu \nabla.\bar{V}\right) \tag{2.2}$$

$$\rho C_{\mathrm{p}}\left(\frac{\partial T}{\partial t} + \bar{V}.\nabla T\right) = \nabla.(k\nabla T) + \dot{Q} + \mu\Phi + \beta T\left(\frac{\partial p}{\partial t} + \bar{V}.\nabla p\right) \tag{2.3}$$

where ρ is density, T is temperature, t is time, $\bar{V}$ is the velocity vector, $\bar{F}$ is body force, p is pressure, μ is dynamic viscosity, C_{p} is specific heat at constant pressure, β is coefficient of volumetric thermal expansion, $\mu\Phi$ is viscous dissipation, and $\dot{Q}$ is a volumetric heat source. The viscous dissipation and pressure work effects are included in the energy equation, Eq. (2.3), with the last two terms multiplied by βT representing the pressure effect. The bulk viscosity is taken as zero, giving the second viscosity coefficient λ as – (2/3) μ and Stokes' relationships are used for the viscous forces in the momentum equation, Eq. (2.2).

The viscous terms in the momentum equation may also be written as $\nabla.\tau$, where the stress tensor τ may then be obtained from the appropriate constitutive equations for Newtonian or non-Newtonian fluids (Batchelor 1967; Tadmor and Gogos 1979; Kundu et al. 2015). The corresponding terms for the flow of materials such as polymers and glass are given later when processes related to these materials are presented. The preceding general equations may be modified for specific materials processing applications to include additional effects and geometries, as needed. As mentioned earlier, analytical solutions are possible in only a few selective or idealized circumstances and are generally impossible or much too involved in most practical cases. Thus, numerical or experimental techniques are generally employed to obtain the desired results. Numerical simulation provides results on the effects of design parameters and operating conditions on the physical variables such as temperature, velocity, heat transfer, and pressure. These are then linked with changes in the material undergoing processing.

2.3 Common Challenges in Mathematical Modeling

Several challenges are encountered in obtaining accurate analytical or numerical solutions for manufacturing and materials processing systems. Some of the important ones are outlined below.

2.3.1 Variable Properties

In most materials processing operations, large material property changes occur with variables such as temperature, pressure, shear rate, and concentration. Since material properties are of direct interest in the process and are essential to obtaining accurate results from the mathematical and numerical modeling, it is necessary to employ accurate, repeatable, and appropriate material property data. However, accurate property data over the range of interest of the independent variables are often not available. In many cases, these data are available at atmospheric pressure and room temperature that could be significantly different from those encountered in the actual process. The dependence on temperature, pressure, and other variables may not be explicitly known. Then, the results of the simulation could be saddled with significant error, making it difficult to base the design and optimization on the numerical results. Sensitivity analysis is often carried out by varying different properties to isolate the dominant ones that could lead to significant error. Properties may also change over time, a common occurrence in polymers, chemicals, and food materials. Therefore, the properties may have to be experimentally measured under realistic and applicable conditions in order to have the necessary confidence in the numerical results. Similar concerns arise for experimental results on the manufacturing process or system.

Data on fluid viscosity are particularly important if material flow arises in the process, for example, in casting and optical fiber drawing. The viscosity data are needed near and above the softening point T_{melt} of silica glass used for optical fiber manufacture. The softening point is around 1900 K, and the variation of viscosity with temperature is critical for a study of the flow, which in turn determines the neck-down as well as the velocity and temperature fields in the fiber. This information can be used to determine possible defects in the fiber and its overall quality. Experimental measurements have indicated that the viscosity varies exponentially with temperature. From a curve fit of the limited data available, an equation for the kinematic viscosity ν is given for silica, in S.I. units, as (Fleming 1964; Myers 1989)

$$\nu = 4545.45 \exp\left[32\left(\frac{T_{\text{melt}}}{T} - 1\right)\right] \tag{2.4}$$

This equation has been used in several numerical simulations of glass flow in optical fiber drawing. Other properties, such as density, thermal conductivity, and specific heat, also vary with temperature, and these variations may also have to be considered, depending on the temperature range in the process. If the temperature range is relatively small, the properties may often be treated as constant at their average values over the temperature range. This considerably simplifies the analysis.

Another important consideration relates to radiation properties that usually depend on temperature, surface condition, composition, and wavelength. However, limited information is often available on the variation of these properties. In optical fiber drawing, for instance, various dopants like GeO_2 and B_2O_3 are often added to the glass to enhance the transmission characteristics of the fiber. The effect of different dopants on viscosity and radiation properties is important in the modeling, but is not accurately known over the temperature range typically encountered in this process.

If the density ρ can be taken as constant, the continuity equation simply becomes

$$\nabla . \bar{V} = 0 \tag{2.5}$$

The last term in Eq. (2.2) drops out and other simplifications arise, as mentioned later. Similarly, if the thermal conductivity k is constant, it can be taken out of the parentheses in the energy equation, making the corresponding terms linear. If the viscosity μ is a constant along with the density, the momentum equation becomes

$$\rho\frac{D\bar{V}}{Dt} = \rho\left(\frac{\partial\bar{V}}{\partial t} + \bar{V}.\nabla\bar{V}\right) = \bar{F} - \nabla p + \mu\nabla^2\bar{V} \tag{2.6}$$

Here, *D*/*Dt* is the particle derivative, which includes the transient variation and the convection term representing the variation along the flow, as given above (Batchelor 1967; Burmeister 1993). If viscosity is not constant, the equation becomes much more involved, as given by the following *x*-component of the momentum equation:

$$\begin{aligned}\rho\left(\frac{\partial u}{\partial t} + \bar{V}.\nabla u\right) &= F_x - \frac{\partial p}{\partial x} + \frac{\partial}{\partial x}\left(2\mu\frac{\partial u}{\partial x} - \frac{2}{3}\mu\nabla\cdot\bar{V}\right)\\ &+ \frac{\partial}{\partial y}\left[\mu\left(\frac{\partial v}{\partial x} + \frac{\partial u}{\partial y}\right)\right] + \frac{\partial}{\partial z}\left[\mu\left(\frac{\partial w}{\partial x} + \frac{\partial u}{\partial z}\right)\right]\end{aligned} \tag{2.7}$$

These equations are considered again later with respect to various materials processing circumstances, along with additional complexities that may arise for different processes.

Usually, the dependence of the properties on temperature *T* is the most important effect. Numerical curve fitting may be employed to obtain a given material property as a function of *T*, as say, $k(T) = k_r\,[1 + a(T - T_r) + b(T - T_r)^2]$, where T_r is a reference temperature at which $k = k_r$. Thus, a continuous function *k*(*T*) is used to replace the discrete data on *k* at different temperatures. This gives rise to nonlinearity since

$$\frac{\partial}{\partial x}\left[k(T)\frac{\partial T}{\partial x}\right] = \frac{\partial k}{\partial x}\frac{\partial T}{\partial x} + k\frac{\partial^2 T}{\partial x^2} = \frac{\partial k}{\partial T}\left(\frac{\partial T}{\partial x}\right)^2 + k(T)\frac{\partial^2 T}{\partial x^2} \tag{2.8}$$

Figure 2.1 shows the temperature dependence of the thermal conductivity *k* of selected solids and liquids. Clearly, depending on the temperature range in a given process, the variation in *k* could be quite significant.

The preceding equations assumed the material to be isotropic, i.e., *k* to be the same in all directions, as in gases, most liquids, and amorphous solids. Then, the heat flux $q'' = -k\nabla T$. However, many materials such as wood, fibrous materials, and many crystalline substances are orthotropic, with different conductivities in different coordinate directions, with the surfaces normal to these directions. Then the conduction equation becomes, for the Cartesian coordinate system,

$$\rho C_p\frac{\partial T}{\partial t} = \frac{\partial}{\partial x}\left(k_x\frac{\partial T}{\partial x}\right) + \frac{\partial}{\partial y}\left(k_y\frac{\partial T}{\partial y}\right) + \frac{\partial}{\partial z}\left(k_z\frac{\partial T}{\partial z}\right) + \dot{Q} \tag{2.9}$$

where the heat flux $q_x'' = -k_x\partial T/\partial x$ and so on for other heat flux components. Similarly, anisotropic materials in which the heat flux in a given direction depends on the temperature gradients in all the directions may be treated.

Besides the properties related to thermal transport, mass transfer, and fluid flow, there are several other properties that are important in materials processing and affect the modeling, as well as the experiments. These include yield point, ultimate stress,

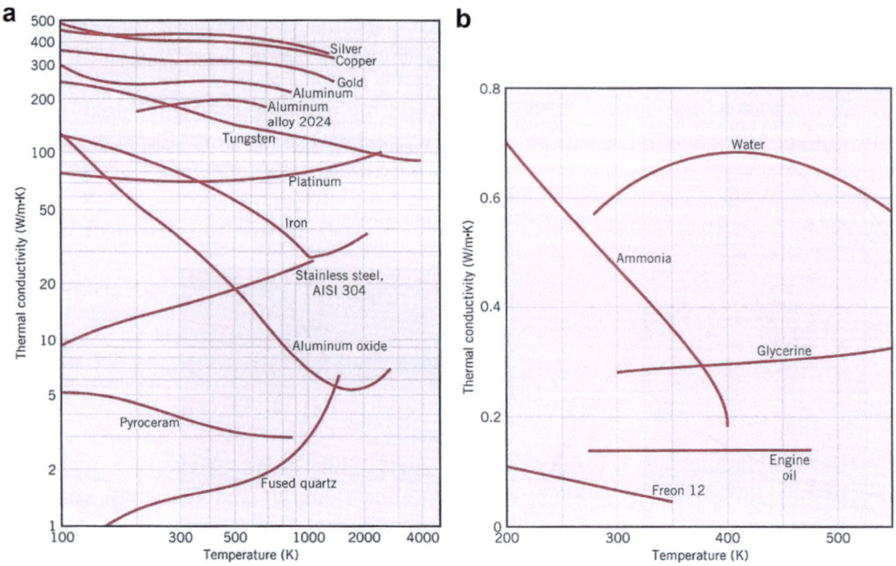

Fig. 2.1 The temperature dependence of the thermal conductivity of various (**a**) solids and (**b**) liquids under saturated conditions (Bergman et al. 2011)

stress-strain relationship, softening point, glass transition temperature, melting point, etc.

The data on other material properties may similarly be represented by appropriate curve fits. Because of the resulting additional nonlinearity, the solution of the equations becomes more involved than that for the constant property circumstance. Iterative numerical procedures are often required to deal with such nonlinear problems. Therefore, average constant property values at appropriate reference conditions are frequently employed to simplify the solution. However, this approach is satisfactory only for small ranges of the process variables. Most manufacturing processes require the solution of the full variable-property problem for accurate predictions of the resulting transport.

Clearly, material properties and their variation with the independent variables like temperature and pressure are important in the modeling and experimental study of materials processing. It is critical to obtain accurate property data over the range of operation and to include the variation of these properties, if significant.

2.3.2 *Viscous Dissipation*

Many materials of engineering interest have large viscosity, typically on the order of a million or more times that of water. These include materials like plastics, polymer composites, and glass. The large viscosity could give rise to significant viscous dissipation in the flow. As seen earlier in the energy equation, Eq. (2.3), an additional term $\mu\Phi$ arises. Since this term represents the irreversible part of work done by the shear stress, it behaves as a thermal source in the flow and affects the temperature distribution. For a Cartesian coordinate system, the viscous dissipation function Φ is given for a two-dimensional flow by the following expression:

$$\Phi = 2\left[\left(\frac{\partial u}{\partial x}\right)^2 + \left(\frac{\partial v}{\partial y}\right)^2 + \left(\frac{\partial w}{\partial z}\right)^2\right] + \left(\frac{\partial v}{\partial x} + \frac{\partial u}{\partial y}\right)^2 + \left(\frac{\partial w}{\partial y} + \frac{\partial v}{\partial z}\right)^2$$
$$+\left(\frac{\partial u}{\partial z} + \frac{\partial w}{\partial x}\right)^2 - \frac{2}{3}(\nabla.V)^2 \tag{2.10}$$

It is evident that the expression is always positive, and it thus provides an energy input to the flow. The corresponding expressions for other coordinate systems may similarly be written. Viscous dissipation is particularly important in flows through narrow regions, such as extrusion dies and microchannels.

Figure 2.2 shows the temperature profiles that arise in the flow between two parallel plates at a separation of $h = 2L$, with temperatures T_1 and T_0 of the upper and lower plates, respectively. In the first case, the upper plate moves at a given velocity U, and, in the second case, the plates are stationary, and the flow is driven by a pressure gradient down the channel. E is the Eckert number, where $E = U^2/[C_p\,(T_1 - T_0)]$ in the first case. The U is replaced by u_m, the average velocity in the second case. Pr is the Prandtl number, $Pr = \nu/\alpha$, where $\alpha = k/\rho C_p$ is thermal diffusivity. Clearly, the temperatures can exceed the imposed temperature at the upper wall, depending on the parameter $E.Pr$. The upper wall may lose energy, be adiabatic, or gain energy from the fluid. Viscous dissipation plays an important role in polymer processing and in optical fiber drawing. In plastic extrusion, for instance, the temperature in the flow can exceed the imposed temperature at the wall due to viscous dissipation, as seen later. Heat may have to be removed in order to maintain the wall at a given temperature (Jaluria 1996). Similarly, in optical fiber drawing, viscous dissipation becomes important as the region narrows sharply to the fiber diameter. In such processes, viscous dissipation must be included in the analysis to predict the temperatures, the associated flow, and material characteristics accurately.

2.3.3 Buoyancy Effects

There are several processes, where buoyancy-induced flows are generated due to temperature or concentration differences. These flows, if significant, can affect the rate of production as well as the quality of the product. Processes such as casting, crystal growing, welding, and soldering involve flows in the melt that could affect the interface between the solid and the liquid. This, in turn, could impact on the product characteristics and the rate at which solidification occurs. Similarly, the flow in a chemical vapor deposition (CVD) reactor is important for the deposition

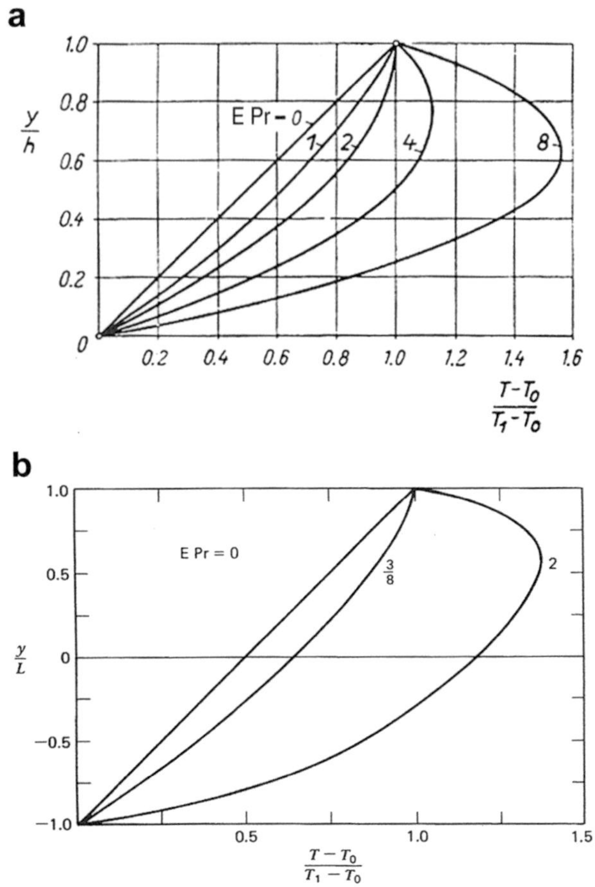

Fig. 2.2 Dimensionless temperature variation in Couette flow and in the flow between two stationary parallel plates, including the effect of viscous dissipation (Burmeister 1993)

process, and if the buoyancy effects due to the heated susceptor, or surface where deposition occurs, are significant, they could affect the uniformity and rate of deposition. Therefore, in such processes, the buoyancy term must be included in the momentum equations.

The buoyancy term is obtained by considering the body force, such as gravity, and the hydrostatic pressure. A coupling of the pressure and body force terms in the momentum equation gives the buoyancy force and a term, known as the motion pressure term, which represents the pressure due to the flow (Gebhart et al. 1988). Figure 2.3 shows the buoyancy force arising in a quiescent ambient medium due to density difference in a gravitational field and examples of buoyancy-driven external and internal flows. The Boussinesq approximations, which neglect density variation in the continuity equation and assume a linear variation of density with temperature or concentration, are applicable for relatively small temperature or concentration differences. For thermal transport, if Boussinesq approximations are invoked, the following terms are obtained for the buoyancy effects:

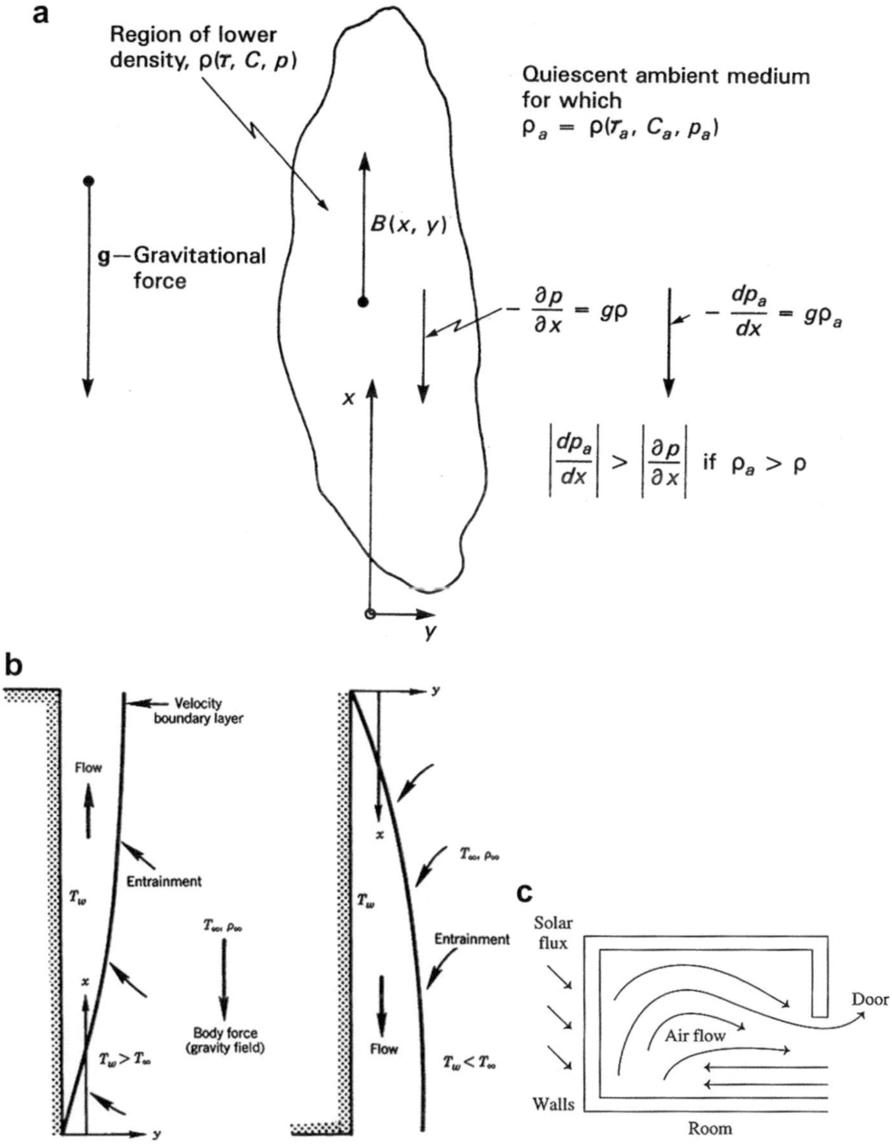

Fig. 2.3 (**a**) Buoyancy force arising due to density difference in a quiescent ambient medium; (**b**) buoyancy-driven flow over a vertical surface for a heated or cooled surface; (**c**) flow driven by buoyancy in a room with an opening

$$\bar{F} - \nabla p = -\bar{e}\, g\rho\beta(\,T - T_{\mathrm{a}}\,) - \nabla p_{\mathrm{d}} \tag{2.11}$$

Here, p_{d} is the motion pressure due to the flow, $\bar{e}$ is unit vector in the direction of the gravitational force, and g is the magnitude of gravitational acceleration. The buoyancy force, which is the first term on the right-hand side, leads to the Grashof number, $Gr = g\beta(T - T_{\mathrm{a}})L^3/\nu^2$, which represents the ratio of buoyancy force to viscous force. Here, L is the height and T_a the ambient temperature. If the temperature or concentration differences are large, the Boussinesq approximations are not applicable. This is frequently the case in materials processing. Then the variable density effect is considered in the continuity equation, and the actual density differences are employed to obtain the buoyancy force. The buoyancy term couples the flow with the energy equation, and the two have to be solved concurrently, unlike common forced convection problems at constant density where the flow is often independent of the temperature distribution.

2.4 Additional Transport Mechanisms and Considerations

2.4.1 Non-Newtonian Fluids

These are fluids whose dynamic viscosity μ is dependent on the shear rate and, thus, on the flow. Materials such as plastics, food, rubber, and several oils that are of interest in a variety of manufacturing processes are non-Newtonian. Figure 2.4 shows the typical behavior of non-Newtonian fluids in terms of the shear stress τ_{yx} versus shear rate *du/dy* graphs for a shear flow such as the flow between two parallel plates with one plate moving at a given speed and the other held stationary. The dynamic viscosity is independent of the shear rate for Newtonian fluids like air and water but increases or decreases with the shear rate for shear thickening or thinning fluids, respectively. These are viscoinelastic (purely viscous) fluids, which may be time-independent or time-dependent, the shear rate being a function of both the

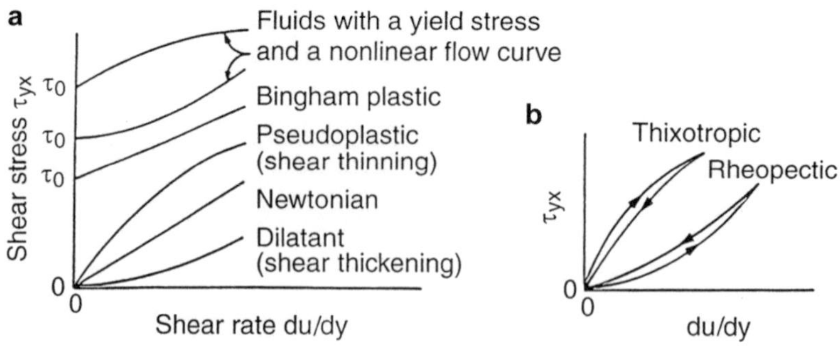

Fig. 2.4 Shear stress variation with shear rate for purely viscous, viscoinelastic, non-Newtonian fluids. (**a**) Time-independent and (**b**) time-dependent fluids. (Adapted from Skelland (1967))

magnitude and the duration of shear in the latter case. The flow dependence of the viscosity significantly complicates the solution for the flow and heat transfer.

In many materials processing applications, the fluid is treated as a Generalized Newtonian fluid, with the viscosity given in terms of the shear rate $\dot{\gamma}$ and temperature T, since the viscosity is also generally a strong function of the temperature (Tadmor and Gogos 1979; Jaluria 1996). A typical constitutive equation in terms of a reference point, given by subscript o, may be written in terms of the power-law index n and temperature coefficient b, as

$$\mu = \mu_o \left[\frac{\dot{\gamma}}{\dot{\gamma}_o}\right]^{n-1} \exp[-b(T - T_0)] \tag{2.12}$$

Other such equations are discussed later for a variety of manufacturing processes.

In addition, viscoelastic behavior, in which the material exhibits both viscous and elastic characteristics during deformation, must also be modeled in some cases, particularly for flow emerging from a die. Viscoelastic fluids show partial elastic recovery on the removal of a deforming shear stress. Food materials are often viscoelastic in nature. Various other approaches have been developed to model different types of non-Newtonian materials. However, despite the critical importance of accurate material property data in manufacturing, usually, only limited data are available in the literature, and the dependence on temperature, concentration, shear rate, etc. is not known accurately.

2.4.2 Surface Tension Effects

Surface tension plays an important role in many materials processing techniques, particularly at small length scales. The flow of molten metal in soldering, as shown in Fig. 1.1c and in Fig. 2.5a, and in welding, as shown in Fig. 2.5b, is largely driven by surface tension. Similarly, surface tension is important in the collapse of the holes in manufacture of hollow and microstructured fibers because of typical diameters of a few microns. Surface tension effects become particularly important in processes such as crystal growing and solidification under microgravity conditions, such as those in space, due to the reduction in the buoyancy force. Marangoni convection that arises due to the variation of surface tension with temperature and concentration is important in the movement of impurities and species. A sketch of such thermocapillary flows is shown in Fig. 2.5c.

For the rectangular container shown in Fig. 2.5c, with its left wall at temperature T_{L}, the right wall at a lower temperature T_{R}, and the bottom insulated, the boundary condition at the free surface is

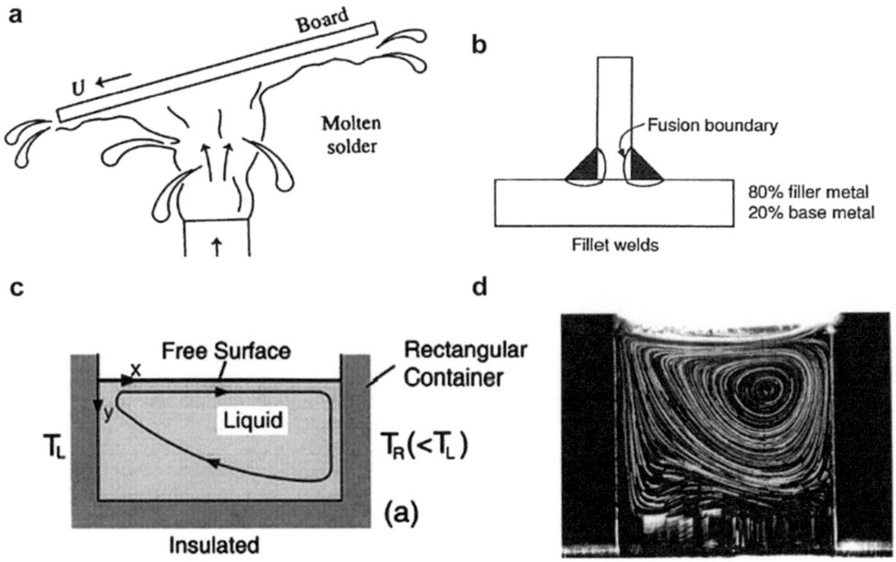

Fig. 2.5 Surface tension effects: (**a**) sketch of the soldering process in the manufacturing of electronic circuitry using wave soldering for joint formation and plating of a moving circuit board; (**b**) a welded joint; (**c**) sketch of thermocapillary convection in a rectangular container; (**d**) observed flow in a $NaNO_3$ melt (Kou 1997)

$$-\mu \frac{\partial u}{\partial y} = \frac{\partial \sigma}{\partial T} \frac{\partial T}{\partial x} + \frac{\partial \sigma}{\partial c_m} \frac{\partial c_m}{\partial x} \tag{2.13}$$

where u is the velocity component along the coordinate axis x. For most pure materials, σ decreases with T, i.e., $\partial\sigma/\partial T < 0$, and since $\partial T/\partial x < 0$ in this case, the fluid is pulled from the left to the right at the surface, resulting in a clockwise circulation, as shown. The flow pattern in a melt of $NaNO_3$ is shown in Fig. 2.5d, indicating the dominance of thermocapillary convection and the vertical flow near the vertical wall due to thermal buoyancy. Similarly, boundary conditions may be written for other geometries. Recent years have seen growing activity in thermocapillary convection due to interest in high product purity and advanced materials processing.

2.4.3 Particulates and Powdery Materials

In various manufacturing processes, the movement and treatment of particles and powders are important. For instance, in spray coating, sputtering, and thin film deposition, the deposition of particles is critical to the fabrication of the final product. Particle motion is also important in mixing and impurity migration in processes like polymer processing and casting. The motion of the particles, which are driven by the flow, is investigated and their trajectories obtained, often using a Lagrangian approach, to characterize the process. Powders are encountered in several materials processing applications, such as powder metallurgy and the processing of food and pharmaceutical materials. Powders are conveyed along channels in these processes with compaction resulting from the imposed or generated pressure and heating due to friction or energy input. The transport processes associated with these materials, as well as the compaction process, are not very well understood. However, recent efforts have focused on these materials because of their practical importance in areas like pharmaceutical engineering.

2.4.4 Phase Change

As mentioned earlier, many materials processing techniques, such as crystal growing, casting, and welding, are based on phase change. Since interest lies in the rate of fabrication as well as in the quality of the product, it is important to study the phase change process and vary the governing parameters to obtain the desired results. There are two main approaches for modeling phase change. The first one treats the two phases as separate, with their own properties and domains, as shown in Fig. 2.6. The location and geometry of the interface between the phases must be determined for each time step or iteration, so that conservation principles may be applied there. For numerical simulation, appropriate discretization of the two regions is carried out (Ramachandran et al. 1982; Bennon and Incropera 1988). This approach is particularly suitable for pure materials such as metals, for which the two phases can be defined clearly and accurately.

The second approach, known as the enthalpy model, essentially smears out the discrete phase transition that occurs in a pure material. However, the numerical modeling is much simpler. The same governing equations are employed over the entire computational domain, and the interface between the two phases is not needed for the application of boundary conditions (Viswanath and Jaluria 1993; Prescott and

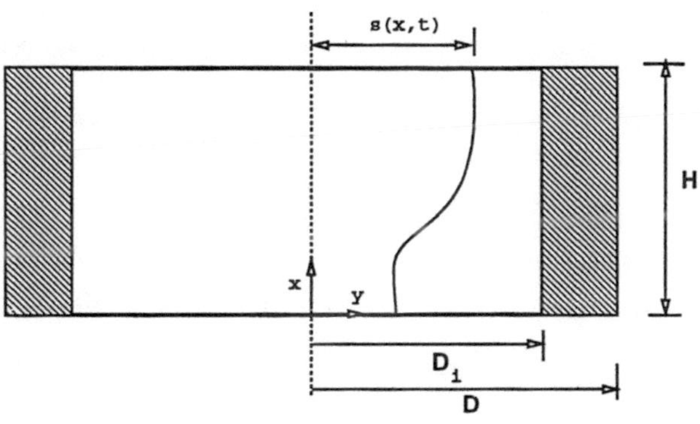

Fig. 2.6 Configuration of a cavity with melting or solidification, indicating solid-liquid interface

Incropera 1996). Also, mixtures such as alloys, impure materials, and polymers that do not have a well-defined melting point can be treated easily by this approach.

2.4.5 Chemically Reactive Materials

Chemical reactions, combined heat and mass transfer, and chemical structural changes, such as chemical conversion, are important in many materials processing circumstances, such as chemical vapor deposition, food processing, manufacturing involving reactive polymers, chemical bonding, and other processes where multiple species arise. For example, single or twin screw extrusion, as sketched in Fig. 1.2d and in Fig. 2.7a, is an important manufacturing technique for food processing, particularly for products like snacks, cereals, pasta, and bread sticks. Materials such as wheat flour, rice flour, and corn meal are fed into the hopper of the extruder along with a chosen amount of water. The dough is kneaded by the rotating screws and cooked or chemically converted due to shear and heat to obtain different extruded products (see, for instance, Harper (1981), Kokini et al. (1992), and Forte and Young (2016)). Chemical reactions occur in such chemically reactive materials to substantially change the structure, properties, and characteristics of the product. These are reflected in the flavor, texture, softness, and other attributes of the extruded food material. Similarly, changes due to chemical reactions are important in the curing of polymers, for example, in surface coating and chemical bonding. The chemical reactions may be triggered by heat, shear, ultraviolet radiation, and other means of energy input.

Chemical reactions are important in many other manufacturing processes. The deposition of a solid material film from the gas phase in chemical vapor deposition systems is the result of chemical reactions in the gases as well as at the surface (Jensen et al. 1991; Mahajan 1996). The temperatures and concentrations of the different chemical species, which include carrier gases as well as precursors for the material to be deposited, affect the chemical kinetics, which in turn affect the deposition. In many cases, the process is limited by the chemical kinetics, implying that the transport processes are quite vigorous and the deposition is largely restricted by the chemical reaction kinetics. In other cases, the process is limited by the heat and mass transfer rates. The chemical kinetics for the precursors employed for depositing several materials are available in the literature. For instance, the chemical kinetics for the deposition of silicon from silane (SiH_4), with hydrogen as the carrier gas, in a CVD reactor is given by the following simple expression (Jensen and Graves 1983):

$$K = \frac{K_o p_{\mathrm{SiH_4}}}{1 + K_1 p_{\mathrm{H_2}} + K_2 p_{\mathrm{SiH_4}}} \tag{2.14}$$

where the surface reaction rate K is in mole of Si/m^{2}s, $K_0 = A \exp(-E/RT)$, E is the activation energy, and A, K_1, and K_2 are constants, which are obtained

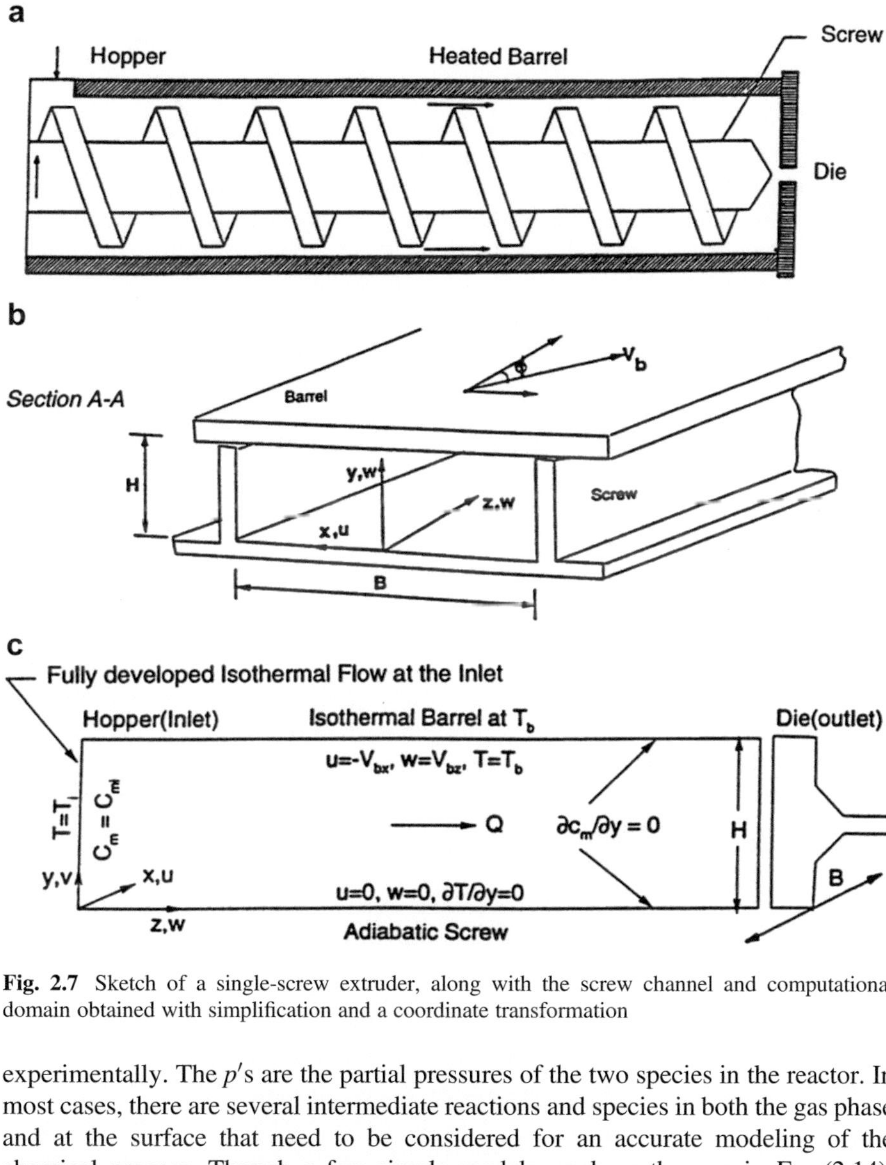

Fig. 2.7 Sketch of a single-screw extruder, along with the screw channel and computational domain obtained with simplification and a coordinate transformation

experimentally. The p's are the partial pressures of the two species in the reactor. In most cases, there are several intermediate reactions and species in both the gas phase and at the surface that need to be considered for an accurate modeling of the chemical process. Though a few simple models, such as the one in Eq. (2.14), have been used for analysis in a few cases, accuracy demands the consideration of the much more complicated problem with intermediate reactions. These aspects are considered later for the CVD deposition process.

2.4.6 Complicated Geometry

In manufacturing and materials processing systems, we frequently encounter complicated geometries and configurations. For instance, in single-screw extrusion, the geometry is usually quite complicated. The modeling is further complicated by the rotation of the screw, as shown in Fig. 2.7a. By locating the coordinate system on the rotating screw and neglecting curvature effects, a steady channel flow with the barrel moving at the pitch angle, as shown in Figs. 2.7b, c, is obtained (Karwe and Jaluria 1990). A simple rectangular screw is shown here, but similar simplifications and transformations can be made for more complicated ones used in several practical screw extruders. Even for relatively simple geometries, large changes in dimensions could arise as is the case in optical fiber drawing where the preform diameter is of the order of 10 cm and the fiber diameter around 125 μm. The liquid region in casting, as solidification proceeds, is quite irregular in shape. Casting and injection molding may also involve complicated shapes and geometries. Sometimes, transformations can be used to simplify the domain. For instance, Landau's transformation (Landau 1950) can be used for the neck-down region in optical fiber drawing to convert it to a cylindrical one. The resulting grid for the two regions, glass and inert gas, obtained from such a transformation is shown in Fig. 2.8. Further discussion on this problem is given in a later chapter. If such transformations cannot be used to simplify the domain, the actually geometry and dimensions have to be employed, with the discretization chosen to capture the complicated geometry. Finite element methods are particularly suitable for the generally intricate configurations encountered in many practical manufacturing problems.

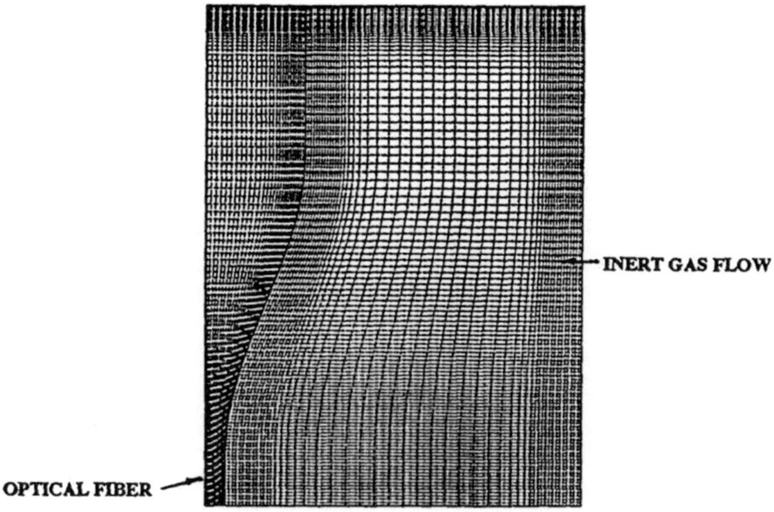

Fig. 2.8 Finite volume grid in actual physical geometry for the optical fiber drawing process obtained from a coordinate transformation, which converts axisymmetric complex shapes to cylindrical ones

2.4.7 Combined Mechanisms

Combined transport mechanisms arise in many cases, such as combined radiation and convection operating at the heated surfaces or in participating media like glass. Combined heat and mass transfer has been mentioned earlier for processes such as chemical vapor deposition, food processing, and drying. For these circumstances, the species conservation equations must be solved for all the species that arise, along with the equations for the flow and temperature, to determine the concentration C_i or the mass fraction ω_i. A simplified form of the species equation may be written in terms of the summation notation as

$$\frac{\partial\left(\rho u_j \omega_i\right)}{\partial x_j} = \frac{\partial}{\partial x_j}\left(\rho D_{ij}\frac{\partial \omega_i}{\partial x_j}\right) \tag{2.15}$$

where x_j represents the coordinate axes and D_{ij} the diffusion coefficient for species i diffusing into the species j. A source or sink term may be included if chemical reactions lead to the species being generated or removed. Energy release and absorbed must also be accounted for. Phase change processes like casting involve convection driven by temperature and concentration differences in the melt, as well as conduction in the mold and in the solidified material. This gives rise to a conjugate heat transfer problem. The governing equations have to be modified to include such combined mechanisms that arise in the process.

2.4.8 Complex Boundary Conditions

The initial and boundary conditions, resulting from the operating conditions in materials processing, are generally quite complicated. A few of these are briefly discussed here. Combined transport mechanisms often operate at the boundaries, as mentioned above. Also, free boundary surfaces are important in processes like optical fiber drawing, where the neck-down is given by a free surface. The profile of this surface is an important part of the solution. The feasibility of the process is determined by the stability of the neck-down profile, which is determined by various forces such as viscous, gravitational, surface tension, and shear forces acting on the fiber. In welding and soldering, free surfaces arise and need to be determined to obtain the shape and characteristics of the final joint. Moving surfaces are also quite common, as is the case in hot rolling and extrusion, and the appropriate boundary conditions must be imposed. Flow through openings need to be considered in some cases as in sprays and dies. Coupled conduction and convection, or conjugate conditions, must be modeled in many cases, such as extrusion due to conduction in the barrel and in the screw. Various other interfaces and moving boundaries occur in phase change processes and have to be addressed (Shyy et al. 1996). It is critically important to impose the boundary conditions accurately since the results obtained are strongly dependent on the transport at the boundaries.

2.5 Other Challenges

In the preceding sections, several complexities and challenges that are encountered in the modeling and simulation of manufacturing processes have been outlined. Many of these considerations also affect the experimental study of these processes. However, there are several other challenges that arise in different processes and need special treatment. Some of these will be considered in detail later, such as multiscale modeling, validation of the mathematical/numerical model, additional mechanisms, coupling of product characteristics with the transport processes, inverse problem to choose the operating conditions for desired product characteristics, uncertainties in the boundary conditions, and system considerations. Other challenges may arise, particularly with new and emerging manufacturing processes. These have to be overcome to obtain accurate results that form the basis for system design, prediction, control, and optimization. The material presented in this book can be used as the basis for meeting these challenges.

References

G.K. Batchelor, *An Introduction to Fluid Dynamics* (Cambridge University Press, Cambridge, UK, 1967)

W.D. Bennon, F.P. Incropera, Developing laminar mixed convection with solidification in a vertical channel. J. Heat Transf. **110**, 410–415 (1988)

T.L. Bergman, A.S. Lavine, F.P. Incropera, D.P. Dewitt, *Introduction to Heat Transfer*, 6th edn. (Wiley, New York, 2011)

L.C. Burmeister, *Convective Heat Transfer*, 2nd edn. (Wiley Interscience, New York, 1993)

J.D. Fleming, Fused silica manual, Final Report for the U.S. Atomic Energy Commission, Oak Ridge (1964) TN, Project B-153

D. Forte, G. Young, *Food and Feed Extrusion Technology: An Applied Approach to Extrusion Theory* (Food Industry Engineering, Brisbane, 2016)

B. Gebhart, Y. Jaluria, R.L. Mahajan, B. Sammakia, *Buoyancy-Induced Flows and Transport* (Taylor and Francis, Philadelphia, 1988)

J.M. Harper, *Extrusion of Foods: Volume I* (CRD Press, Boca Raton, 1981)

Y. Jaluria, Heat and mass transfer in the extrusion of non-Newtonian materials. Adv. Heat Transf. **28**, 145–230 (1996)

Y. Jaluria, *Design and Optimization of Thermal Systems*, 2nd edn. (CRC Press, Boca Raton, 2008)

Y. Jaluria, Thermal issues in materials processing. J. Heat Transf.. **135**, 061701-1–14 (2013)

K.F. Jensen, D.B. Graves, Modeling and analysis of low pressure CVD reactors. J. Electrochem. Soc. **130**, 1950–1957 (1983)

K.F. Jensen, E.O. Einset, D.I. Fotiadis, Flow phenomena in chemical vapor deposition of thin films. Annu. Rev. Fluid Mech. **23**, 197–232 (1991)

M.V. Karwe, Y. Jaluria, Numerical simulation of fluid flow and heat transfer in a single-screw extruder for non-Newtonian fluids. Numer. Heat Transf. **17**, 167–190 (1990)

J. L. Kokini, M. V. Karwe, C. T. Ho (eds.), *Food Extrusion Science and Technology* (Marcel Dekker, New York, 1992)

S. Kou, *Transport Phenomena and Materials Processing* (Wiley, New York, 1997)

P.K. Kundu, I.M. Cohen, D.R. Dowling, *Fluid Mechanics*, 6th edn. (Academic Press, New York, 2015)

H.G. Landau, Heat conduction in a melting solid. Appl. Math. Q. **8**, 81–94 (1950)

R.L. Mahajan, Transport phenomena in chemical vapor-deposition systems. Adv. Heat Transf. **28**, 339–425 (1996)

M.R. Myers, A model for unsteady analysis of preform drawing. AICHE J. **35**, 592–602 (1989)

P.H. Oosthuizen, D. Naylor, *Introduction to Convective Heat Transfer Analysis* (McGraw-Hill, New York, 1998)

P.J. Prescott, F.P. Incropera, Convection heat and mass transfer in alloy solidification. Adv. Heat Transf. **28**, 231–338 (1996)

N. Ramachandran, J.P. Gupta, Y. Jaluria, Thermal and fluid flow effects during solidification in a rectangular enclosure. Int. J. Heat Mass Transf. **25**, 187–194 (1982)

W. Shyy, H.S. Udaykumar, M.M. Rao, R.W. Smith, *Computational Fluid Dynamics with Moving Boundaries* (Taylor & Francis, Philadelphia, 1996)

A.H.P. Skelland, *Non-Newtonian Flow and Heat Transfer* (Wiley, New York, 1967)

Z. Tadmor, C. Gogos, *Principles of Polymer Processing* (Wiley, New York, 1979)

R. Viswanath, Y. Jaluria, A comparison of different solution methodologies for melting and solidification problems in enclosures. Numer. Heat Transf. **24**, 77–105 (1993)

Chapter 3
Numerical Modeling and Experimentation

In the preceding chapter, the transport mechanisms that form the basis for materials processing were discussed. The governing equations, based on the conservation principles, were presented, along with many additional effects that are of particular interest in various manufacturing processes. The solution to these equations is needed in order to provide information on the behavior of the system and its dependence on the important parameters in the problem. The results also provide guidelines for choosing the various boundary conditions to obtain the desired product and the inputs needed for the design and optimization of the process. Though experimental results can be obtained in a few selected cases, analytical and numerical approaches, particularly the latter, are extensively used to obtain the desired information and inputs. This chapter considers analytical and numerical, as well as experimental, methods for studying materials processing systems.

3.1 Analytical Solution

As mentioned earlier, because of the complexity of typical manufacturing processes, analytical results are usually not obtained, and one has to resort to numerical modeling to obtain the desired results. However, analysis may sometimes be used with various simplifications and idealizations. In some cases, the model for a particular component or element of a manufacturing system can be simplified so that analysis may be used to study its behavior, while numerical methods are used for other more complex conditions and components. Analysis may also be used to indicate the general trends under limiting conditions and thus guide the numerical scheme. Analytical results are particularly useful for validating the model, this being a critical and necessary component of any modeling effort. Thus, analytical solutions, whenever possible, are valuable in the overall modeling process since they provide:

Y. Jaluria, *Advanced Materials Processing and Manufacturing*, Mechanical Engineering Series, https://doi.org/10.1007/978-3-319-76983-7_3

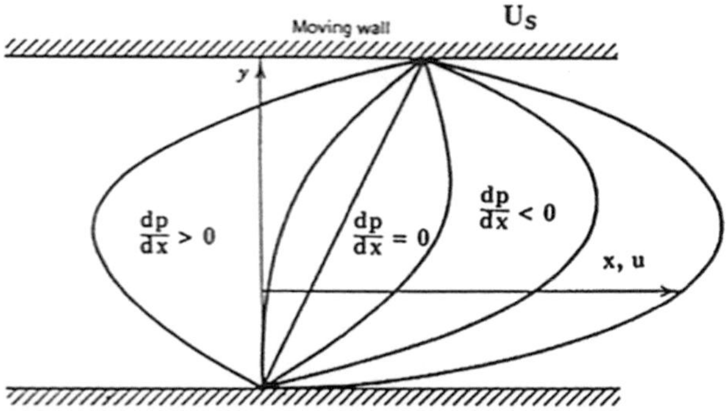

Fig. 3.1 Velocity profiles for developed flow of a Newtonian fluid in a channel of height H with combined effects of shear due to the upper wall moving at velocity U_s and an imposed pressure gradient dp/dx

1. Results that can be used for validating mathematical and numerical models
2. Physical insight into the basic mechanisms and expected trends
3. Limiting or asymptotic conditions
4. Quantitative results for certain simple components

The validation of the mathematical and numerical models, expected physical characteristics of the process, and limitations on important variables are all important aspects in the development of a numerical or experimental approach to study the process.

A few examples of analytical solutions may be given here. The complex flow in a screw extruder, such as the one shown in Fig. 2.7, was simplified to shear- and pressure-driven flow in a channel. The simplest case is that of fully developed flow for which the velocity field is assumed to remain unchanged downstream. Analytical solutions can be obtained for such channel flows driven by pressure and shear, as shown in Fig. 3.1 for Newtonian fluids. When the pressure gradient is zero, the flow is only due to the viscous effect of the moving upper wall and is termed drag flow. For Newtonian flow, the velocity profile is linear, and the dimensionless flow rate, or throughput, q_V, which is the volume flow rate divided by the product of wall speed and cross-sectional area, is simply 0.5. This flow is the well-known Couette flow between two parallel plates, with the upper wall moving at constant speed, in the absence of a pressure gradient. For a favorable pressure gradient, $dp/dx < 0$, the throughput exceeds 0.5, and for an adverse pressure gradient, $dp/dx > 0$, it is less. Similar trends are expected for non-Newtonian fluids though the profiles and the q_V value for drag flow would be different.

Another analytical solution that has been used to model the extrusion process is that of flow through a die. For a Newtonian fluid, the developed flow through a circular pipe is the well-known Poiseuille flow. Then, the axial velocity v and the relationship between the pressure drop Δp across a cylindrical region of length L and radius R, with mass flow rate $\dot{m}$, are given by (Burmeister 1993)

$$v = 2v_{av}\left(1 - \frac{r^2}{R^2}\right) \quad \text{and} \quad \Delta p = \frac{8m\dot{\nu}L}{\pi R^4} \tag{3.1}$$

where v_{av} is the average velocity. For a non-Newtonian fluid, the corresponding equation for Δp was obtained by Kwon et al. (1986) as

$$\Delta p = \frac{2L}{R}\tilde{C}(T)\left[\frac{3n+1}{4n}\frac{4\dot{m}}{\rho\pi R^3}\right]^n \tag{3.2}$$

where $\tilde{C}$ is a temperature-dependent coefficient in the viscosity expression, which is given as $\mu = \tilde{C}(T)(\dot{\gamma})^{1-n}$, n being the power-law index and $\dot{\gamma}$ the shear rate. For a Newtonian fluid, $n = 1$, this equation yields the results given in Eq. (3.1).

Another common example is the heat treatment of a small metal part, which may be modeled as lumped, i.e., no temperature variation within the component. The temperature variation with time for an object at initial temperature T_0 cooling or heating in a fluid at temperature T_a due to convection is then given by (Bergman et al. 2011)

$$T - T_a = (T_0 - T_a)\exp\left(-\frac{hAt}{\rho CV}\right) = (T_0 - T_a)\exp\left(-\frac{t}{t_r}\right) \quad \text{where} \quad t_r = \frac{\rho CV}{hA} \tag{3.3}$$

The temperature of a lumped mass cooling due to convection is sketched in Fig. 3.2. In this equation, h is the convective heat transfer coefficient, A the surface area, V the volume of the object, ρ the material density, and C the specific heat. The parameter t_r is known as the characteristic or response time. The response time is important in the mathematical modeling of time-dependent processes. With increasing t_r, the temperature variation of the body with time becomes slower, indicating a slower response.

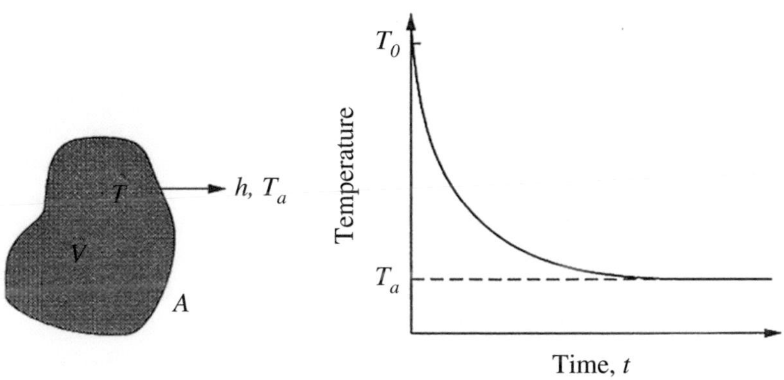

Fig. 3.2 Lumped mass approximation of a heated body undergoing convective cooling

If, in addition to convective transport, the mass is subjected to a heat flux input q'' at the surface, the temperature variation is obtained as

$$T - T_a = (T_o - T_a)\left\{\exp\left(-\frac{t}{t_r}\right) + \frac{q''}{h}\left[1 - \exp\left(-\frac{t}{t_r}\right)\right]\right\} \tag{3.4}$$

Thus, a steady-state temperature of $T_a + (T_o - T_a)q''/h$ is obtained. Similarly, other boundary and initial conditions may be considered, including energy generation in the lumped mass. In some cases, the material may be treated as semi-infinite, so that a one-dimensional, transient, problem is considered. This approximation can often be made for short times in manufacturing involving, for instance, laser heating and thermal processing at the surface.

Similarly, the flow velocity in a channel may be assumed to be uniform, thus allowing a simple analysis to be carried out to determine the temperature rise due to heat input. Thus, the axial variation of the bulk temperature T_m is given for constant heat flux q'' or for convective heating as

$$\frac{dT_m}{dx} = \frac{q''P}{\dot{m}C_p} \quad \text{or} \quad \frac{dT_m}{dx} = \frac{P}{\dot{m}C_p}h(T_a - T_m) \tag{3.5}$$

Here, P is the perimeter and C_p the specific heat at constant pressure. Therefore, the variation of T_m with the axial distance x may be determined. This approximation is often made for channel flows and for moving rods and plates undergoing materials processing. Similarly, several other assumptions and simplifications are often made in order to obtain analytical solutions, which are then coupled with numerical and experimental results for other components to complete the model for the entire system.

3.2 Numerical Modeling

It is necessary to develop a numerical model for components and processes that cannot be studied by using an analytical model, which is a common situation in most cases. A numerical model of the manufacturing system refers to a computational or numerical representation of the system on a computer, which may be used to approximate the behavior and characteristics of the system. It consists of a numerical scheme or procedure that would yield a solution to the governing mathematical equations, with numerically imposed boundary and initial conditions, relevant property data, component characteristics, and other inputs needed for representing the entire system. The numerical algorithm, as well as its implementation on a computer, constitutes the numerical model.

In the preceding chapter, the general governing equations were presented, along with a wide range of challenges that typically arise in materials processing (Jaluria 2013). However, a few additional aspects that are of particular relevance to the numerical modeling of materials processing systems are outlined here.

3.2.1 Accurate Imposition of Appropriate Boundary Conditions

The accuracy of any numerical simulation is strongly dependent on the imposition of realistic, appropriate, and accurate boundary conditions. Even though isothermal and uniform heat flux conditions at given surfaces are commonly used in fundamental studies, these conditions are seldom employed in practical circumstances. Also, the coupling between different components or parts of the system must be taken into account for a realistic model. For example, in modeling the solidification process in an enclosed region, using the multiple-domain or the enthalpy-porosity model, the coupled conduction, or conjugate transport, in the walls of the mold is an important consideration. The boundary conditions for heat removal are imposed at the outer surface of the mold, which must be taken as an integral part of the computational domain. The effect of the applied boundary conditions, which may vary with location and time, on the solidification process can be obtained by solving this conjugate problem, which yields the temperature distribution in the mold as well as that in the solid and the liquid. Figure 3.3 shows the effect of conduction in the mold on the resulting temperature and velocity distributions, as well as on the solidification. Thus, it is important to model the conjugate transport in the walls, insulation, and other boundary regions in many manufacturing systems in order to obtain realistic and accurate simulation results.

Other examples can be given where the proper imposition of boundary conditions is critical to an accurate numerical modeling of a practical system. In order to provide the correct energy transport at a given surface, it is preferable to use energy balance in the finite region that is representative of the surface, rather than the differential form of the boundary condition, which is based on the dimensions of this finite region approaching zero (Jaluria and Torrance 2003; Bergman et al. 2011). Figure 3.4 shows a sketch of typical finite regions that represent different grid points in the computational domain. Similarly, a force balance at a free surface, such as a meniscus, is appropriate for specifying the conditions at the surface. The numerical form of the energy, force, or mass balance allows one to ensure that the boundary conditions are realistic, accurate, and appropriate for finite regions and grids used for the numerical model. Further details on these aspects are given in later chapters.

In some cases, the boundary conditions may not be accurately known, and the solution of the entire system, with all its components, to obtain the relevant boundary condition may be prohibitive in terms of cost and effort. One such circumstance is the optical fiber drawing furnace, where the wall temperature distribution is a critical input to the process (see Fig. 3.5). But this distribution is not easily determined experimentally because of limited access to the furnace and modeling is complicated by the presence of many control and traverse subsystems in the draw furnace. An inverse calculation, using the limited temperature data obtained from a graphite rod immersed in the furnace, was used by Issa et al. (1996) to determine the wall temperature, which can then be used to accurately simulate the draw process. This approach is discussed in detail in a later chapter. Other cases where experimental data are used to obtain the relevant boundary conditions and thus proceed to the numerical solution are also common.

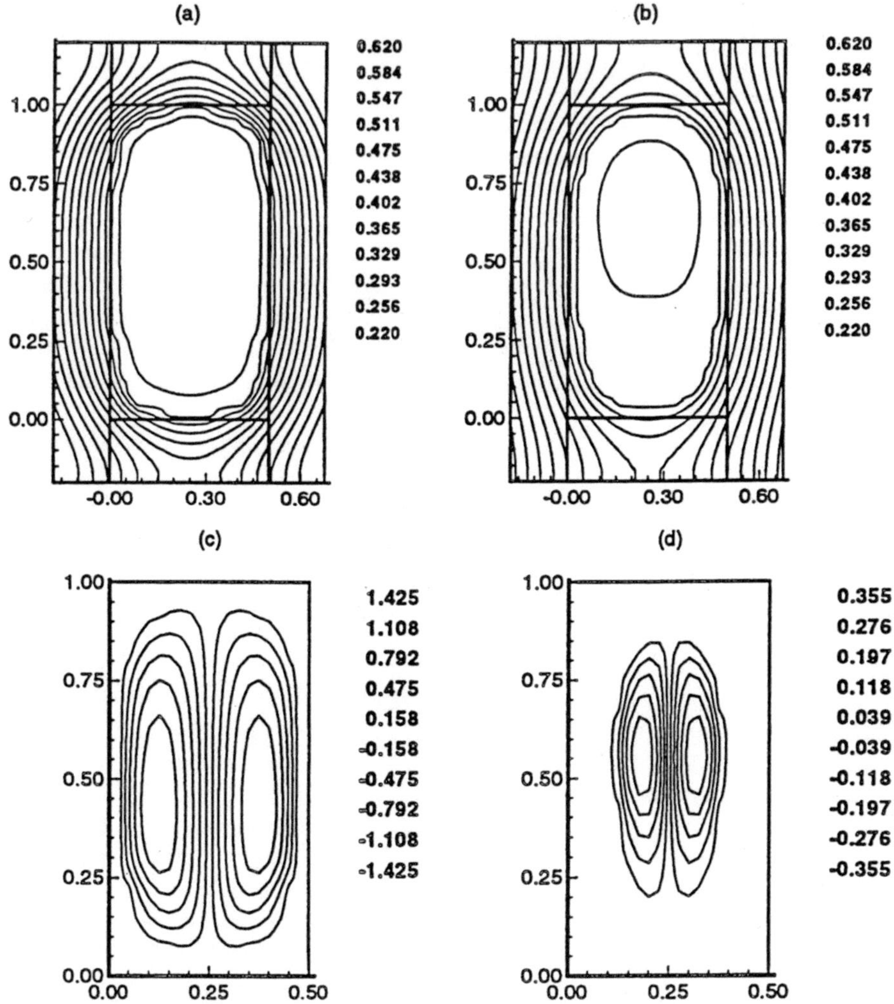

Fig. 3.3 Effect of conjugate boundary conditions on the flow and heat transfer in the solidification process in mold casting, shown in terms of isotherms in the top figures and streamlines in the bottom figures

Frequently, the system may involve openings through which flow may occur either due to an imposed pressure difference or due to flow driven by buoyancy. The former is modeled by a given flow rate or imposed pressure. As an example of the latter case, consider the numerical modeling of flow in an enclosure with an opening. The computational domain is generally extended, and the boundary conditions are imposed far away from the region of interest to ensure accurate simulation results, as sketched in Fig. 3.6 (Abib and Jaluria 1988). Such domain extensions are also useful to ensure accurate imposition of mass and energy inflow into given regions. Other

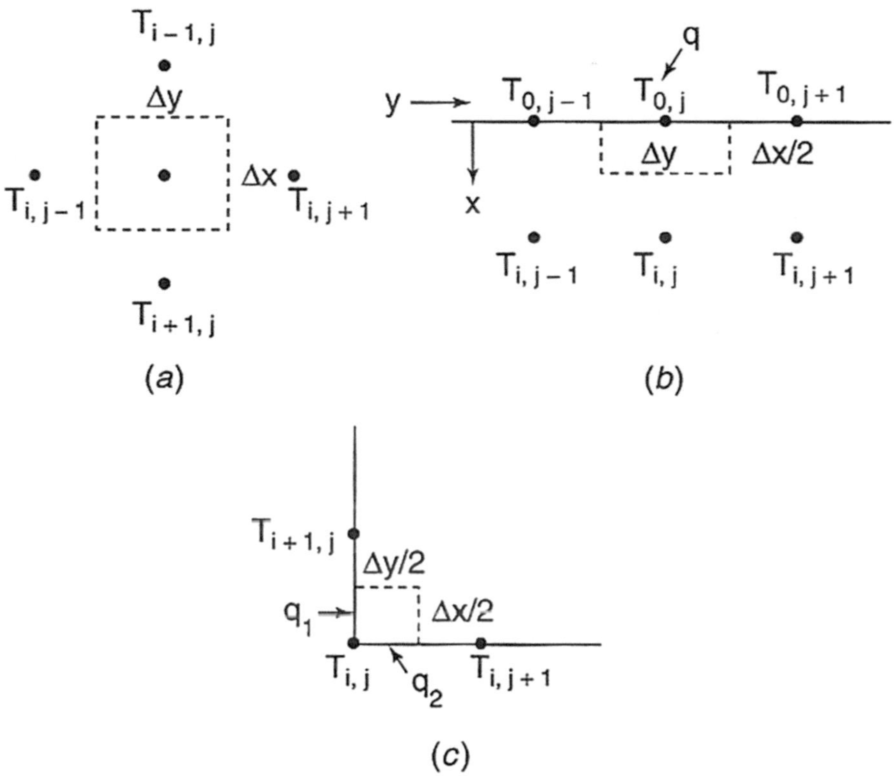

Fig. 3.4 Finite regions represented by (**a**) interior, (**b**) surface, and (**c**) corner grid points

such approaches have been used in practical manufacturing systems to obtain the relevant boundary conditions and thus accurately simulate the process.

3.2.2 Numerical Modeling of Complicated Regions

As mentioned earlier, complicated domains and geometries are frequently encountered in materials processing. The modeling of such regions is quite a challenge, and various approaches have been developed to obtain accurate results. Some of the common ones are:

Nonuniform discretization
Multizone models
Adaptive methods
Complex boundaries
Transformations
Finite volume and finite element methods

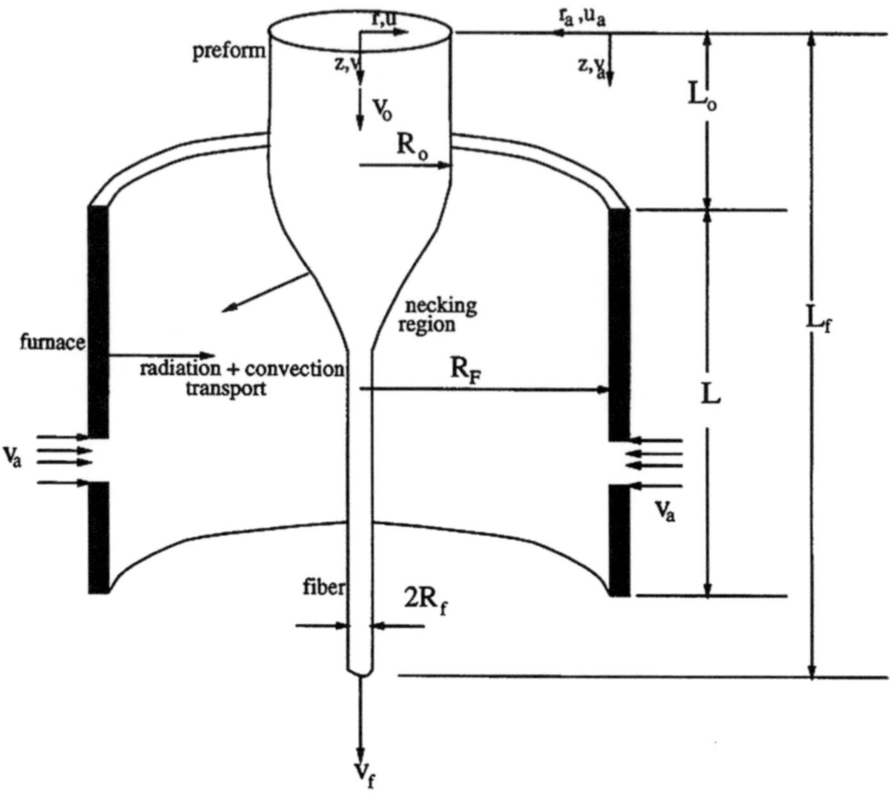

Fig. 3.5 Neck-down region in optical fiber drawing, indicating the important transport mechanisms

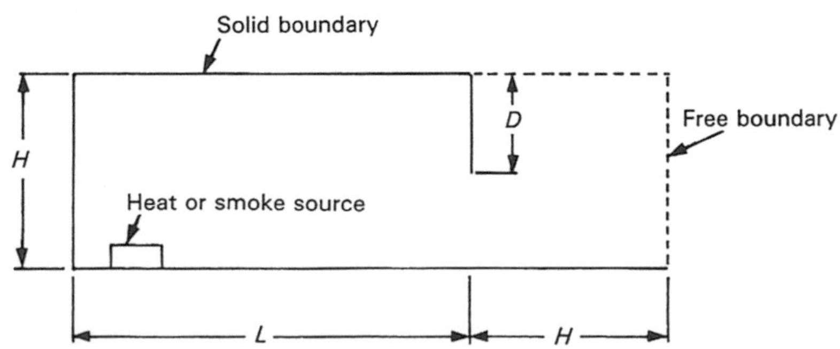

Fig. 3.6 Domain extension to impose the boundary conditions for buoyancy-induced flow in an enclosure with an opening

All these approaches will be discussed at various stages in the book with respect to different processes. Therefore, just a brief outline is given here. Because of large variations in the computed variables in different directions and subdomains,

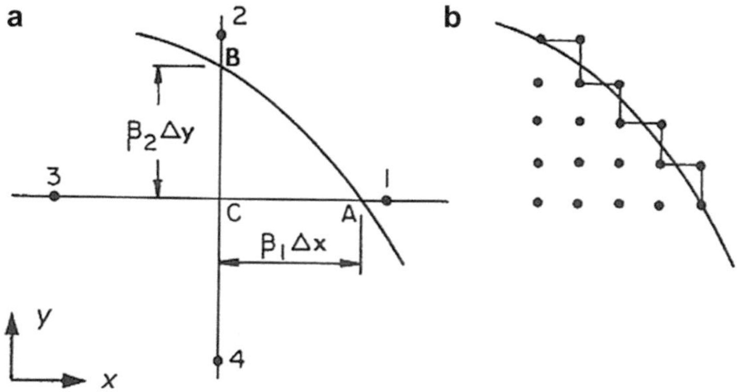

Fig. 3.7 (**a**) Grid points near a curved boundary, with a square or rectangular mesh in the conduction region. The boundary conditions are treated by deriving special equations for these points; (**b**) replacement of a curved boundary by steps formed from the grid employed

nonuniform, multizone, and adaptive approaches are employed. The grid could be nonuniform to capture the large gradients near the walls; for instance, different zones are discretized differently depending on expected variations due to, say, changes in material properties, and adaptive methods allow the discretization to be varied as the solution proceeds in order to focus on regions where large changes occur. Complex boundaries may be handled by deriving special equations at the grid points near the boundary, as shown in Fig. 3.7a, or by approximating the boundary in terms of the chosen grid, as sketched in Fig. 3.7b. Transformations allow complex domains to be converted to simpler domains, as sketched in Fig. 3.8 and as discussed in the previous chapter in terms of Landau's transformation. Then the transformed domain, being much simpler, may be handled with regular grids that yield body-fitted coordinates in the physical domain (Tannehill et al. 1997). This approach is of particular interest in processes like optical fiber drawing where the dimensions change substantially over a relatively short distance. Finally, finite volume and finite element methods allow considerable flexibility in approximating complex domains and are useful for many practical manufacturing problems. Figure 3.9 shows a sketch of the finite element grid employed for the complex shape of a particular single screw polymer extruder.

3.2.3 Modeling of Combined Mechanisms and Complex Transport Phenomena

In most thermal processing systems, coupled transport mechanisms generally arise and complicate the modeling. Conjugate transport was considered in the preceding section, where the effects of combined conduction and convection were discussed.

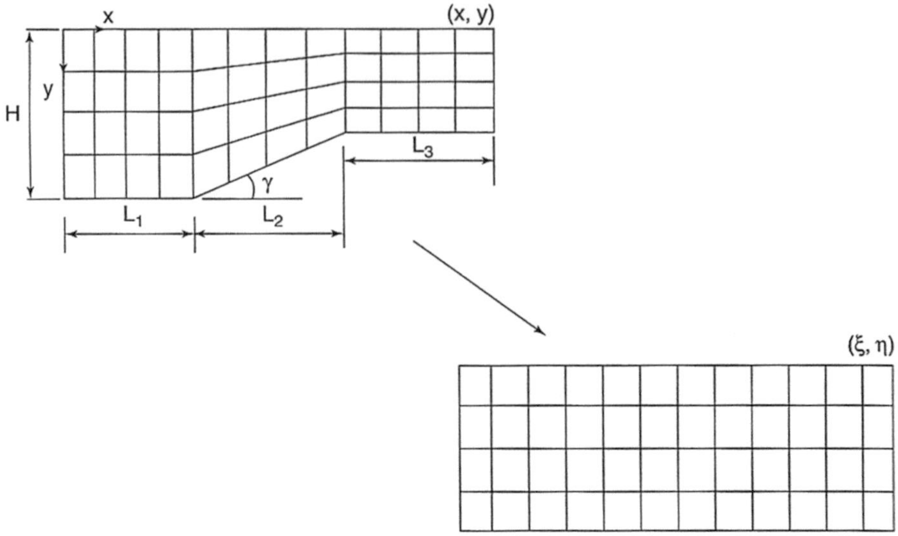

Fig. 3.8 Coordinate transformation to convert a two-dimensional region with varying cross-sectional area to a rectangular one

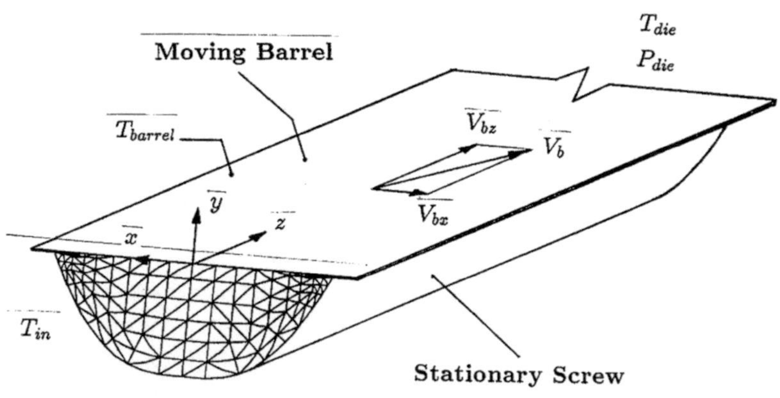

Fig. 3.9 Finite element grid for the complicated domain a single screw polymer extruder

Similarly, in the furnace for optical fiber drawing, thermal radiation and convection arise as coupled mechanisms, as shown in Fig. 3.5. Convection arises both in the inert gas environment and in the silica glass, which is a subcooled liquid. Beyond the softening point, T_{melt}, the glass is treated as a highly viscous liquid, with viscosity obtained from an equation such as Eq. (2.4). Below the softening point, the viscosity is very high, and the glass behaves essentially as a solid. Radiation is the dominant mode of transport, and the glass is largely heated up by radiation. Using radiation models such as the zonal method, the radiation transport in the glass as well as in the furnace is determined to obtain the energy absorbed. The modeling must include the combined mechanisms to determine the resulting transport, temperature variation, and the flow, as well as the free surface profile as the fiber is drawn from a cylindrical perform, which is several centimeters in diameter, to the fiber diameter of 125 μm (Li 1985).

Another manufacturing process that was mentioned earlier is the screw extrusion of reactive polymers like food materials. Figure 2.7 showed a sketch of the extruder, a simple mathematical model to obtain the domain as a channel and the boundary conditions for channel flow. The process involves convective combined heat and mass transfer, and the resulting product depends on the inlet and imposed concentration C as well as the temperature T. For food materials, the concentration of moisture is of particular importance (Kokini et al. 1992). The governing equations thus involve the flow equations along with the energy and mass transfer equations. Chemical reactions occur and give rise to source terms in the energy and mass conservation equations. The properties also vary with concentration, besides the temperature and the shear rate for non-Newtonian materials. In such combined transport mechanisms, it is necessary to model the different mechanisms accurately and to ensure that the coupling effects are not neglected. For instance, combined heat and mass transfer processes arise in chemical vapor deposition, as shown in Fig. 1.3c and described earlier. The coupling between the thermal and mass transport are given in terms of the Soret and Dufour effects that may not be negligible and need to be included (Mahajan and Wei 1991).

Also, as discussed earlier, most practical manufacturing processes involve significant additional effects, such as buoyancy, surface tension, and viscous dissipation that considerably complicate the transport phenomena being modeled. Free surfaces arise during processes like fiber drawing and coating. Figure 3.10 shows the upper and lower menisci that arise in wire or fiber coating by passing it through an applicator containing the coating liquid. The resulting shape of such free surfaces is governed by a balance of the forces due to shear, tension, gravity, and surface

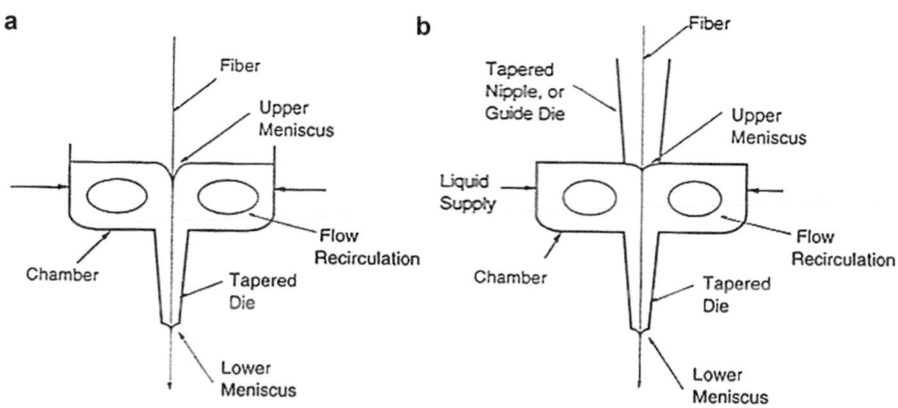

Fig. 3.10 Sketches of (**a**) open cup and (**b**) pressurized coating applicators

tension. Similarly, a force balance is used at interfaces in multilayered fibers, along with the conservation principles, to determine the resulting profiles. In hollow optical fiber drawing, a major concern is the collapse of the central core, which is needed for applications such as power delivery, sensors, and infrared radiation transmission. Because of the size of the core, surface tension is an important parameter and plays a very significant role in the collapse. The numerical model must address the combined and complex transport phenomena associated with the process.

Thus, a wide variety of additional effects often arise in practical processes and complicate the transport phenomena being mathematically and numerically modeled. It is generally best to consider all the additional effects that may arise and to carry out a detailed scale analysis to determine which ones need to be retained. Sensitivity analysis, based on a parametric study, can also be used to determine the dominant effects and parameters. Then the complex process, with the important additional effects, can be modeled.

3.2.4 Multiscale Modeling

A common challenge faced in the modeling of manufacturing processes is the presence of transport mechanisms and subsystems at different length or time scales. The governing equations for different scales in the problem may be different, and thus the numerical approaches used in different regions may be quite different. As an example, consider the numerical simulation of pressure-driven nitrogen slip flow in long microchannels, with conjugate heat transfer in the walls under a uniform heat flux wall boundary condition, as shown in Fig. 3.11. For the gas phase, the momentum and energy equations are solved, considering variable properties, rarefaction, which involves velocity slip, thermal creep and temperature jump, compressibility, and viscous dissipation. For conduction in the solid region, on the other hand, the energy equation is solved with variable properties. Thus, the two regions are treated with different approaches. The effect of substrate material and thickness

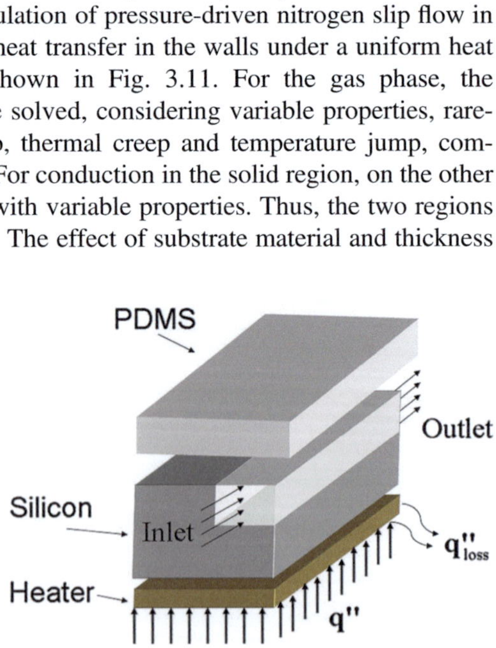

Fig. 3.11 Sketch showing different length scales due to flow in a microchannel embedded in a solid region for heat removal or delivery

on the thermal transport and on the temperature distributions is substantial. Therefore, a multiscale problem must be solved to obtain accurate results on this problem.

In materials processing, the characteristics and quality of the material being processed are generally determined by the transport processes that occur at the micro- or nanometer scale in the material, for instance, at the solid-liquid interface in crystal growing or at sites where defects are formed in an optical fiber. However, engineering aspects are generally concerned with the commercial or macroscale, involving practical dimensions, systems, and appropriate boundaries. Therefore, different scales arise and need to be solved by different methodologies, ultimately coupling the two to obtain the overall behavior (Jaluria 2014). Several examples that involve these considerations are seen later.

Similarly, in reactive thermal processing, such as food and reactive polymer extrusion, the microscopic changes in the material are linked with the operating conditions that are imposed on the system. The chemical conversion process is then represented by the chemical kinetics, which is dependent on the local temperature and concentration (Wang et al. 1989). These microscale conversion mechanisms are coupled with the flow and heat transfer in a screw extruder to obtain the conversion, pressure, and other important quantities.

3.3 Numerical Solution

The basic governing equations given earlier are the ones usually encountered in fluid flow and heat and mass transfer. Though additional complexities due to the geometry, boundary conditions, material property variations, combined mechanisms, etc., arise in materials processing, as discussed earlier, the numerical solution of the governing equations is based on the extensive literature on computational fluid mechanics and heat transfer (see, for instance, Roache (1976), Minkowycz and Sparrow (1997, 2001), and Minkowycz et al. (2006)). Among the most commonly employed techniques for solving these equations is the SIMPLER algorithm, given by Patankar (1980) and the several variations of this approach. This method employs the finite volume formulation with a staggered grid, so that the value of each scalar quantity such as pressure and temperature is associated with the grid node and the vector quantities like velocity are displaced in space relative to the scalar quantities and generally located on the faces of the control volume. This grid system has an advantage in solving the velocity field since the pressure gradients that drive the flow are easy to evaluate and the velocity components are conveniently located for the calculation of the convective fluxes. A pressure correction equation is used during the iteration or time marching to converge to the solution.

For two-dimensional and axisymmetric problems, the governing equations are often cast in terms of the vorticity and stream function by eliminating the pressure from the two components of the momentum equation and by defining a stream function to take care of the continuity equation (see Jaluria and Torrance (2003)). This reduces the number of equations by one, and pressure is eliminated as a variable, though it can be calculated after the solution is obtained. The solution yields the stream function, which is used for obtaining the velocity field and plotting streamlines; the temperature, which is used for plotting the isotherms and calculating heat transfer rates; and the vorticity. Because the stream function is specified on the boundaries, i.e., a Dirichlet problem is to be solved. Convergence of the stream function equation is usually faster than that of the pressure equation which involves gradient conditions. Thus, this approach is generally advantageous, as compared to the methods based on the primitive variables of velocity, pressure, and temperature, for two-dimensional and axisymmetric flows. The latter approach is more appropriate for three-dimensional circumstances.

Both transient and steady-state solutions are of interest, though the material undergoing processing is generally subjected to a time-dependent phenomenon. In the transient case, time marching is used with convergence at each time step to obtain the time-dependent variation of the flow, temperature field, heat and mass transfer rates, and transformation in the material, such as chemical conversion. For steady problems also, time marching may be used to obtain the desired results at large time. However, the problem can also be solved by iteration or by using false transients with large time steps, as presented by Mallinson and de Vahl Davis (1973). Though central differences are generally desirable for the approximations, numerical instability with the convection terms is often avoided by the use of upwind, exponential, or power-law differencing schemes, as discussed by Patankar (1980). Because of the inaccuracy due to false diffusion, second-order upwind differencing and third-order QUICK schemes have become quite popular for discretizing the convection terms (Leonard 1997). Under-relaxation is generally needed for convergence due to the strong nonlinearities that arise in these equations mainly due to convection and property variations. Several methods are available to solve the vorticity transport and energy equations. The alternating direction implicit (ADI) method of Peaceman and Rachford (1955) and modifications of this time-splitting method are particularly efficient for two-dimensional problems. Similarly, cyclic reduction, successive overrelaxation, and other standard methods may be used for the stream function or the pressure equation.

A wide range of numerical methods are available, based on finite difference, finite volume, finite element, spectral, and other approaches, to solve the differential equations that describe given materials processing techniques. These are generally solved using nonuniform grids, with finer grids employed in regions where large gradients are expected. Other discretization methods, including adaptive grids, meshless methods, boundary-fitted coordinates, and multigrid methods may be employed to solve the problem accurately and efficiently, depending on the problem at hand. The strong variation in material properties, like viscosity, necessitates fine grids, linearization, and iterative procedures, using a variety of available and modified numerical techniques. Finite element methods are particularly suited to complicated domains and complex boundary conditions. A wide range of commercial software, as well as those in the public domain, are available for numerical analysis. These include general purpose software for computational fluid dynamics (CFD) like ANSYS Fluent, CFD-ACE+, FIDAP, and OpenFOAM, as well as software to solve the governing mathematical equations like MATLAB, COMSOL Multiphysics, and Maple. Specialized software like Moldflow for injection molding are also available for various processes. The code may be developed, modified, or augmented by the user as well. For further details on the extensive work done on numerical methods for many different circumstances, the references mentioned earlier may be consulted.

3.4 Verification and Validation of the Model

As discussed previously, many simplifications and idealizations are employed in the modeling of practical manufacturing processes and systems because of the complexities that arise. Therefore, it is critical to verify and validate the mathematical and numerical models to ensure that the results obtained are applicable, realistic, and accurate. Unless the models are satisfactorily validated, the simulation results cannot be used as the basis for choosing appropriate operating conditions and for design and optimization. It is important to verify that the numerical model accurately solves the given equations. This aspect, known as verification, involves grid refinement and varying the arbitrarily chosen parameters, such as convergence criteria and computational domain boundaries, to ensure that the results are essentially independent of the values chosen. Validation involves ensuring that the mathematical model is an accurate representation of the physical problem (Roache 1998). Among the approaches used for validation are a consideration of the physical behavior of the results obtained, comparisons with available analytical and numerical results, particularly benchmark solutions, and comparisons with available experimental data. Validation is often obtained over a limited region, and numerical models are then used to obtain results over a much larger domain.

Because of the critical importance of verification and validation, extensive efforts have to be made to obtain experimental data, whenever possible, for comparison with numerical predictions. In several cases, a separate, well-designed, experimental setup may be needed to achieve this. For instance, in the modeling and simulation of single and twin screw polymer extruders, a specially designed cam-driven thermocouple system was employed to obtain the temperature profile in the rotating screw, and two rotating cylinders were used to study the mixing phenomena and thus validate the model for twin screw extrusion (Esseghir and Sernas 1991; Sastrohartono et al. 1990). More is said on this in a later chapter.

In the polymer coating of optical fibers, typical coating thicknesses of the order of 40–50 μm are applied to the uncoated fiber or as secondary coating to a coated fiber. The basic coating process involves drawing the fiber of diameter around 125 μm through a reservoir of coating fluid, with inlet and outlet dies. This is immediately followed by curing of the polymer coating material around the fiber. At the die exit, the coating material is drawn out with the fiber, forming a downstream meniscus, which influences the coating characteristics. Thus, an important consideration in the coating process is the exit meniscus, which represents the profile of the free surface as the fluid exits from the die due to the viscous drag from the moving fiber. The forces due to gravity, polymer shear, surface tension, and external shear due to air are included to determine the overall force balance at the free surface. The force imbalance is used to generate an iteration scheme, starting with a guessed profile, till the force balance is satisfied, and a converged meniscus is obtained (Yoo and Jaluria 2008). The numerical results are compared with experimental data. As shown later in Chap. 8, a good agreement on the exit meniscus profile is observed, indicating the validity and accuracy of this approach. Overall, it is necessary to make all possible efforts to validate the mathematical/numerical models, even if it means spending considerable time and effort in developing an experimental arrangement to obtain the data needed for comparisons.

3.5 Modeling of Manufacturing Systems

In the preceding sections, we have discussed the mathematical and numerical modeling of various processes and components that are of interest in materials processing and manufacturing. However, there is another important aspect that must be considered and that relates to the numerical modeling of the overall manufacturing system, which usually consists of several components, since the process undergone by the given material or part results from the energy and mass exchange with the various constituents of the system.

Consider, for instance, a typical furnace, which consists of an electrical or gas heater, walls, insulation, enclosed gases, and the material or part undergoing heat treatment. The transport mechanisms in all these components are coupled through the boundary conditions. Thus, the heater exchanges thermal energy with the walls, the gases, and the material. Similarly, the material undergoing heat treatment is in energy exchange with the heater, the walls, and the gases. The gas flow is driven by an externally imposed pressure difference, such as that due to a fan, a moving conveyor, and buoyancy. Figure 3.12 shows a sketch of a simple system for this process and the information flow diagram that indicates the coupling between the different parts.

Each individual component may first be mathematically modeled and numerically simulated as uncoupled from the others, by employing prescribed boundary conditions. The mathematical and numerical model is then verified and validated, as discussed in the preceding section. Then, these individual simulations are combined, employing the appropriate coupling through the boundary conditions. This procedure provides a systematic approach to the numerical model of the system, which may be a simple one or a complicated practical one, such as a blast furnace for making steel. The numerical model of the system may again be validated by the use of available data on similar existing systems or on experimental systems and prototypes fabricated for the purpose. Similar approaches may be used for other manufacturing systems, such as casting and crystal growing systems, by considering the different components to develop the overall model for the system. Due to the material and dimensions of the different components, different governing equations

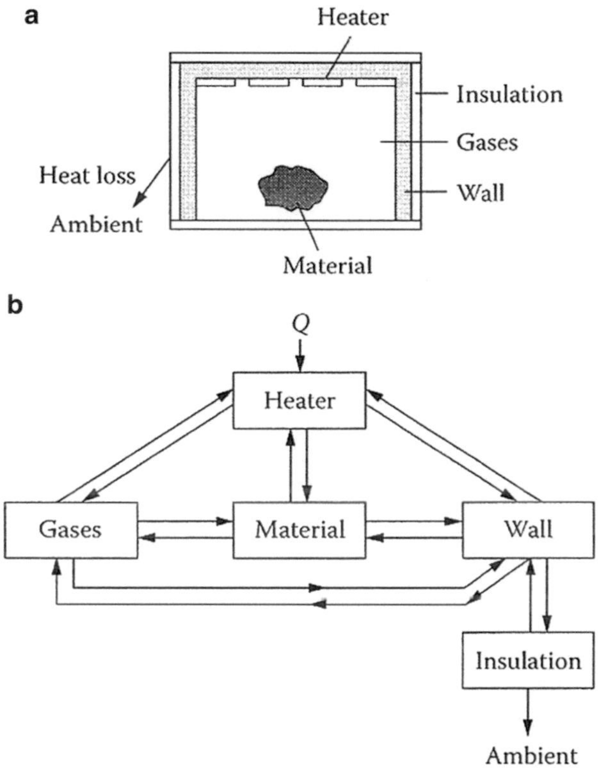

Fig. 3.12 An electric furnace for thermal processing of an object and an information flow diagram showing the coupling between different parts of the system

may be applicable for different parts of the system, and these different numerical models are finally coupled to yield the model for the complete system.

3.6 Numerical Simulation

Once the model is confirmed to be a valid and accurate representation of the manufacturing system, the model is subjected to changes in the design parameters and operating conditions. This process of studying the behavior of the system by means of a model, rather than by fabricating a prototype, is known as *simulation*. The results obtained allow us to consider many different design possibilities as well as a variety of operating conditions. Therefore, the choice of operating conditions to achieve desired product characteristics and production rates may be based on the simulation. Different designs may be evaluated to choose an acceptable design, and safe levels may be established for the operating conditions. These results are also used for optimization of the system. Therefore, the success of the design and optimization process is strongly dependent on the numerical modeling and simulation of the system (Jaluria 2003).

System simulation is one of the most important elements in studying the behavior of a given manufacturing process or system; determining the appropriate operating conditions; determining the temperature, shear, pressure, and other physical variables that a given object is subjected to during processing; and undertaking the design and optimization of the system. Since experimentation on a prototype of the actual system is generally expensive and time consuming, simulation is critical to obtain the desired information on the system behavior and response under different conditions. A one-to-one correspondence is established between the model and the physical system by validating the model, as discussed earlier. Then the results obtained from the simulation are indicative of the behavior of the actual system.

In summary, numerical simulation can be used to:

1. Study system behavior under various conditions and for different design parameters
2. Choose operating conditions for desired productivity and product characteristics
3. Evaluate different designs for selection of an acceptable design
4. Determine safety limits for the system
5. Determine effects of different design variables for optimization
6. Improve or modify existing systems
7. Investigate sensitivity of the design to different variables

All these are important considerations. The study of system behavior and choice of operating conditions have been mentioned earlier. Evaluation of different designs is important because several designs are typically generated for a given application. If each of these were to be fabricated and tested for acceptability, the cost would be prohibitive. System simulation is employed effectively to investigate each design and to determine if the given requirements and constraints are satisfied, thus yielding an acceptable or workable design. Predicting the behavior of the system under off-design conditions, i.e., values beyond those used for the design, is another important use of system simulation. Such a study provides valuable information on the operation of the system and how it would perform if the conditions under which it operates were to be altered, as under overload or fractional-load circumstances. The deviation from design conditions may occur due to many reasons, such as variations in energy input, differences in raw materials fed into the system, changes in the characteristics of the components with time, changes in environmental conditions, and shifts in energy load on the system. These outputs also indicate the safety limits of the system. It is important to determine, for instance, the maximum temperatures in a polymer processing system without damage to the material. Safety features can then be built into the system.

System simulation plays an important role in optimization, which may involve minimization of parameters such as cost per item and energy consumption per unit output or maximization of quantities such as output and product quality. Whatever the criterion for optimization, it is essential to change the variables over the design domain, determined by physical limitations and constraints, and to study the system behavior. Then, using various techniques available, the optimal design is determined. Simulation can also be used for correcting a problem in an existing system or for modifying the system for improving its performance. Rather than changing a particular component in order to correct the problem or improve the system, simulation is first used to determine the effect of such a change. Since the simulation closely represents the actual physical system, the usefulness of the proposed change can be determined without actually carrying out the change.

A question that arises frequently in materials processing is the effect of a given variable or component on the product and the system performance. This relates to the sensitivity of the process to the design variables and operating conditions. A substantial reduction in the cost of the system or improvement in the product may be obtained by changes in these. System simulation is used to determine the sensitivity of the system and the process to the variables and to decide if slight alterations can be made in the interest of reducing the cost without significant sacrifice in output.

3.7 Experimentation

An experimental study of the process or the system is generally much more involved and expensive than numerical modeling. However, it is of particular importance in materials processing, because of the need for better physical understanding, the complexity of the problem in many practical circumstances, and characterization of the product and its properties. Also, in some cases, numerical techniques are not available, are difficult to implement, or need experimental inputs to proceed with the simulation. In short, experimentation is needed for the following main reasons:

1. Providing better physical understanding of the basic phenomena
2. Validation of the mathematical and numerical models
3. Providing inputs to simplify or initiate the simulation
4. Providing results in cases where modeling is difficult, unavailable, or inaccurate
5. Obtaining property data for the materials involved
6. Determining the properties and characteristics of the product

Experimentation is often needed to understand the basic transport phenomena that govern the process because of the typical complexity of practical systems. For example, flow visualization in a channel or die, made of transparent material, can be used to understand the flow pattern and mixing mechanisms, allowing an improved design to avoid recirculation and stagnation. We have already discussed the importance of experimental data in validating the model. Experimental data can also be used to simplify the analysis or allow the modeling to proceed by providing inputs on boundary conditions and on various components. For instance, the heat transfer by convection at a boundary may be based on empirical correlations available in the literature, particularly for turbulent transport. The temperature distribution at the wall of a furnace may be obtained by solving the conjugate convection-conduction problem. But the problem can be considerably simplified if the temperature distribution is measured and supplied as an input to the model.

Similar considerations apply for inflow conditions in ovens and temperature of rollers or dies.

In several cases, modeling is very complicated and inaccurate, or the basic mechanisms are not fully understood. The modeling of turbulent flows is such a circumstance. Similarly, coating processes often involve dynamic contact angles, which result from a balance of surface tension, pressure, viscous, and gravitational forces. This could lead to a fairly complicated problem for which the surface properties must be well known to obtain accurate results. However, if experimentation is used to obtain the contact angle or the meniscus under a range of operating conditions, considerable simplification is obtained in the modeling (Ravinutala and Polymeropoulos 2002).

Experimental data are critical to the numerical simulation of manufacturing processes since the modeling is strongly dependent on the material properties used. As mentioned earlier, properties vary with the local conditions like temperature, concentration, and pressure. Changes in the material during the process also affect the properties. In several practical processes, the properties also vary with the source of the raw material and with time. This is often the case with polymers and other raw materials. Thus, it is important to measure the properties over the expected ranges of temperature, pressure, and other variables in the process and use these data in the simulation.

Finally, since the obvious focus is on the product, it is important to experimentally study the characteristics and quality of the product and determine its important properties. The characteristics and properties vary with the product and application under consideration. For instance, thickness and uniformity of a coating or a deposited thin film are of main interest in these processes. In polymer and food extrusion, the degree of conversion is of main concern. Similarly, in a soldered or welded joint or in casting, voids are important. A wide range of diagnostic tools is available to determine the relevant properties, defects, and other characteristics. For instance, a scanning electron microscope (SEM) can be used to study a given product closely to study its shape and dimensions. Figure 3.13 shows the SEM images of microchannels fabricated in silicon by wet and dry etching. Clearly, the latter results in much better defined microchannels at the desired dimensions. More is said on this in a later chapter.

As an example of defects in the product, Fig. 3.14 shows the results from a numerical simulation to determine the normalized concentration, n_d/n_{do}, of a particular type of defects, E', in an optical fiber at two fiber drawing speeds and different furnace wall temperature distributions. These defects are measured by a study of the transmission characteristics of the fiber. TP I-IV refers to uniform, Gaussian, parabolic, and parabolic profile with uniform temperature in the lower region of the furnace. The defects approach a uniform value near the end of the draw process. A higher draw speed reduces the time in the furnace, resulting in lower defect concentration. The wall temperature has a substantial effect on the defects, with larger temperatures leading to more defects. This problem is considered in greater detail in a later chapter. But this figure is included to indicate the importance of defects in the product and the need to quantify these through numerical modeling and experimentation.

Fig. 3.13 Microchannels in silicon fabricated by (**a**) wet etching and (**b**) plasma etching

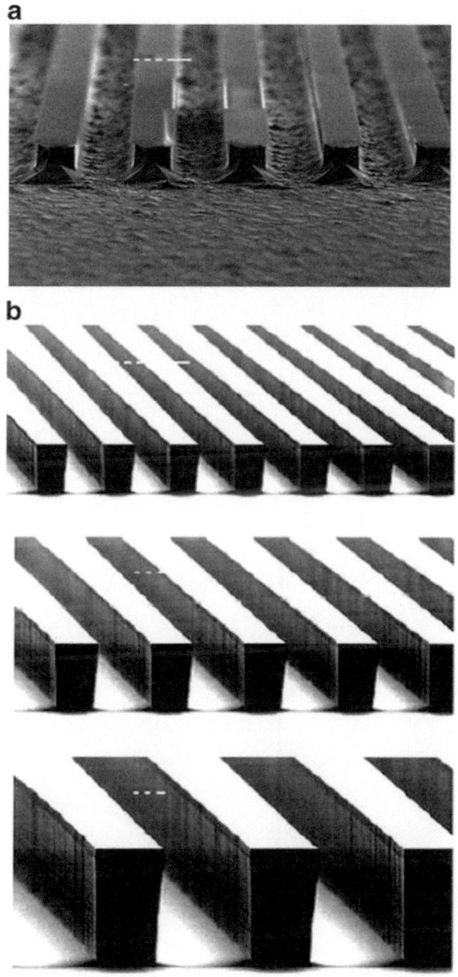

Similarly, other techniques based on optical and x-ray scanning may be used to study the product. A differential scanning calorimeter (DSC) may be used to determine the amount of conversion or transformation in a product due to the process. An atomic force micrometer (AFM) may be used to study the surface and determine irregularities. Also, the product may be employed in the desired application, such as optical data transmission, to determine if the desired performance is achieved. Strength, hardness, ductility, etc. may also be measured to determine if the given materials processing technique such as heat treatment is satisfactory.

Available experimental results are generally quite limited in manufacturing processes due to complexities like high temperature and pressure, inaccessibility of the probe into the region of interest, changes in the material characteristics as it undergoes the given process, inconsistency of the raw materials, uncertainties in the

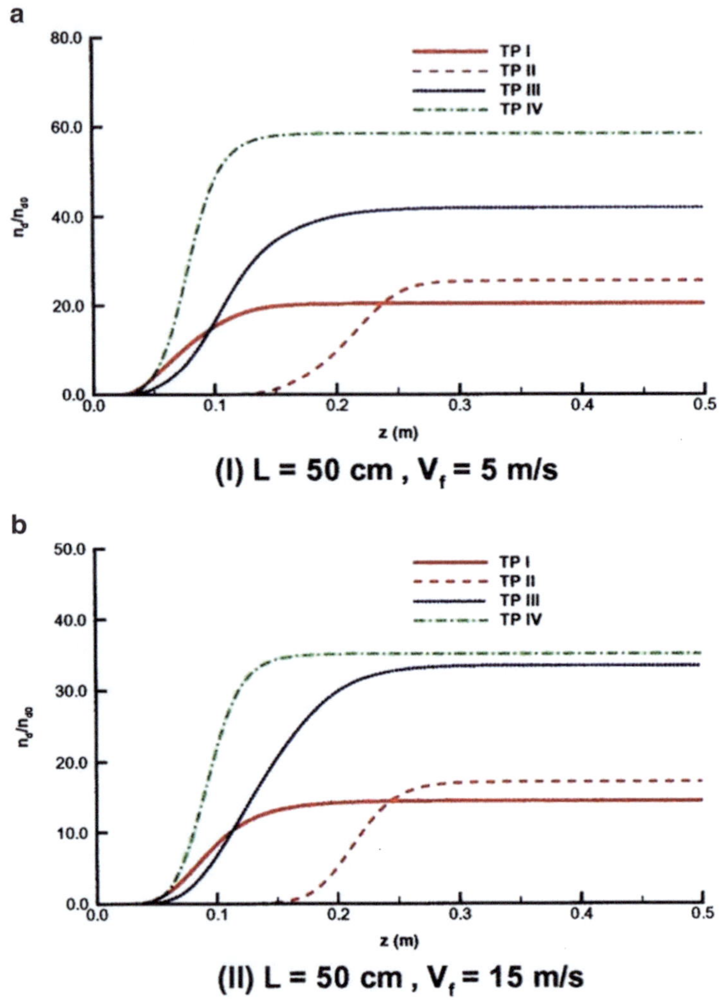

Fig. 3.14 Axial variation of the normalized concentration of E' defects in optical fiber drawing for different furnace wall temperature distributions and draw speed V_f in a furnace of length L of 50 cm

imposed conditions, measurement difficulties in a production environment, and so on. Consequently, experiments on laboratory systems are often substituted for measurements on the full-scale systems, thus limiting the usefulness of the data. Scale-up is then an important consideration and must be investigated in detail. Overall, experimental data are crucial in materials processing systems and, despite the effort and cost, are obtained whenever possible. Only a brief outline of experimentation is given here, and further details are included in later chapters with respect to various manufacturing processes.

3.8 Link Between Transport Processes and Product Characteristics

One of the most crucial aspects in the study, analysis, modeling, and simulation of materials processing and manufacturing systems is the dependence of the product characteristics and properties on the system design and operating conditions. This is based on the link between the transport processes that are being modeled or studied experimentally and the final product. The properties of the material undergoing thermal processing must be measured and appropriately investigated to accurately quantify this dependence, which would then lead to the proper choice of variables to obtain the desired result. However, this is an area in which there is acute lack of data and critical work is needed in the future.

Numerical and experimental investigations can lead to the prediction of the conditions, in terms of temperature, pressure, stress, and other variables, undergone by the material in the manufacturing process. The next and particularly critical step is to determine the changes in the structure or composition of the material as it goes through the system. Thus, a study of the transport processes can, in principle, lead to a determination of the physical and chemical properties of the final product. But this requires a detailed information on material behavior and how structural or chemical changes occur in the material as a consequence of the temperature, pressure, and other conditions to which it is subjected. Additional diagnostics may also be undertaken to link important properties like strength, porosity, defects, and ductility of the product with the transport processes.

There are thus two main aspects in linking the product with the system parameters and operating conditions. The first involves predicting the conditions undergone by the material in the process. The second one is the link between the transport processes to which the material is subjected and the properties of the final product. For instance, the thermal cycle needed for a particular process, such as heat treatment of steel and plastic thermoforming, may be determined on the basis of material considerations. The material has to be heated beyond its recrystallization temperature, maintained at this value for a given time for the transformations or process to be completed, and then cooled at a rate determined again by material considerations. Figure 3.15 shows a sketch of such a desired thermal variation in the material to be processed, indicating heating, cooling and soaking, or maintaining, at the desired temperature for homogeneity in the material. In metals, a gradual cooling would lead to annealing that restores the ductility of the material. Different cooling rates lead to different material characteristics. Numerical modeling and simulation, along with selective experimental data, can be used to study the transport processes for a variety of operating conditions and thus determine methods of controlling the heating/cooling rates or the soaking time over which a given temperature is maintained. Thus, the conditions that lead to desired temperature, pressure, or stress variations with time in the material can be determined, with uniformity in the product determined by the uniformity of these variables in the object.

Finally, the properties and characteristics of the product are obtained through various diagnostic means. The performance of the product in the given application is

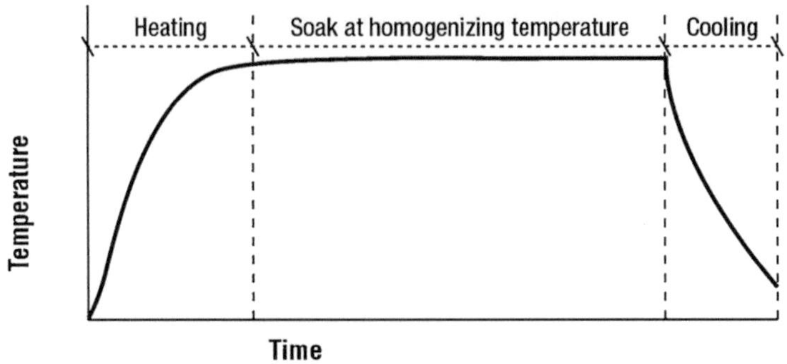

Fig. 3.15 Sketch of the temperature cycle for a heat treatment or thermoforming process

also investigated. In some cases, the product may be sampled by a group of users to ensure that it is satisfactory. This is a common procedure in processed food products to determine flavor, taste, crispiness, etc. Similar approaches are used for other consumer products. Thus, the characteristics and quality of the final product are determined. One then proceeds to link this with the transport processes and ultimately with the operating conditions. The choice of the operating conditions is an inverse problem, which is generally complicated and nonunique. All these considerations are discussed later with respect to different manufacturing processes.

References

A. Abib, Y. Jaluria, Numerical simulation of the buoyancy-induced flow in a partially open enclosure. Numer. Heat Transf. **14**, 235–254 (1988)

T.L. Bergman, A.S. Lavine, F.P. Incropera, D.P. Dewitt, *Introduction to Heat Transfer*, 6th edn. (Wiley, New York, 2011)

L.C. Burmeister, *Convective Heat Transfer*, 2nd edn. (Wiley Interscience, New York, 1993)

M. Esseghir, V. Sernas, A cam-driven probe for measurement of the temperature distribution in an extruder channel. SPE ANTEC Tech. Papers **37**, 54–57 (1991)

J. Issa, Z. Yin, C.E. Polymeropoulos, Y. Jaluria, Temperature distribution in an optical fiber draw tower furnace. J. Mater. Process. Manuf. Sci. **4**, 221–232 (1996)

Y. Jaluria, Thermal processing of materials: From basic research to engineering. J. Heat Transf. **125**, 957–979 (2003)

Y. Jaluria, Challenges in the accurate numerical simulation of practical thermal processes and systems. Int. J. Numer. Methods Heat Fluid Flow **23**, 158–175 (2013)

Y. Jaluria, Numerical modeling of multiple length scales in thermal transport processes. Int. J. Numer. Methods Heat Fluid Flow **24**, 781–796 (2014)

Y. Jaluria, K.E. Torrance, *Computational Heat Transfer*, 2nd edn. (Taylor & Francis, New York, 2003)

J. L. Kokini, C.-T. Ho, M. V. Karwe (eds.), *Food Extrusion Science and Technology* (Marcel Dekker, New York, 1992)

T.H. Kwon, S.F. Shen, K.K. Wang, Pressure drop of polymeric melts in conical converging flow: experiments and predictions. Polym. Eng. Sci. **28**, 214–224 (1986)

B.P. Leonard, Bounded Higher-order upwind multidimensional finite-volume convection-diffusion algorithms, in *Advances in Numerical Heat Transfer*, ed. by W. J. Minkowycz, E. M. Sparrow (Eds), vol. 1, (Taylor & Francis, Philadelphia, 1997), pp. 1–57

T. Li (ed.), *Optical Fiber Communications, Vol. 1: Fiber Fabrication* (Academic Press, New York, 1985)

R.L. Mahajan, C. Wei, Buoyancy, Soret, Dufour and variable property effects in silicon epitaxy. J. Heat Transf. **113**, 688–695 (1991)

G.D. Mallinson, G. de Vahl Davis, G., 1973, The method of false transient for the solution of coupled elliptic equations. J. Comput. Phys. **12**, 435–461 (1973)

W. J. Minkowycz, E. M. Sparrow (eds.), *Advances in Numerical Heat Transfer*, vol 1 (Taylor & Francis, Washington, DC, 1997)

W. J. Minkowycz, E. M. Sparrow (eds.), *Advances in Numerical Heat Transfer*, vol 2 (CRC Press, Boca Raton, 2001)

W. J. Minkowycz, E. M. Sparrow, J. Y. Murthy (eds.), *Handbook of Numerical Heat Transfer*, 2nd edn. (Wiley, New York, 2006)

S.V. Patankar, *Numerical Heat Transfer and Fluid Flow* (Taylor and Francis, Washington, DC, 1980)

D.W. Peaceman, H.H. Rachford, Numerical solution of parabolic and elliptic differential equations. J. Soc. Ind. Appl. Math. **3**, 28–41 (1955)

S. Ravinutala, C.E. Polymeropoulos, Entrance meniscus in a pressurized optical fiber coating applicator. Int. J. Exp. Heat Transfer Fluid Mech. **26**, 573–580 (2002)

P.J. Roache, *Computational Fluid Dynamics* (Hermosa Publishers, Albuquerque, 1976)

P.J. Roache, *Verification and Validation in Computational Science and Engineering* (Hermosa Publishers, Albuquerque, 1998)

T. Sastrohartono, M. Esseghir, T.H. Kwon, V. Sernas, Numerical and experimental studies of the flow in the nip region of a partially intermeshing co-rotating twin screw extruder. Polym. Eng. Sci. **30**, 1382–1398 (1990)

J.C. Tannehill, D.A. Anderson, R.H. Pletcher, *Computational Fluid Mechanics and Heat Transfer*, 2nd edn. (Taylor & Francis, Washington, DC, 1997)

S.S. Wang, C.C. Chiang, A.I. Yeh, B. Zhao, I.H. Kim, Kinetics of phase transition of waxy corn starch at extrusion temperatures and moisture contents. J. Food Sci. **54**, 1298–1301 (1989)

S.Y. Yoo, Y. Jaluria, Numerical simulation of the meniscus in the non-isothermal free surface flow at the exit of a coating die. Numer. Heat Transf. **53A**, 111–131 (2008)

Part II
Different Manufacturing Processes

In the preceding chapters, the basic considerations that arise in materials processing have been discussed. These included the governing equations and boundary conditions for common processes, various challenges posed by the complexities that arise in typical materials processing systems, and additional mechanisms that are often encountered. Some of the common approaches for analysis, numerical simulation, and experimentation were also outlined, along with a few examples. The basic aspects that link different processes were discussed in terms of the underlying mechanisms, governing equations, and important parameters. However, major differences exist between different materials processing techniques and special treatment is needed for each process. The results that are of interest and that are obtained from numerical or experimental investigations are also often quite different. This section considers several important processes in greater detail in order to illustrate the application of the approaches outlined earlier, as well as to present characteristic results and their practical implications. Additional processes that may be of interest, particularly those that are new and emerging, are also briefly discussed.

Chapter 4
Manufacturing Processes with Phase Change

Processes with phase change are extensively used in manufacturing. In most cases, the change occurs between the liquid and the solid phases. The phase change from liquid or solid to vapor and vice versa is also of interest in processes such as drying, condensation, and deposition. Solidification and melting have been studied extensively because of their importance in a wide range of manufacturing processes, as mentioned in earlier chapters. Some of the important ones may be listed as:

Casting
Crystal growing
Welding and soldering
Injection molding
Thermal sprays
Additive manufacturing

In each of these applications, different systems and approaches have been employed, for instance, mold casting, which is a batch process, and continuous casting, which is a continuous process. Pure materials such as metals, as well as mixtures like alloys and polymers, have been considered. The heat transfer in the solid is coupled with that in the liquid and thus with the flow that arises in the liquid due to buoyancy, rotation, surface tension, or other mechanisms. As a consequence, the problems involving phase change are generally quite involved, and numerical methods are employed for most practical situations.

Processes such as casting have come down the ages and continue to be important today due to different materials and applications that are of interest. The demands on product quality and on production rates have resulted in substantial improvements in the classical process and the development of new systems. Advances in materials, such as specialized alloys and polymers, have also resulted in considerations of defects, impurities, and characteristics of the product. This chapter focuses on such phase change processes with respect to a variety of manufacturing techniques. It discusses the various analytical, numerical, and experimental approaches that may be adopted to obtain results that may be used to develop systems for new and

Y. Jaluria, *Advanced Materials Processing and Manufacturing*, Mechanical Engineering Series, https://doi.org/10.1007/978-3-319-76983-7_4

emerging materials while enhancing the current systems and processes. Though the rate of melting or solidification and possible voids that affect product quality are of major interest, recent focus has also been on the microstructure, uniformity, and imperfections in order to obtain the desired product quality by varying the operating conditions and system parameters.

There are several important aspects in manufacturing processes such as casting. These include rate of temperature rise and decrease that affects the thermal stresses introduced in the material, transient effects in the flow with respect to oscillations and instabilities, density change in phase change, melting and solidification of multicomponent materials and mixtures, absorbed and released gases, and the nature and behavior of the solid-liquid interface. Though melting is also important in the overall process, much of the focus in the literature has been on solidification since the quality and characteristics of the product are generally directly impacted by the solidification process. Consequently, substantial work has been done over the years considering different aspects of the process as well as different materials and manufacturing systems (see, for instance, Hughel and Bolling (1971), Flemings (1974), Kurtz and Fisher (1998), and Dantzig and Rappaz (2017)). This chapter presents some of the main aspects of the process and the approaches that may be adopted for analysis, design, and optimization.

4.1 Analysis

For the simple case of mold casting, a molten material is poured into a cavity or mold, and the solid/liquid interface moves with time, as shown in Fig. 4.1a, due to heat removal, which leads to solidification. Similarly, heat addition results in melting. In a typical mold casting system, a riser is provided in order to supply the liquid to fill in any gaps left due to volume reduction, or shrinkage, during solidification, as sketched in Fig. 4.2. In continuous casting, Czochralski crystal growing process, or other continuous processes, the interface remains essentially stationary, as shown in Fig. 4.1b. Similarly, in one-dimensional solidification in a planar or long cylindrical region, shown in Fig. 4.1c, the interface moves parallel to the boundaries as time elapses. The location of the moving boundary is not known and must be obtained from the solution, taking the transport processes in the various regions into account. There are two main approaches for analyzing these processes, these being the interface tracking approach and the enthalpy method.

4.1.1 Interface Tracking

If a change of phase occurs at a boundary, such as the solid-liquid interface, the energy absorbed or released due to the change of phase must be taken into account. Thus, the boundary conditions at the moving interface between the two phases must

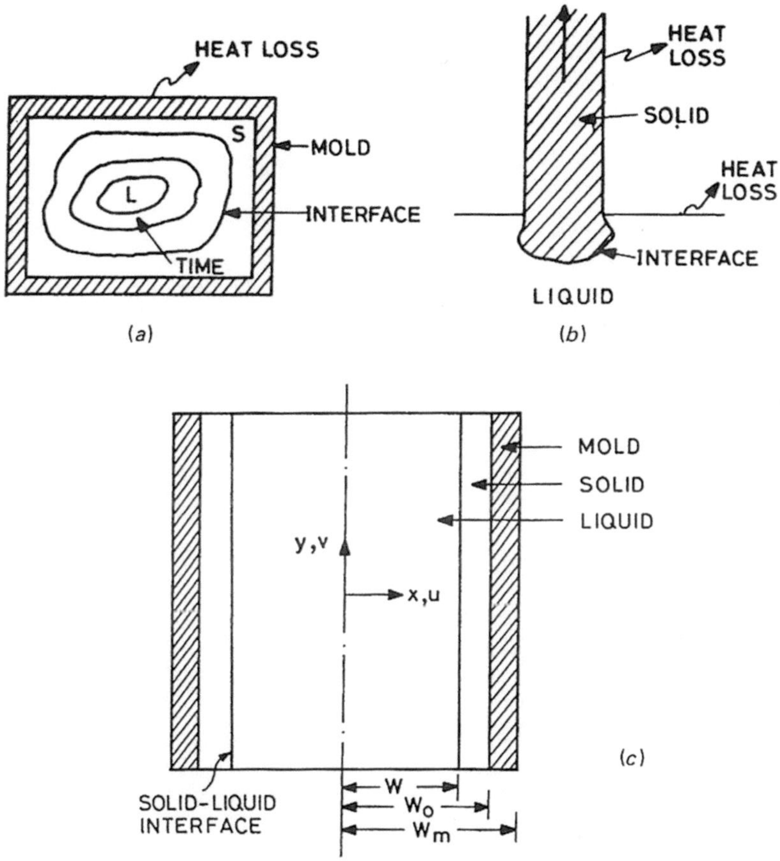

Fig. 4.1 (**a**) Movement of the solid/liquid interface during solidification in an enclosed region; (**b**) interface in a Czochralski crystal growing process; (**c**) solidification in a long cylindrical or planar mold

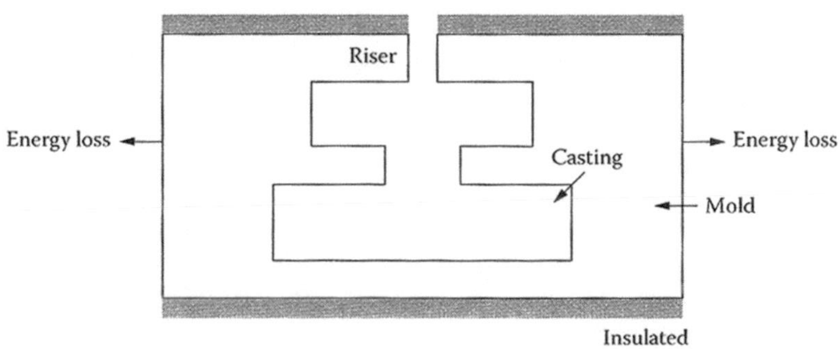

Fig. 4.2 Sketch of a simple mold casting system

be given if the interface location and geometry are to be determined. For one-dimensional solidification, with the interface moving in the y-direction, this boundary condition is given by the equation

$$k_{\mathrm{s}}\frac{\partial T_{\mathrm{s}}}{\partial y}-k_{\mathrm{l}}\frac{\partial T_{\mathrm{l}}}{\partial y}=\rho L_{\mathrm{h}}\frac{d\delta}{dt} \tag{4.1}$$

where $y = \delta$ is the location of the interface; subscripts s and l represent solid and liquid, respectively; ρ is the liquid density; and L_{h} is the latent heat of fusion per unit mass. This implies that the energy released due to solidification is conveyed by conduction in the two regions. In melting, the energy is absorbed for phase change to occur. Similarly, for two-dimensional solidification, the boundary condition is written as (Sparrow et al. 1977; Ramachandran et al. 1982; Viskanta, 1988)

$$\left(k_{\mathrm{s}}\frac{\partial T_{s}}{\partial y}-k_{\mathrm{l}}\frac{\partial T_{\mathrm{l}}}{\partial y}\right)\left[1+\left(\frac{\partial\delta}{\partial x}\right)^{2}\right]=\rho L_{\mathrm{h}}\frac{d\delta}{dt} \tag{4.2}$$

Similarly, the boundary conditions may be written for other coordinate systems and for 3D problems.

For a stationary interface, as shown in Fig. 4.1b, with the material moving at velocity U, the boundary condition is (Siegel 1978, 1984)

$$\left(-k\frac{\partial T}{\partial n}\right)_{\mathrm{l}}+\rho U L_{\mathrm{h}}\frac{dy}{ds}=\left(-k\frac{\partial T}{\partial n}\right)_{\mathrm{s}} \tag{4.3}$$

where ds is a differential distance along the interface and n is coordinate normal to it. Also, the temperature at the interface in all these cases is the melting point T_{m}, which is assumed to be a fixed value. The transport processes in the two regions, solid and liquid, are treated separately by applying the conservation principles.

The analysis of these problems is fairly involved since the interface location and shape must be determined for each time step or iteration. The governing equations are the same as those given earlier for the solid and the liquid. Appropriate discretization of the two regions may be carried out, with their own individual properties like thermal conductivity and density. A coordinate transformation, such as the Landau transformation discussed earlier, may be employed to simplify the computational domains. Several other techniques have been developed to treat such moving boundary problems and the complicated domains that arise. In continuous casting and crystal growing, the interface between the solid and the liquid is essentially stationary, but it is not known a priori, and an iterative procedure may be adopted to determine its shape and location. Transformations and body-fitted coordinates may be employed to approximate the irregular-shaped computational domains.

4.1.2 Enthalpy Method

In the second approach, the conservation of energy is considered in terms of the enthalpy *H*, yielding the governing energy equation as

$$\rho\frac{DH}{Dt} = \rho\frac{\partial H}{\partial t} + \rho\bar{V}.\nabla H = \nabla.(k\nabla T) \tag{4.4}$$

where each of the phase enthalpies H_i is defined as

$$H_i = \int_0^T C_i dT + H_i^o \tag{4.5}$$

C_i being the corresponding specific heat and H_i^o the enthalpy at 0 *K*. Then, the solid and liquid enthalpies are given by, respectively,

$$H_s = C_s T \qquad H_l = C_l T + [(C_s - C_l)\,T_m + L_h] \tag{4.6}$$

The continuum enthalpy and thermal conductivity are given, respectively, as

$$H = H_s + f_l\,(H_l - H_s) \qquad k = k_s + f_l\,(k_l - k_s) \tag{4.7}$$

where f_l is the liquid mass fraction, obtained from equilibrium thermodynamic considerations as

$$f_l = \begin{cases} 0 & h \le C_s T_m \\ (h - C_s T_m)/L_h & C_s T_m < h < (C_s T_m + L_h) \\ 1 & h \ge (C_s T_m + L_h) \end{cases} \tag{4.8}$$

The dynamic viscosity μ is expressed as the harmonic mean of the phase viscosities in the limit as $\mu_s \to \infty$, i.e., $\mu = \mu_l/f_l$.

This model smears out the discrete phase transition in a pure material. But the numerical modeling is much simpler since the same equations are employed over the entire computational domain and there is no need to keep track of the interface between the two phases (Bennon and Incropera 1988; Prescott and Incropera 1996). In addition, impure materials, mixtures, and alloys, which do not have a fixed melting point, can be treated very easily by this approach. Figure 4.3 shows the two approaches outlined here for numerical modeling, indicating a single domain for the enthalpy method and the interface between the two regions for the two-region approach. Both the approaches have been used to study materials processing with phase change, and results from the two have been compared for pure materials with a fixed melting point (see, for instance, Viswanath and Jaluria (1993) and Voller (1997)).

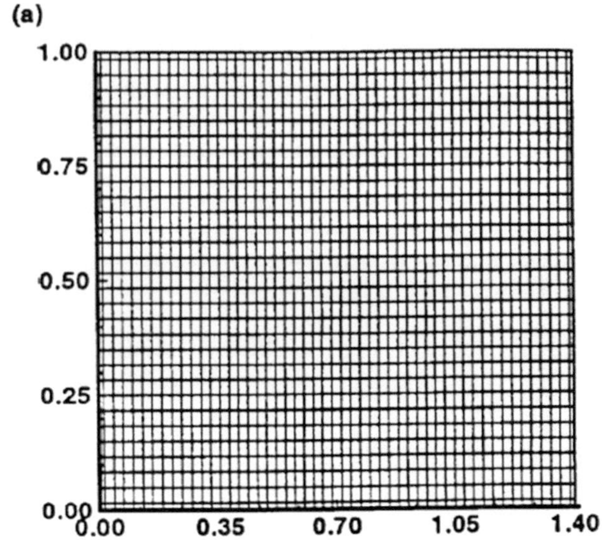

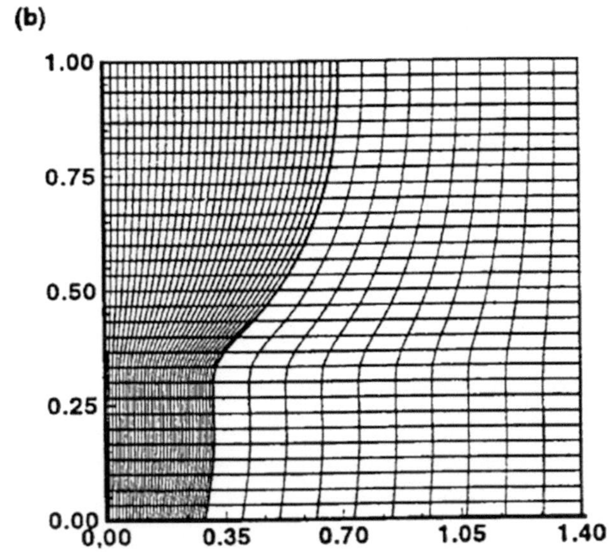

Fig. 4.3 Numerical grids used for the (**a**) enthalpy method (single region) and (**b**) the two-phase interface tracking (two-region) method

4.2 Simple Conduction Models

For the casting process in an enclosed region, a large number of design parameters arise, such as materials, initial melt pour temperature, cooling fluid and its temperature and flow rate, geometry, and dimensions. The quality of the casting is determined by grain size, composition, directional strength, concentration of defects, voids, and thermal stresses. It is necessary to carry out a thermal analysis of the solidification process, using modeling and simulation, to obtain inputs for design and to evaluate the nature of the casting. As mentioned earlier, solidification and melting have been studied extensively, considering pure materials as well as mixtures such as alloys. The buoyancy-driven flow due to temperature and concentration differences in the liquid or melt region is coupled with the conduction in the solid. However, earlier studies neglected the convection effects.

Several models can be developed for the study of solidification under different conditions. Considering mold casting, some of these may be listed as:

1. *Steady conduction in solid model:* The mold is taken at fixed temperature, and steady conduction in the solid is assumed.
2. *Chvorinov model:* Entire thermal resistance is assumed to be due to the mold, and energy balance is used.
3. *Lumped mold model:* Temperature in the mold is assumed to be uniform and time-dependent. Steady conduction in the solid is assumed.
4. *One-dimensional conduction model:* Transient 1D temperature distribution is assumed in the mold, solid, and melt.
5. *Two- and three-dimensional models:* Natural convection flow in the melt may be included.
6. *More sophisticated models:* Needed for alloys, generation of voids and defects, complicated geometries, etc.

In the last three models, buoyancy-driven flow in the liquid may be included for more accurate and realistic results. The first three models neglect the flow in the melt. The liquid may be taken at or above the melting point. The simplest model is the one with steady conduction in the solid, with the boundary at a fixed temperature of T_{a} and the liquid at the melting temperature. Then the energy equation is obtained for density ρ being the same for solid and liquid as

$$\rho L_{\mathrm{h}} \frac{d\delta}{dt} = k \frac{T_{\mathrm{m}} - T_{\mathrm{a}}}{\delta} \tag{4.9}$$

which gives the solid region thickness as a function of time as

$$\delta = \sqrt{\frac{2k(T_{\mathrm{m}} - T_{\mathrm{a}})t}{\rho L_{\mathrm{h}}}} \tag{4.10}$$

This expression indicates that the solid thickness δ increases with time t as $t^{1/2}$, starting with zero. It also shows the dependence of δ on the thermal properties and temperatures. Even though this is an idealized situation, the result is important in starting the numerical scheme for solidification in a region for various other conditions since the solid layer is of zero thickness at the beginning and this expression may be used for short times to obtain a finite thickness before proceeding to more complicated and appropriate numerical models.

If the density changes due to the phase change, the movement of the interface is affected due to expansion or contraction on solidification. A parameter b, where

$b = (\rho_s - \rho_l)/\rho_l$, may be used to represent the density change on solidification. Then the energy equation and the expression for δ are obtained as

$$\rho L_{\mathrm{h}}(1+b)\frac{d\delta}{dt} = k\frac{T_{\mathrm{m}} - T_{\mathrm{a}}}{\delta} \qquad \delta = \sqrt{\frac{2k(T_{\mathrm{m}} - T_{\mathrm{a}})t}{\rho L_{\mathrm{h}}(1+b)}} \tag{4.11}$$

Here, b is positive for a contraction on solidification and negative for an expansion. Similarly, modifications in other equations for δ for different cases may be made to take into account the density change on solidification.

The next three models from the preceding list are sketched in Fig. 4.4, with the liquid taken at the melting point. Each model has its own level of accuracy and

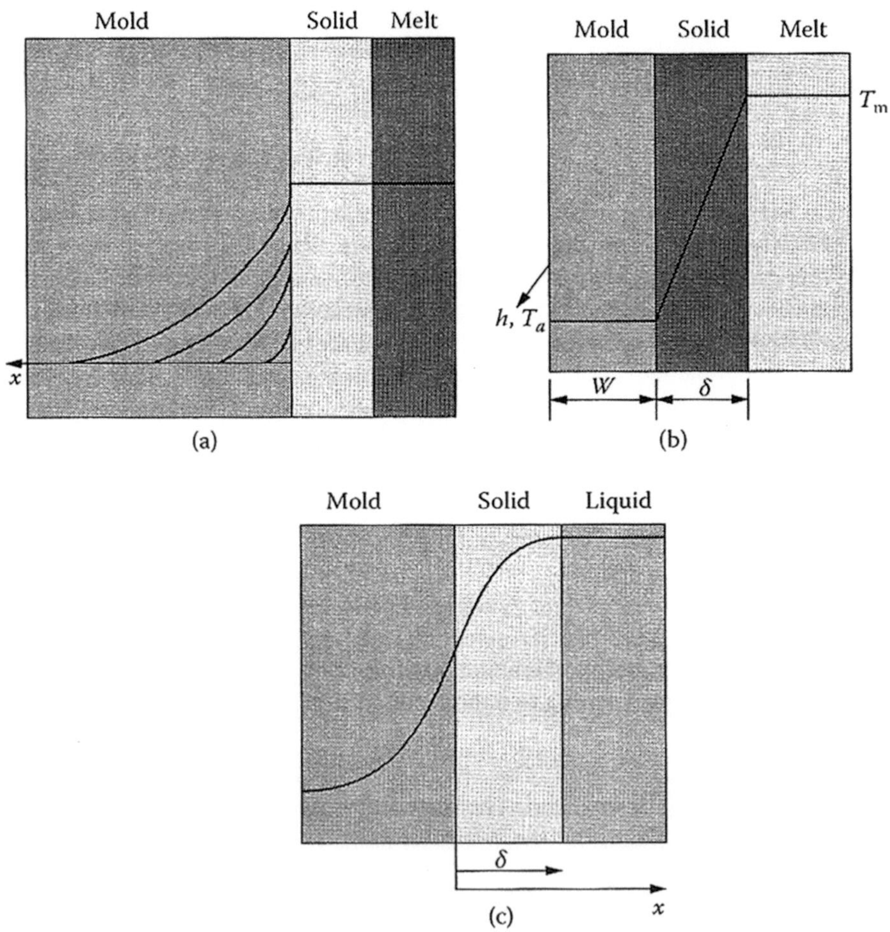

Fig. 4.4 Different conduction models for mold casting, with the melt at the melting point T_{m}. (**a**) Chvorinov model; (**b**) lumped mold model; (**c**) semi-infinite model

validity. Different models may be chosen, depending on the application and materials involved. For instance, if an insulating material such as ceramic or sand is used for the mold, the Chvorinov model may yield good results since most of the thermal resistance is in the mold. For the planar case shown, if the mold is treated as semi-infinite, the heat flux q and the energy balance are given by

$$q = \frac{k(T_{\mathrm{m}} - T_{\mathrm{a}})}{\sqrt{\pi \alpha t}} \qquad \rho V_{\mathrm{o}} L_{\mathrm{h}} = \frac{2A_{\mathrm{o}} k(T_{\mathrm{m}} - T_{\mathrm{a}}) t_{\mathrm{s}}^{1/2}}{\sqrt{\pi \alpha}} \tag{4.12}$$

where T_{a} is the initial temperature of the mold and t_{s} is the solidification time and V_{o} the volume and A_{o} the surface area of the casting. Then the solidification time may be written as

$$t_{\mathrm{s}} = \gamma \left(\frac{V_{\mathrm{o}}}{A_{\mathrm{o}}} \right)^2 \tag{4.13}$$

where γ is known as the Chvorinov constant. Similarly, expressions for cylindrical and spherical configurations may be derived.

If the mold is treated as lumped, the fluid is taken at T_{m}, and energy storage in the solidified region is neglected, energy balance leads to Eq. (4.9) and the following equation:

$$(\rho C)_c \frac{dT_c}{dt} = k \frac{T_{\mathrm{m}} - T_c}{\delta} - h(T_c - T_{\mathrm{a}}) \tag{4.14}$$

where subscript c refers to the mold and h is the convective heat transfer coefficient at the outer surface of the mold. These two coupled ordinary differential equations can be easily solved numerically, though analytical solutions may be possible in some cases. Figure 4.5 shows typical results for the solidification of a 20 cm thick plate of metal with given properties, $T_{\mathrm{m}} = 1200$ K and $h = 20$ W/m^{2}K. The effect of mold thickness on the temperature distribution in the mold and on the thickness of the solidified region is shown. Additional results are given by Jaluria (2008).

Similarly, analytical solutions may be obtained for cases where the semi-infinite approximation is made, an example of which is shown in Fig. 4.4c (Jiji 2009). Many such conduction models for a variety of conditions and parameters are discussed in most manufacturing books (see, for instance, Ghosh and Mallik (1986)).The focus is on solidification time, movement of the interface, and possible voids left due to density change on solidification. The interface movement is used to determine where cavities may be left, so that a suitable mold system, with an appropriate riser, may be designed. For a given practical circumstance, one could start with a relatively simple model and move to more complicated models if the desired information is not

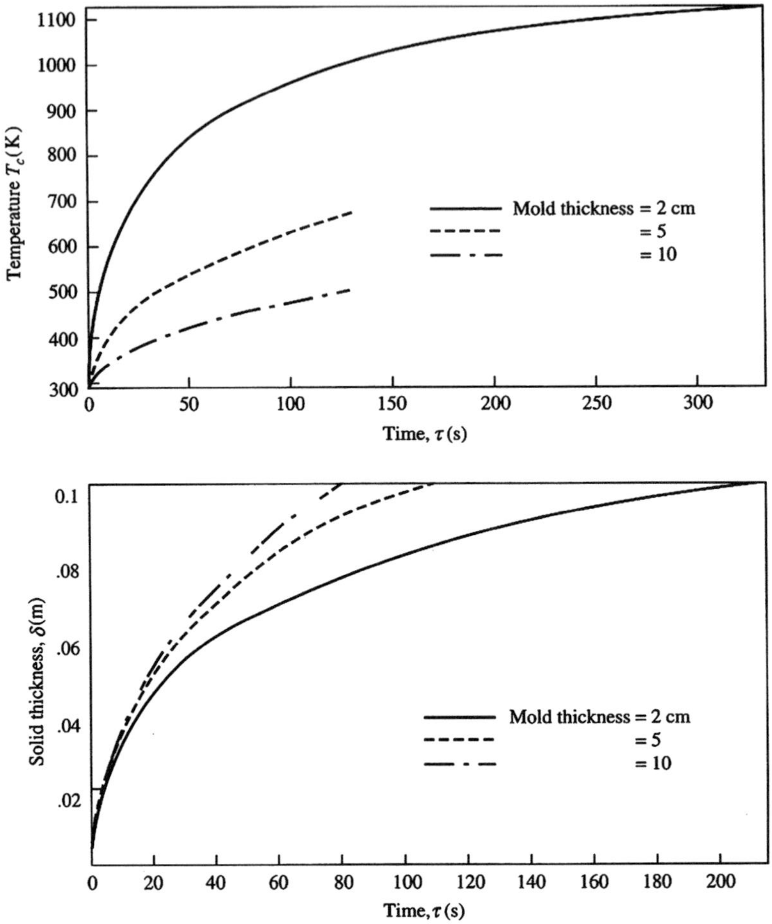

Fig. 4.5 Mold temperature T_c and solid region thickness δ as functions of time t for different values of the mold wall thickness W, at $h = 20$ W/m^2 K

obtained. It is also possible to start with the simplest model and keep on improving the model till the results remain essentially unchanged from one model to the next. As an example, in a typical design session, the cooling parameters are first varied to reduce the solidification time. If the solidification time does not reach the desired value, the pour temperature of the melt may be varied. If even this does not satisfy the requirements, the thickness of the mold wall may be changed. The material of the wall may also be varied, if needed. Thus, by first varying the operating conditions and then the dimensions and materials, the solidification time may be minimized or brought below a desired value. Knowledge-based design methodology may also be used to improve the model automatically to obtain an acceptable or optimal design, as discussed later in Chap. 10.

4.3 One-Dimensional Solidification

Let us consider one-dimensional solidification in a planar region, as shown in Fig. 4.1c. If the subscripts 1, 2, and 3 refer to the liquid, solidified region, and the mold, respectively, the governing equations for the configuration shown are

$$\frac{1}{\alpha_1}\frac{\partial T_1}{\partial t}=\frac{\partial^2 T_1}{\partial x^2} \tag{4.15a}$$

$$\frac{1}{\alpha_2}\frac{\partial T_2}{\partial t}=\frac{\partial^2 T_2}{\partial x^2} \tag{4.15b}$$

$$\frac{1}{\alpha_3}\frac{\partial T_3}{\partial t}=\frac{\partial^2 T_3}{\partial x^2} \tag{4.15c}$$

With the initial conditions given, for instance, as

$$\text{At}\quad t=0: T_1=T_r \text{ and } T_3=T_\text{a} \tag{4.16}$$

Here, T_1 is at T_r, which is greater than the melting point T_m, and T_a is the ambient temperature and the initial temperature of the mold. Convective heat transfer occurs at the outer surface of the mold with a heat transfer coefficient h. There is no solidified region at $t = 0$. But the approximation given in Eq. (4.10) may be used for short times to obtain the solid region thickness $W_\text{o} - W$, with T_m at the solid-liquid interface and T_a at the mold-solid interface. A linear temperature variation is initially taken within this solid layer. In actual practice as well, there is a chilling effect that gives rise to a thin layer of solid adjacent to the mold, and then the solid-liquid interface moves, as given by the preceding governing equations.

The boundary conditions for the three coupled partial differential equations, given here, are obtained from symmetry conditions at the center plane, heat flux and temperature continuity, and the phase change energy balance as

$$\begin{aligned}
&x=0: \frac{\partial T_1}{\partial x}=0\\
&x=W: T=T_2=T_\text{m};\quad k_2\frac{\partial T_2}{\partial x}-k_1\frac{\partial T_1}{\partial x}=\rho_1 L_\text{h}\frac{dW}{dt}\\
&x=W_\text{o}: T_2=T_3;\quad k_2\frac{\partial T_2}{\partial x}=k_3\frac{\partial T_3}{\partial x}\\
&x=W_\text{m}: -k_3\frac{\partial T_3}{\partial x}=h(T_3-T_\text{a})
\end{aligned} \tag{4.17}$$

The last equation can easily be modified to include radiative transport, if significant. A constant uniform temperature of T_a may be used for very large values of h. Since the equations are coupled and have a moving boundary, the problem is quite involved to solve analytically, and a numerical solution may be obtained by using finite-difference, finite volume, or other methods.

The moving interface can be treated numerically by employing a suitable transformation, as discussed by Sparrow et al. (1977) and Ramachandran et al. (1981a). Thus, transformed coordinates are defined as

$$
\begin{aligned}
x_1 &= \frac{x}{W(t)} \\
x_2 &= \frac{x - W(t)}{W_o - W(t)} \\
x_3 &= \frac{x - W_o}{W_m - W_o}
\end{aligned}
\tag{4.18}
$$

The temperature and time are nondimensionalized as

$$
\theta = \frac{T - T_a}{T_r - T_a}; \ \bar{t} = \frac{\nu t}{W_o^2} \tag{4.19}
$$

where ν is the kinematic viscosity of the liquid and T_r is a reference temperature, which could be taken as the liquid pour temperature. The thermal diffusivity α may also be used instead of ν, if the flow in the liquid is neglected. The solid-liquid interface is held stationary over the chosen time step. The solution marches in time, employing inputs from the previous time step. Though explicit methods are simpler and more convenient, implicit methods may be used to avoid instability.

The temperature differences in the liquid give rise to a time-dependent buoyancy-induced flow. The convection terms drop out due to the one-dimensional approximation, and thermal field is independent of the flow. The flow is important in a consideration of the movement of impurities in the liquid and of the quality of the casting obtained, as discussed by Szekely and Stanek (1970). Since an infinitely long planar circumstance is considered, the enclosure is simulated by satisfying the condition of zero net mass flow in the vertical direction, the transverse velocity being zero in this case. The momentum equation becomes

$$
\frac{\partial v}{\partial t} = \nu \frac{\partial^2 v}{\partial x^2} + g\beta(T - T_p) \tag{4.20}
$$

where v is the vertical velocity and T_p is a temperature that is adjusted iteratively for each time step in the numerical computation, so that the condition of zero flow in the vertical direction is satisfied. The initial and boundary conditions for this equation are

$$
\begin{aligned}
t = 0 : v = 0 & \\
t > 0; x = 0 : & \quad \frac{\partial v}{\partial x} = 0 \\
x = W : & \quad v = 0
\end{aligned}
\tag{4.21}
$$

The temperature distribution of the previous time step is employed to determine the buoyancy term, which is then fed into the momentum equation to obtain the velocity field at the next time step.

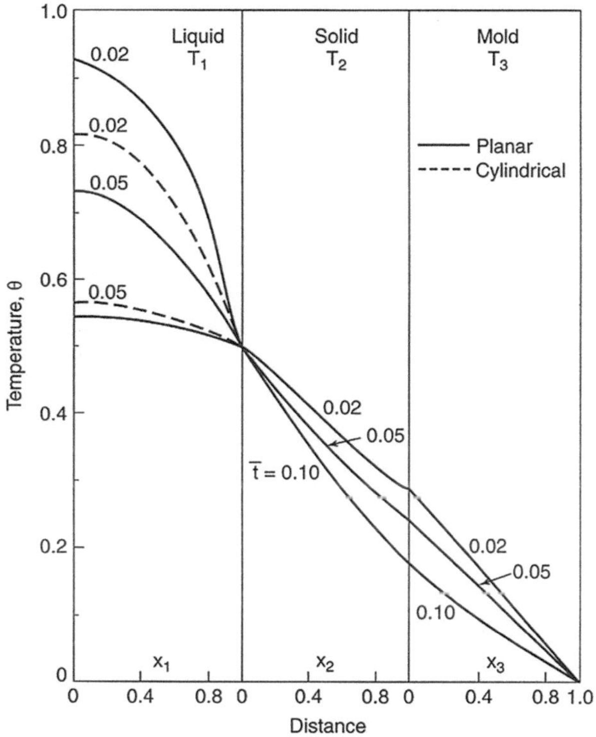

Fig. 4.6 Temperature distributions in the three regions, liquid, solid, and mold, for solidification with a constant temperature being maintained at the outer surface of a tall vertical mold

A few results obtained by Ramachandran et al. (1981a) for this problem are shown in Figs. 4.6 and 4.7 at two Prandtl number Pr ($= \nu/\alpha$) values. Results for both planar and cylindrical cases are shown. The temperature distributions in all three regions and the resulting velocity in the liquid are shown at different times. Due to the normalization, all the regions vary from 0 to 1. At low Pr, the thermal effects penetrate at a faster rate, and, therefore, the maximum velocity is attained earlier. A decay of the flow occurs as temperature differences and thickness of the liquid region decrease with increasing time. The flow rises in the central region of the liquid and flows downward near the solid-liquid interface. The density is assumed to remain unchanged due to phase change. But this effect can easily be built into the calculations, as seen earlier. Also, variable properties, different cooling conditions, and different casting and mold materials can be considered. This is a relatively simple problem. But it brings out most of the important aspects in the modeling and simulation of the mold casting process. Also, such one-dimensional models are adequate for solidification near the boundaries, and the results may be used to approximate two- and three-dimensional problems by considering separate transient solidification processes near the different boundaries. For greater accuracy, the multidimensional problem, with buoyancy-driven flow in the liquid, must be considered, as outlined in the next section.

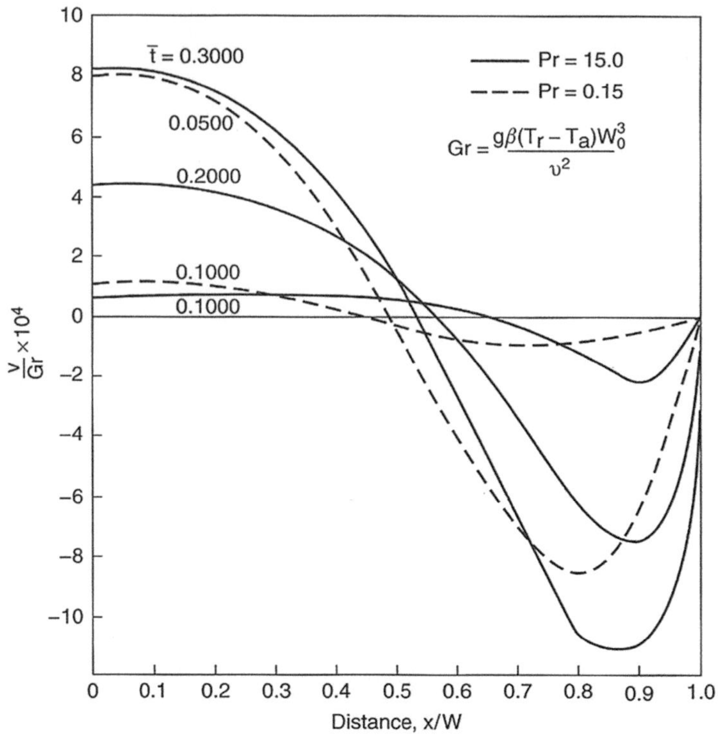

Fig. 4.7 The velocity distribution in the liquid region as a function of the Prandtl number for solidification in a tall vertical rectangular enclosure

4.4 Results for Multidimensional Problems

4.4.1 Interface Tracking: Transformed Grid Method

The two- or three-dimensional problem must be considered if the one-dimensional approximation is inapplicable, which is frequently the case. The problem is fairly involved and has often been solved without initial solid subcooling ($T_s < T_m$), during melting, or liquid superheat ($T_f > T_m$), during solidification. The effect of subcooling has been considered by Sparrow et al. (1978). A coordinate transformation may be employed with a boundary immobilization technique of the phase front, based on one-dimensional Landau transformation, as outlined earlier. Various other approaches for moving boundaries with multidimensional grid movements have been presented (Shyy et al. 1996; Jana et al. 2007). A buoyancy-induced flow arises in the liquid due to the temperature differences, and thus the flow is coupled with the thermal transport. The governing momentum equations for the two-dimensional case, for constant properties in the Cartesian coordinate system, are obtained from the general equations given in Chap. 2 as

$$\frac{\partial u}{\partial t}+u\frac{\partial u}{\partial x}+v\frac{\partial u}{\partial y}=\nu\left(\frac{\partial^2 u}{\partial x^2}+\frac{\partial^2 u}{\partial y^2}\right)-\frac{1}{\rho}\frac{\partial p}{\partial x} \tag{4.22a}$$

$$\frac{\partial v}{\partial t}+u\frac{\partial v}{\partial x}+v\frac{\partial v}{\partial y}=\nu\left(\frac{\partial^2 v}{\partial x^2}+\frac{\partial^2 v}{\partial y^2}\right)-\frac{1}{\rho}\frac{\partial p}{\partial y}+g\beta(T-T_{\text{ref}}) \tag{4.22b}$$

where T_{ref} is a reference temperature that may be taken as the melting point T_{m}. An adjustment in T_{ref}, as done for the one-dimensional case for each time step, is not needed since the continuity equation is explicitly satisfied in the enclosure. The energy equation that governs the temperature in the liquid is

$$\frac{\partial T}{\partial t}+u\frac{\partial T}{\partial x}+v\frac{\partial T}{\partial y}=\alpha\left(\frac{\partial^2 T}{\partial x^2}+\frac{\partial^2 T}{\partial y^2}\right) \tag{4.22c}$$

The convection problem without phase change has been investigated extensively to yield the buoyancy-driven flow in enclosures for a wide range of boundary conditions, aspect ratios, and parametric values (see, for instance, the benchmark solution given by de Vahl Davis (1983)). With phase change, we are concerned with a moving boundary. The boundary conditions are given by Eq. (4.2), with the interface temperature at T_{m}. The numerical scheme is applied to the transformed grid, and time marching is used, as discussed earlier. Figure 4.8 shows the numerically computed interface at various dimensionless time intervals, from Ramachandran et al. (1981b, 1982). The effect of natural convection flow in the liquid is seen in the inclination, with the vertical, of the interface. Additional results are given in the references mentioned.

This problem is a typical one among phase change problems and, when complexities due to property variations, heat conduction in the mold, combined modes operating at the outer surface of the mold and complicated geometries are included, the problem becomes quite involved. Unfortunately, such complexities are important in most practical problems. The three-dimensional problem is generally of interest. Sometimes, simplifications may be used to reduce the problem to a rectangular or cylindrical configuration. Density change on solidification or melting is another aspect that has to be considered in most cases and is incorporated in the interface movement at each time step.

4.4.2 *Enthalpy Formulation*

As mentioned earlier, this approach leads to a single domain, and interface tracking is not needed. It is also valuable in the simulation of impure materials and mixtures, such as plastics and alloys, that do not have a fixed melting point, and the phase change occurs over a finite temperature range. The basic approach and the governing equations were given earlier for a pure material with a fixed melting point.

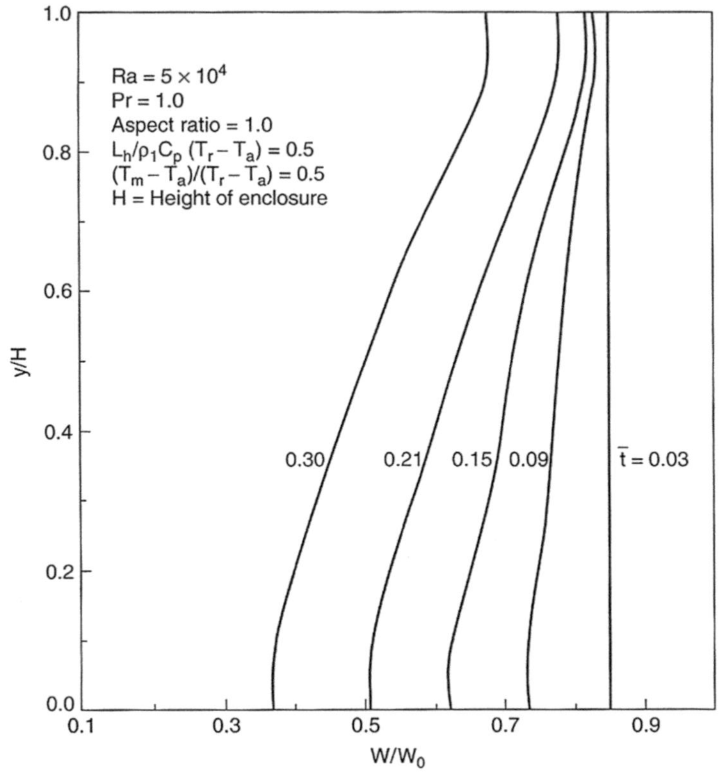

Fig. 4.8 Solid-liquid interface as a function of time for the 2D solidification process in a rectangular enclosure of height *H*, with natural convection effects included in the analysis

Figure 4.9 shows the numerical results for melting in an enclosed region using the enthalpy model. Streamlines and isotherms are shown for four different times during the melting of pure gallium. This is a benchmark problem in which melting is initiated by a step change in the temperatures at the left and right boundaries, the left being at temperature higher than the melting point and the right lower. At very early times, before the buoyancy-driven flow is initiated, the interface is parallel to the heated wall, and the heat transfer in the melt is dominated by conduction.

As the heating progresses, natural convection in the melt sets in. The streamlines indicate the effect of thermal buoyancy, which causes the interface between the solid and the liquid to bend rather than remain parallel to the vertical boundaries. The fluid rises at the hot wall and travels across the top of the cavity to impinge on the solid-liquid interface, thereby producing greater melting in this region. Since the hotter fluid stays near the top, melting is accelerated in the upper region. The flow cools down as it returns along the interface and moves along the bottom of the cavity toward the hot wall. The amount of material melted increases with time till it reaches a steady state for this problem. The recirculation in the liquid is clearly seen. These results are found to agree well with experimental results available in the literature

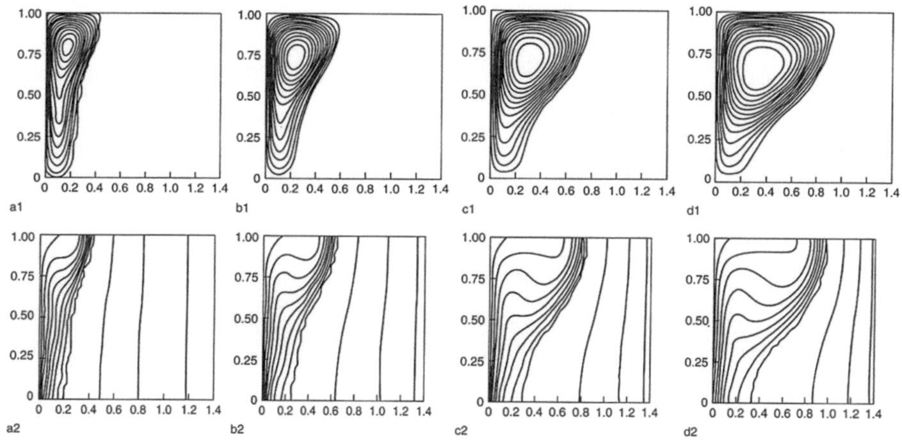

Fig. 4.9 Streamlines (top) and isotherms (bottom) for melting of gallium in an enclosed region, with the left vertical boundary at temperature higher than the melting point, the right vertical boundary at a temperature lower than the melting point, and the two remaining boundaries insulated. The enthalpy method is used, and the results are shown at different dimensionless time, following the onset of melting, t of (**a**) 0.5248, (**b**) 1.0416, (**c**) 1.5622, (**d**) 1.9789

(Viskanta 1988). The two-region approach can also be used for modeling this problem. For pure metals, the two-phase, two-region approach leads to more accurate results, whereas the enthalpy method does not yield a sharp interface, as expected, but smears the velocities over a finite region. A lot of work has been done on such melting and solidification problems, as reviewed by Voller (1997, 2006).

4.4.3 Conjugate Transport

The conduction heat transfer in the mold is also important in these problems. Figure 4.10 shows typical numerical results when conduction in the mold is coupled with heat transfer in the liquid and the solid (Viswanath and Jaluria 1995). With increasing time, the liquid region shrinks due to solidification, whereas the solidified region increases. The effect of the imposed conditions at the outer surface of the mold on the solidification process can be investigated by solving this conjugate problem, which yields the temperature field in the mold along with that in the solid and the liquid, as shown. The interface movement is dependent on the mold material and thickness, the cavity geometry, and the cooling conditions. All these aspects need to be considered for the design of a practical system. For instance, the mold material and thickness may be varied to control the solidification time. The cooling conditions may be varied to affect the interface movement and appropriately design the riser for the casting.

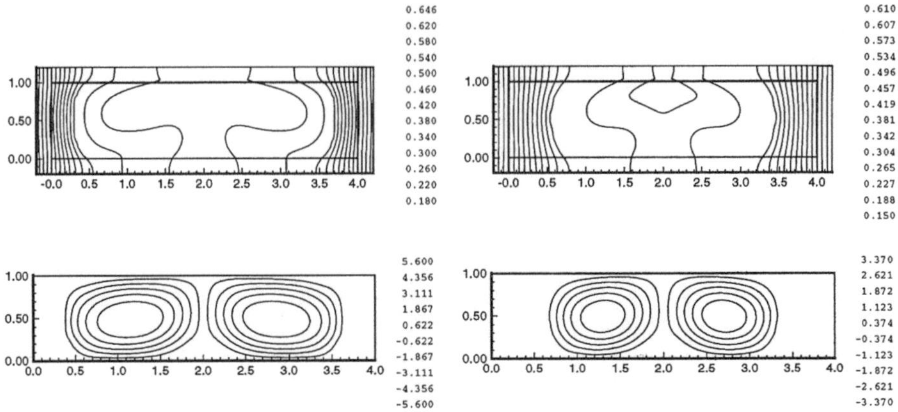

Fig. 4.10 Isotherms (top) and streamlines (bottom) for solidification of tin in a cavity of aspect ratio 4.0 with conjugate transport to the mold. Figures on left, $t = 1.0$, and, figures on right, $t = 2.0$

Figure 4.11 shows some typical results on solidification of tin with dimensionless time t^* for different values of the aspect, or width to height, ratio A of the cavity and for different values of the mold wall thickness W. Here, the main dimensionless parameters are the Prandtl number, taken as 0.011 here, the Rayleigh number, and the Stefan number, given by

$$Ra = \frac{g\beta(T_\mathrm{i} - T_\infty)D^3}{\nu\alpha} \qquad Ste = \frac{L_\mathrm{h}}{C_\mathrm{p}(T_\mathrm{i} - T_\infty)} \tag{4.23}$$

where D is the width of the cavity, T_i is the initial temperature of the liquid, and T_∞ is the ambient temperature. A larger mold wall thickness results in greater heat removal rate and faster solidification due to energy storage in the mold. A thinner wall has lower heat removal rate and higher temperatures, leading to increased natural convection flow. Similarly, heat removal rate is higher for lower aspect ratio.

Many such results have been obtained in the literature to determine the resulting flow and the interface movement with time, thus yielding the solidification time. Generally, the entire casting does not have to solidify before removing it from the mold. If a thick crust has been formed, the casting may be removed and the interior region allowed to solidify more gradually. In summary, the inclusion of conjugate transport; variable properties, particularly density change on solidification; and a variety of cooling conditions at the outer surface of the mold are important for practical processes.

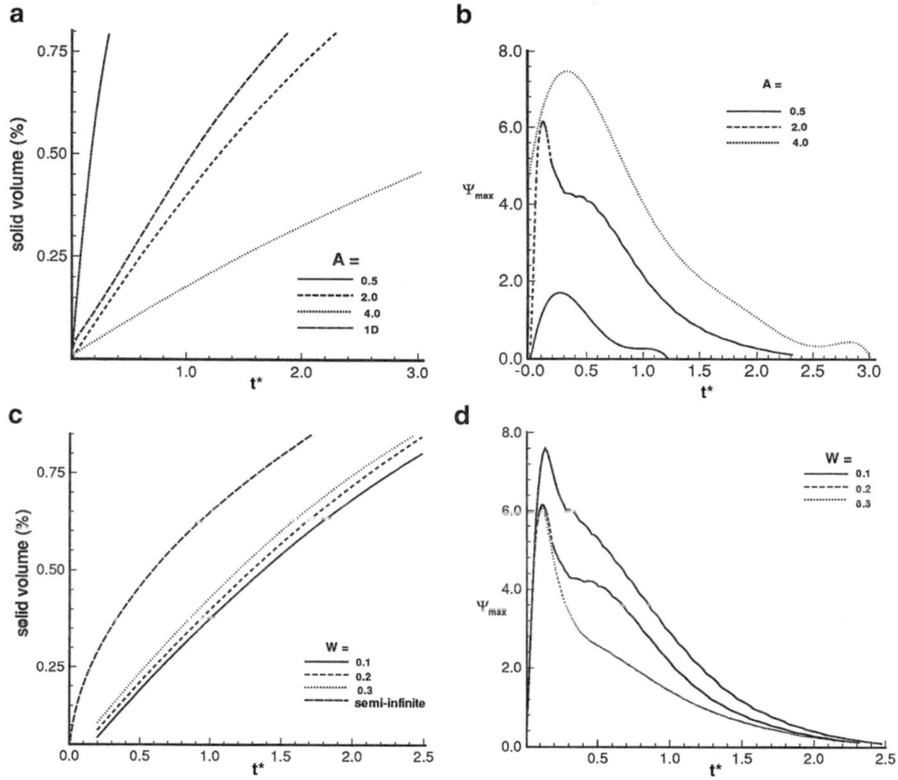

Fig. 4.11 Effect of aspect ratio *A* and mold wall thickness *W* on (**a**, **c**) rate of solidification, and on maximum stream function ψ_{max} (**b, d**) for the solidification of molten tin in a rectangular cavity at Stefan number, *Ste*, of 2.4 and Rayleigh number *Ra* of 10^6

4.4.4 Experimental Studies

Experimental results are valuable not only for validating the models but also for providing better physical insight and obtaining realistic simulation results. Experimental data are used for the material properties and for the practical limitations on the process. However, experimental studies have been relatively few because of the complexity of the process arising from a moving interface and time-dependent flow (Bathelt et al. 1979; Gau and Viskanta 1986). Detailed experimental results are particularly important for critically evaluating the various models employed for simulation as well as for providing information on the characteristics of the interface for development of microscale models. Figure 4.12 shows experimental results, along with numerical predictions, for the melting and solidification of pure tin (Wolff and Viskanta 1987, 1988). These are benchmark problems, as mentioned earlier, and have been used extensively for model validation. The comparisons are fairly good, though the differences at small time indicate the need to improve the model or to more closely approximate the experimental conditions.

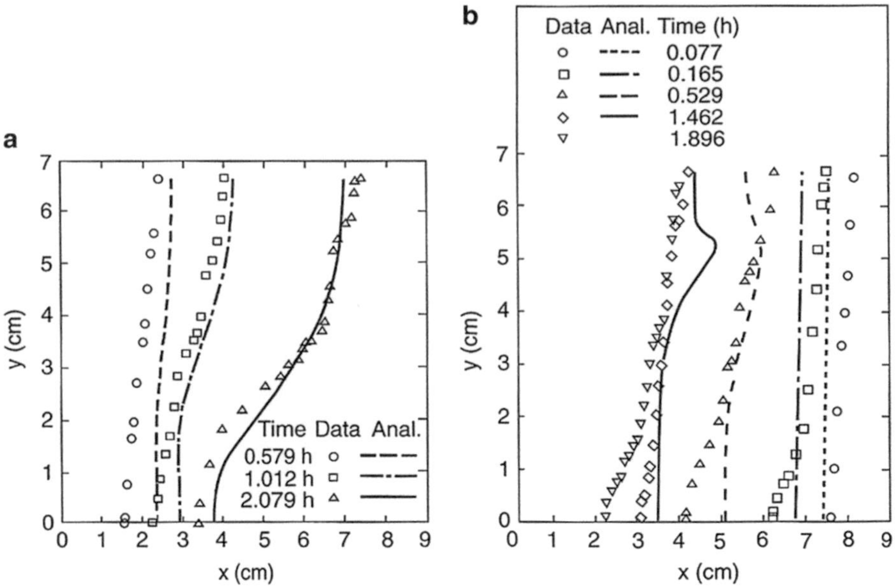

Fig. 4.12 Comparison between measured and predicted interface locations during (**a**) melting and (**b**) solidification of pure tin from a vertical surface (Wolff and Viskanta 1987, 1988)

Similarly, Banaszek et al. (1999) carried out experiments and numerical simulations to demonstrate the importance of conduction in the wall, as shown in Fig. 4.13 for solidification of water. Again, the comparison between experimental and numerical results is quite good, lending support to the enthalpy model used for this problem. Such numerical and experimental studies can be used to determine the progression of the solidification front and thus monitor the generation of voids and other defects in the casting.

Because of the importance of solidification in materials processing and manufacturing, experimental studies have been carried out on a wide range of materials, different operating conditions, and a variety of casting problems. Much of the recent effort has focused on alloys, mixtures, polymers, and various other new and emerging materials. Also, of particular interest has been the microstructure resulting from solidification, as discussed later. Several of these experimental studies have been discussed in the references given earlier.

4.5 Continuous Casting

As mentioned earlier and sketched in Fig. 1.1d, solidification or melting may occur over an essentially stationary interface with the material moving across a mold or a heating region. The energy balance at the interface determines the rate of phase

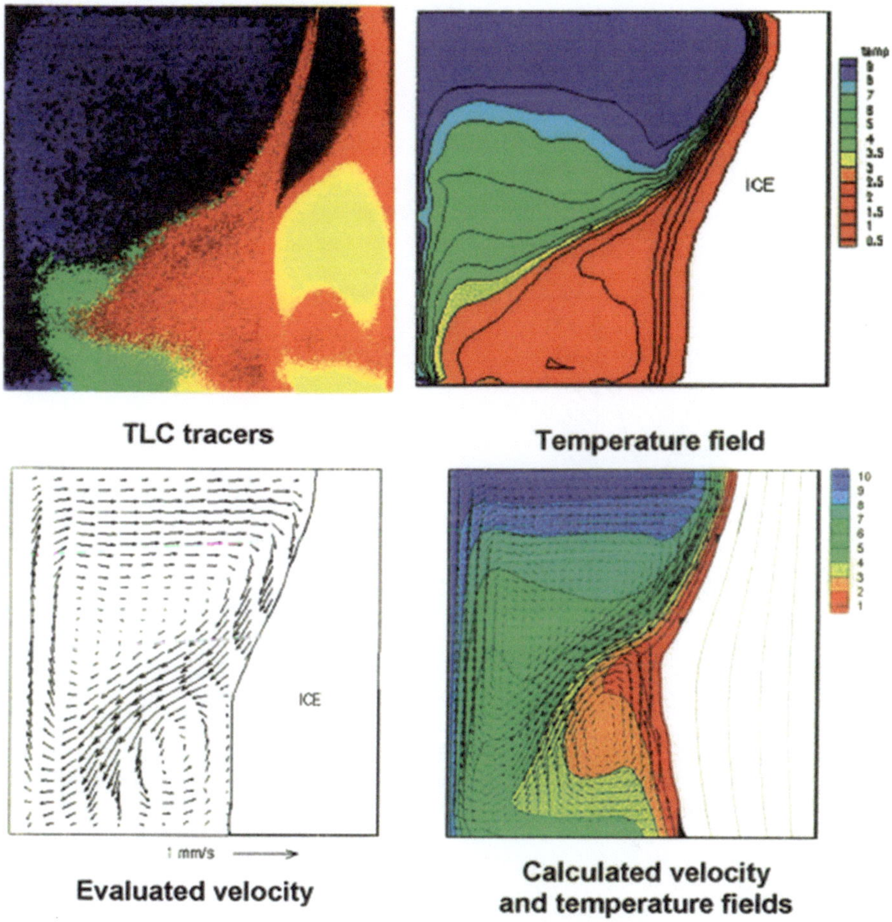

Fig. 4.13 Experimental and numerical results for water solidification driven by convection and conduction (Banaszek et al. 1999)

change, as given by Eq. (4.3). For solidification, the energy released due to phase change is removed by thermal transport in the regions away from the interface. A simple conduction model may be developed for the liquid, phase change, and solid regions, shown in Fig. 4.14a, and given by the equations

$$\rho C_{\mathrm{p}} U_{\mathrm{s}} A \frac{dT}{dx} = kA \frac{d^2T}{dx^2} - h_1 P(T - T_\infty) \tag{4.24a}$$

$$\rho U_{\mathrm{s}} A L_{\mathrm{h}} \frac{dm}{dx} = h_2 P(T_{\mathrm{m}} - T_\infty) \tag{4.24b}$$

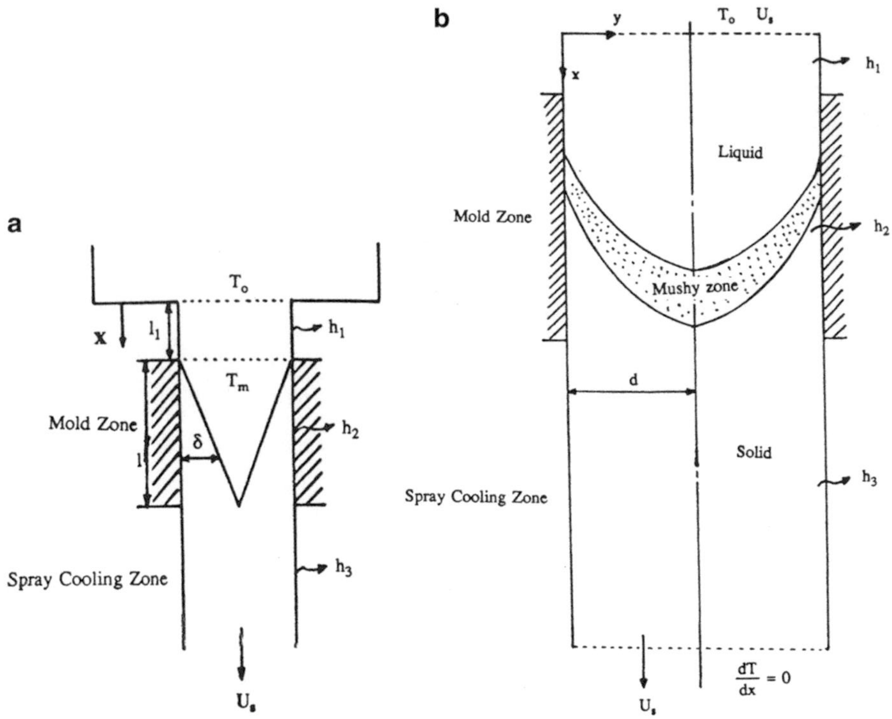

Fig. 4.14 (**a**) Schematic of the one-dimensional, three-zone model for the continuous casting process; (**b**) schematic of a continuous casting process for constant heat transfer coefficient at the mold and spray cooling, with the boundary conditions

$$\rho C_p U_s A\frac{dT}{dx} = kA\frac{d^2T}{dx^2} - h_3 P(T - T_\infty) \tag{4.24c}$$

where h_1, h_2, and h_3 are the convective heat transfer coefficients in the three regions, A is the area of cross section, P is the perimeter, and m is the mass fraction of solid formed. The problem can be solved for different operating conditions and different lengths of the three regions. The results from the first region feed into the second region and the results from the second region into the third. The temperature may be assumed to remain constant at T_m in the mold, where the solid fraction goes from zero to 1.

Typical results on the dimensionless temperature θ are shown in Fig. 4.15 for different Peclet number Pe and Biot number Bi, where

$$Pe = \frac{U_s d}{\alpha} \quad Bi = \frac{hd}{\alpha} \quad 1^* = \frac{l}{d} \quad \theta = \frac{T - T_\infty}{T_m - T_\infty} \tag{4.25}$$

The temperature decreases to the melting point, remains constant at this value during phase change, and then decreases to the ambient temperature. Clearly, the

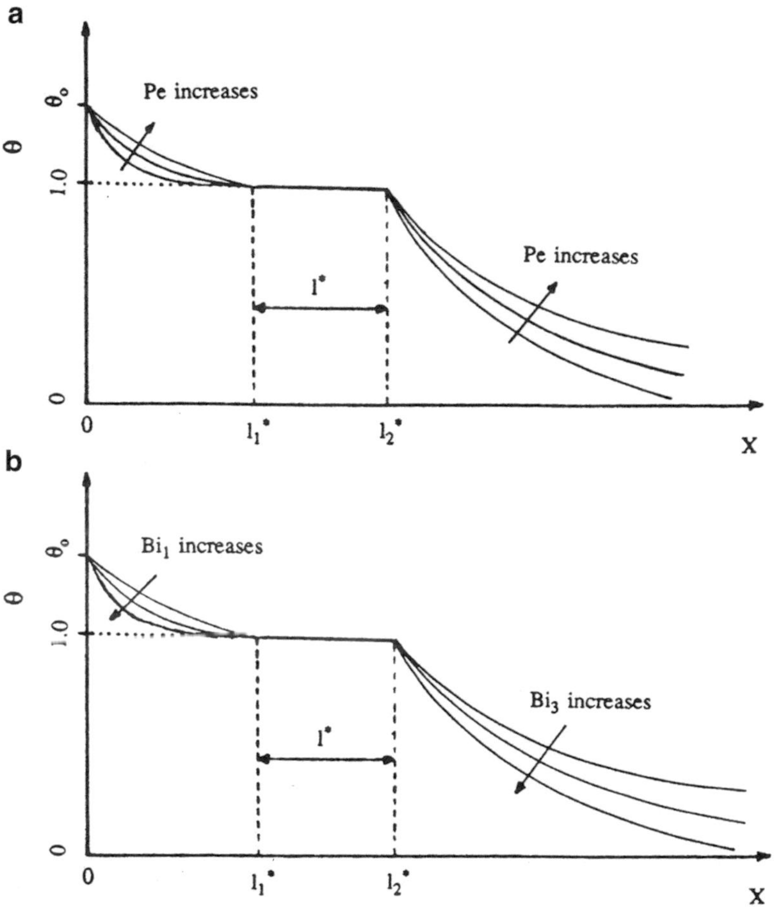

Fig. 4.15 Temperature profiles from the three-zone model for different Peclet and Biot numbers

temperature variations depend on *Pe*, Bi_1, and Bi_3, which may be adjusted to obtain the desired characteristics. Similarly, Bi_2 represents cooling in the mold and may be chosen to vary the region over which the phase change occurs. This is basically a design problem where the parameters may be chosen to obtain the desired solidification behavior and temperature distribution.

Similarly, numerical results may be obtained for continuous casting considering the flow in the liquid region. The results obtained by using the enthalpy model are shown in terms of isotherms in Fig. 4.16 (Kang and Jaluria 1993). The material is n-octadecane which starts as a liquid at the top and solidifies as it flows through a mold. The buoyancy effects in the flow are found to be small in this case. The shaded region indicates the demarcation between pure liquid and pure solid. Therefore, the liquid fraction f_l is 1.0 at the top of the shaded region and zero at the bottom of this region. A value of 0.5 may be taken to represent the liquid-solid interface, but the

Fig. 4.16 Temperature distributions in continuous casting. (**A**) Effect of Pe on the results for n-octadecane, (a), (b), (c) Pe = 1.0, 0.5, 0.3, (d) aluminum at *Pe* = 0.3; (**B**) effect of Bi_2, (a), (b), (c) Bi_2 = 0.05, 0.1, 0.15

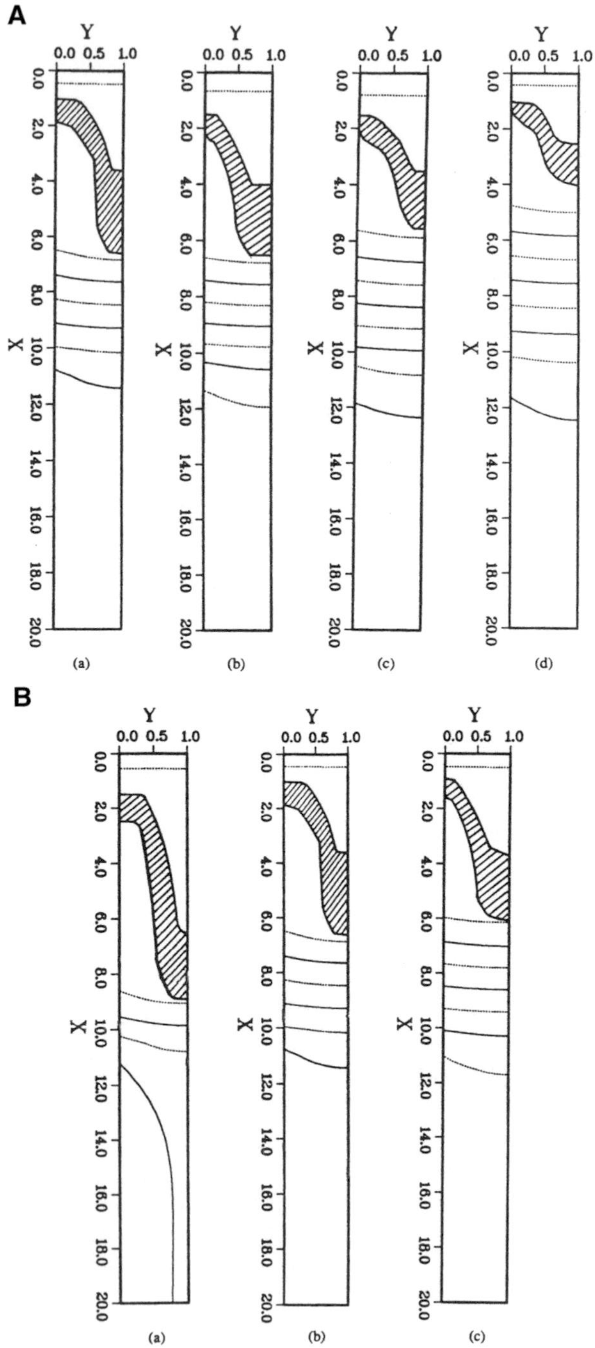

enthalpy method only yields a finite region over which solidification is predicted to occur. It is seen that the material solidifies over a shorter distance at a larger value of the Biot number, as expected. The effect of the Peclet number is to extend the phase change region due to the increased speed U_s. Similarly, results can be obtained for different materials, geometries, and operating conditions. The results may then be used for the design and control of the process.

4.6 Alloys

There has been a growing interest in the phase change processing of mixtures, particularly alloys, because of their extensive applications. Steel, bronze, duralumin, nichrome, constantan, chromel, solder, inconel, and stainless steel are some of the commonly used alloys. Substantial work has been done to understand the melting and solidification processes in order to obtain the desired quality and characteristics. Though the basic concerns are similar to pure materials, many additional complexities arise in alloys, and these have been considered in a wide range of experimental and numerical studies. An excellent review is presented by Prescott and Incropera (1996).

In alloys, solidification occurs over a range of temperatures, rather than a fixed melting point obtained in pure materials. Figure 4.17 shows a typical phase diagram of a binary mixture and the cooling curve. A distinct interface separating the liquid and solid phases is not obtained. Instead the solid is formed as a permeable, fluid-saturated, crystalline matrix over a multiphase region often known as the *mushy zone*. Latent energy is released or absorbed due to phase change between the interfaces that separate the phases in the mushy zone. The distribution of this energy depends on the specific structure of the multiphase region. This latent energy is

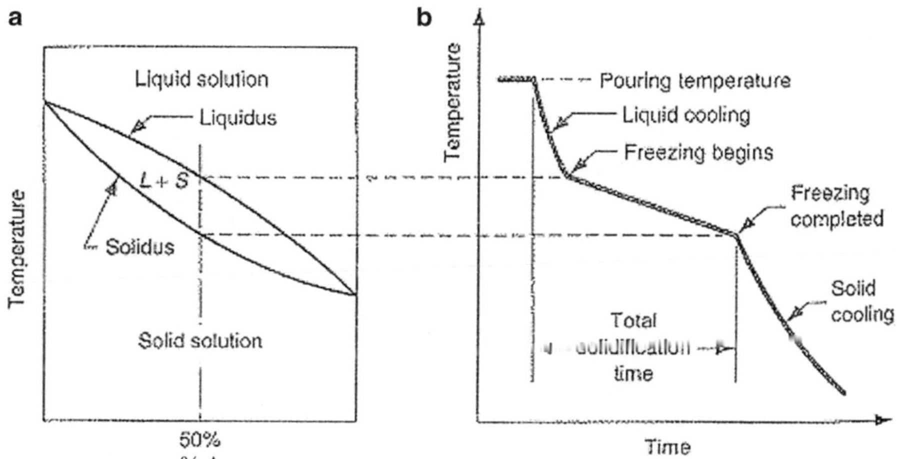

Fig. 4.17 A typical (**a**) phase diagram and (**b**) cooling curve for a mixture of components A and B

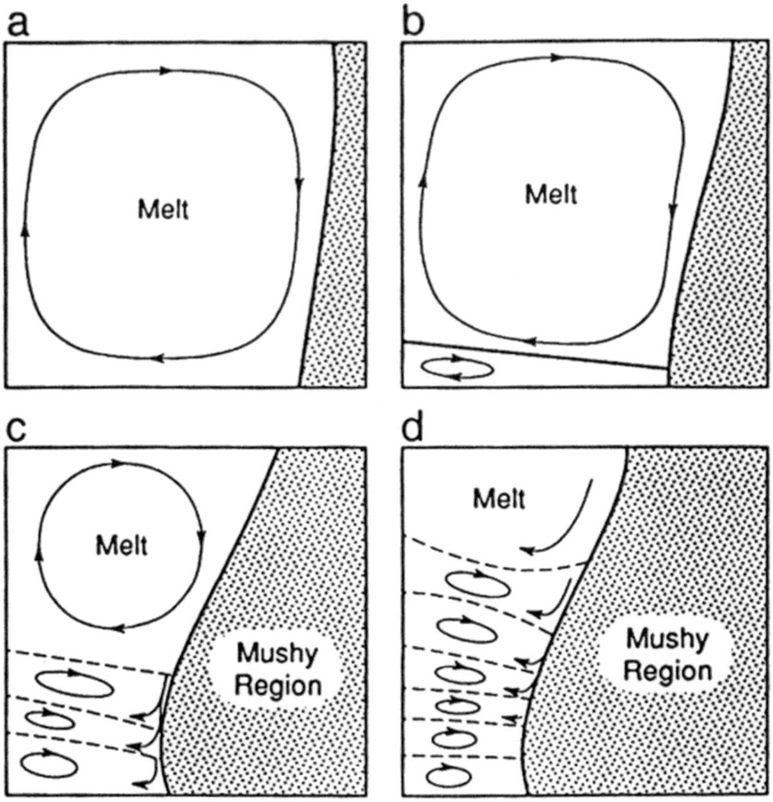

Fig. 4.18 Schematic of double-diffusive convection during solidification of aqueous Na_2CO_3 solution at various times following start of the solidification process. (**a**) 10 min, (**b**) 30 min, (**c**) 75 min, and (**d**) 150 min (Thompson and Szekely 1988)

transferred by conduction through the solid phase and by convection in the liquid phase. Also, in most systems, the chemical components have different solubilities in the different phases. Thus, a chemical species may be preferentially incorporated or rejected at the solid/liquid front. Combined heat and mass transfer processes arise in this case and significantly affect the flow. In general, the driving forces for solidification are the concentration and thermal gradients at the interface. Figure 4.18 shows a schematic of the double-diffusive convective flow that arises and increases in intensity with time. The left wall is heated and solidification occurs on the right wall. As the salt-enriched liquid is ejected from the mushy zone, it forms a layer at the bottom, and subsequently additional layers arise, with recirculation in each layer driven by horizontal temperature gradients. Lower heat transfer at the bottom results in larger mushy region thickness (Thompson and Szekely 1988). Many interesting experiments on the solidification of aqueous NH_4Cl solutions have been carried out to study the flow structure.

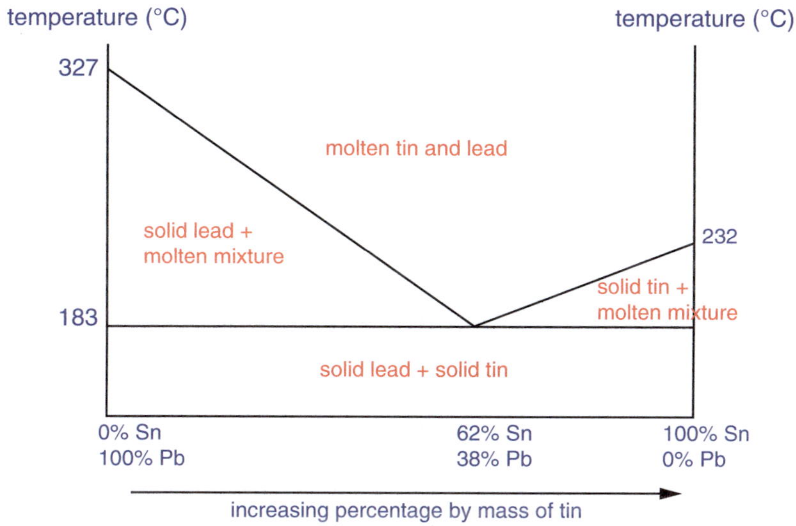

Fig. 4.19 A simplified phase diagram of a lead-tin alloy, showing the eutectic point

The enthalpy method is the most extensively used approach to model phase change in alloys. The phase diagram could be of the form shown in Fig. 4.17 or may be more complicated, depending on the different phases and solid solutions formed in the process. It may also have a eutectic point, which is the lowest possible melting temperature over all of the mixing ratios for the component species. A eutectic mixture has a fixed melting point, and, thus, the cooling curve is similar to that for a pure material. Figure 4.19 shows a simplified version of the real tin-lead phase diagram. Since this alloy is of interest in soldering, detailed property data are available along with more realistic phase diagrams, as discussed in a later chapter. But this figure shows the eutectic point and the basic characteristics of such alloys. Segregation during conventional solidification results in precipitation of a weight fraction f_e of the eutectic. For a binary phase diagram, such as the Al-Cu system, the weight fraction f_e is determined using the Scheil equation as

$$f_e = \left(\frac{C_e}{C_0}\right)^{1/(k_e-1)} \tag{4.26}$$

where C_e and C_0 are the eutectic and initial solute concentrations and k_e is the equilibrium partition coefficient. The flow in the mushy zone may be modeled as flow through porous media as given by the Carman-Kozeny equation, expressed as

$$\nabla p = -C\frac{(1-f)^2}{f^3+b}u_l \tag{4.27}$$

It can be seen that $f = 0$ (fully solid) results in a large value for the coefficient of the fluid velocity u_l and thus forces the velocities to become zero in that region. The

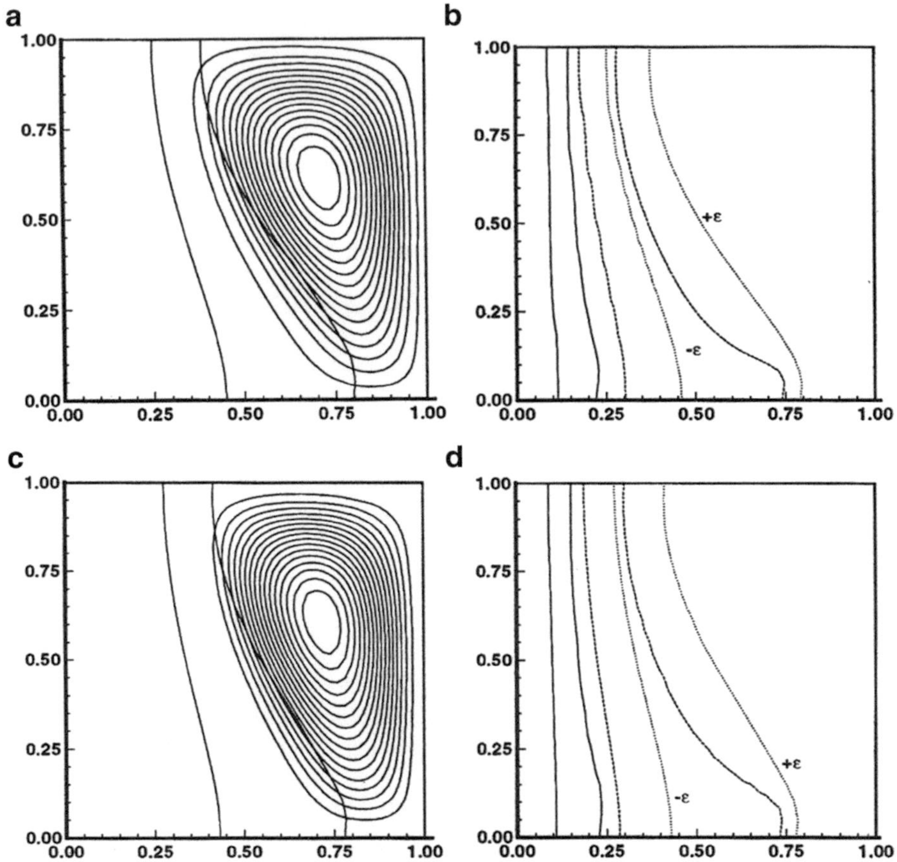

Fig. 4.20 Solidification of an alloy for $Ra = 10^4$, $\varepsilon = 0.1$, $Pr = 10^3$, $Ste = 5$, $C = 10^7$ for (**a**, **b**) $b = 0.1$ and (**c**, **d**) $b = 0.01$. (**a**, **c**) Steady-state streamlines and interface positions; (**c**, **d**) transient development of the mushy zone

constants C and b are suitably chosen to simulate flow characteristics in porous media. The validity of this equation in describing the flow in a dendritic mushy region was examined experimentally by Poirier (1986) and is of practical significance. Other models and approaches have also been developed and applied to a variety of phase change problems.

A few typical results from this approach are shown in Fig. 4.20 for a tin-lead alloy for both the steady-state case of the test problem mentioned earlier and for the transient development of the mushy zone. Here the parameter ε represents the mushy region, with $+\varepsilon$ being the liquidus and $-\varepsilon$ being the solidus temperatures. For small values of C, the convective field was seen to penetrate more into the mushy region. However, for really small values of C, significant flow was seen to exist in the solid region, which is not acceptable. A value greater than around 10^7 was found to be acceptable, since larger values did not affect the results significantly. For small

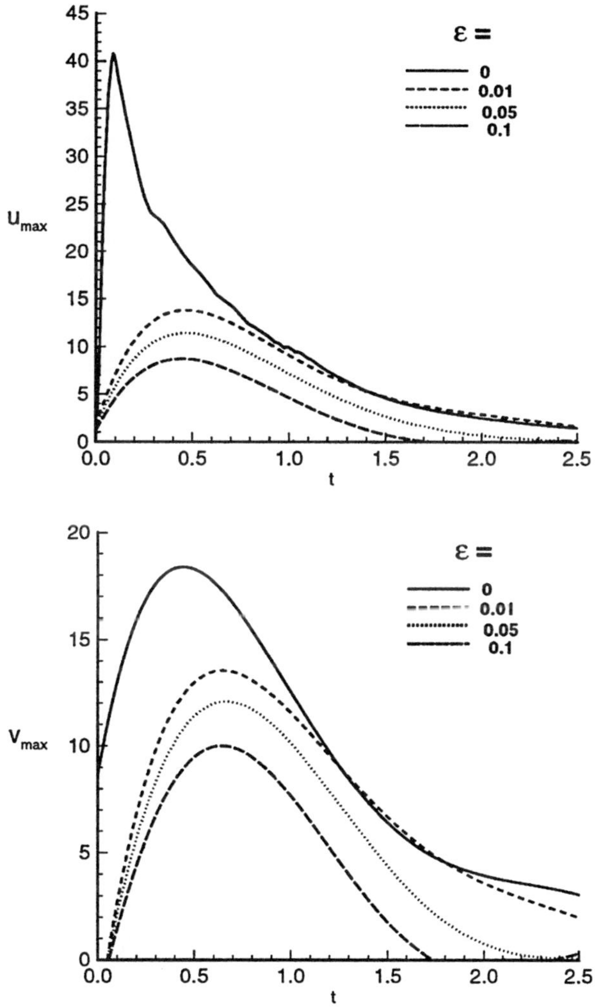

Fig. 4.21 Variation of the maximum horizontal velocity u_{max} and vertical velocity v_{max} in the cavity for different phase change temperature ranges ε at $Ra = 10^6$, $Ste = 5$, and $A = 2.0$

values of b, the mushy zone is advanced more to the right, resulting in a smaller liquid region. However, for values less than around 0.01, the results are unaffected by a further decrease. Obviously, these parameters can be chosen appropriately for different materials on the basis of their phase change characteristics. The calculated transient variation of the maximum velocities is shown in Fig. 4.21. A well-defined maximum is obtained in the horizontal velocity for the pure metal case. For alloy phase change, the velocities are gradually extinguished over a large region. A smaller phase change temperature range ε results in higher velocity in the melt region, and the time at which the maximum occurs is delayed as ε increases. Similarly, temperature profiles and other physical quantities of interest may be obtained. References such as Voller and Prakash (1987), Brent et al. (1988), and others given earlier may be consulted for further details.

Interest also lies in understanding microscopic phenomena associated with solidification. The solidification front is not always smooth, and, under practical circumstances, a variety of microscopically complicated growth structures will develop. This phenomenon, known as microsegregation, involves a chemical separation and concentration of the impurities and alloy segments at the microscale. The front can be divided into various morphological forms such as planar, cellular, and dendritic, depending on the temperature gradient and front velocity. In columnar dendritic growth, long aligned dendritic stalks are produced due to unidirectional heat flow. Equiaxed dendritic crystals grow to the same extent in several directions in multidirectional heat flow. These structures arise in mixtures as well as in pure materials, though solubilities and different species complicate the phenomena in mixtures. These structures exist because they are the most efficient morphologies for the diffusion of solute and dissipation of heat.

Various models have been proposed and experiments carried out to characterize such structures and growth (Wang and Beckermann 1993; Beckermann and Wang 1995; Voller, 2000, 2008; Rappaz et al. 2004). For instance, Fig. 4.22 shows equiaxed and columnar dendritic crystals. Averaging volumes and dendrite envelopes that may be used for modeling of the microscopic phenomena are shown. These considerations clearly lead to a multiscale problem, and the coupling between

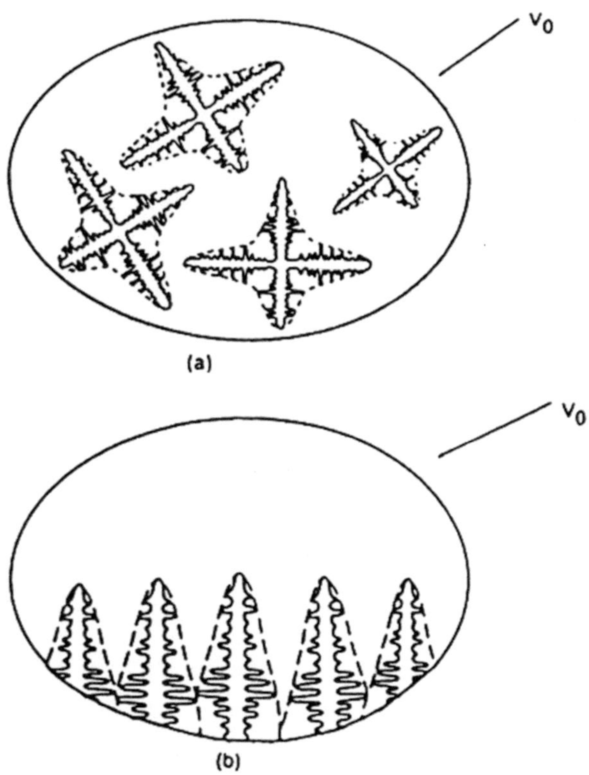

Fig. 4.22 Schematic illustration of the averaging volume and the dendrite envelopes for (**a**) equiaxed growth and (**b**) columnar growth (Beckermann and Wang 1995)

the microscale processes at the interface with the overall macroscopic system is critical to accurate modeling and predictions on the process. Similarly, spatial variations in composition occur in metal alloy castings over much larger length scales, from millimeters and centimeters to meters. This phenomenon is known as macrosegregation and negatively affects the quality of the casting (Voller 2014). It arises due to the relative flow of segregated liquid and solid during solidification and is seen in essentially all casting processes, being of particular concern in large castings and ingots. Much of the recent effort has focused on micro- and macrosegregation in order to improve the quality of the castings and obtain better control of the solidification process.

4.7 Crystal Growing

Crystal growing is an important process since most semiconductor devices are fabricated from single crystals grown from the vapor phase or from the melt. The former generally involves sublimation and chemical transport in a sealed enclosure (Rosenberger 1980). The latter was mentioned earlier, and sketches of two important techniques, Czochralski and floating zone, are shown in Fig. 4.23. The former was

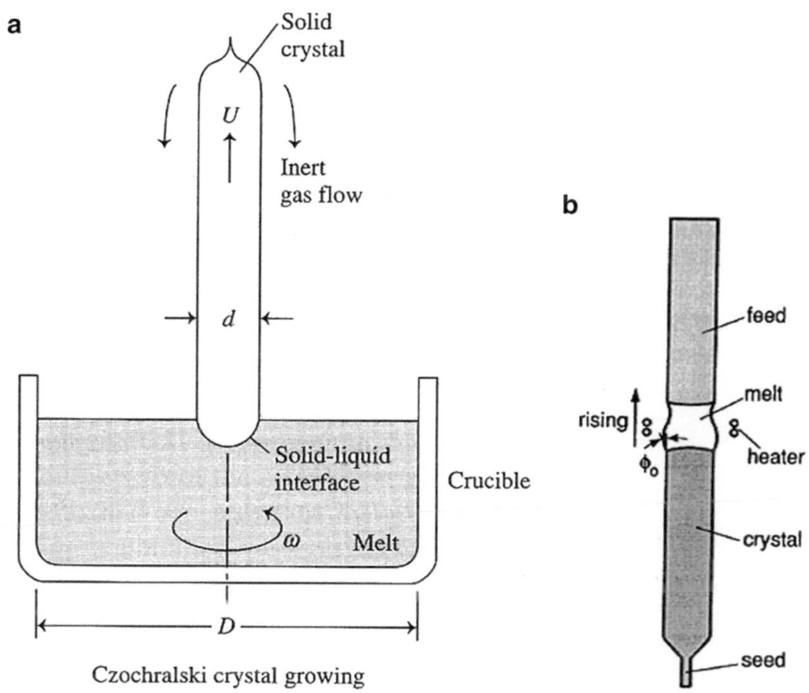

Fig. 4.23 (**a**) Czochralski crystal growing; (**b**) floating-zone method for crystal growth

shown earlier in Figs. 1.3b and 4.1b as well. In Czochralski crystal growing, molten material such as silicon is allowed to solidify across an interface as a seed crystal is withdrawn and in the floating-zone method, a molten zone is established between a polycrystalline charge rod and a crystalline rod. The process is similar to the continuous casting process discussed earlier, though several other complexities arise and particularly stringent requirements on the quality of the crystal are imposed. Several other crystal growth techniques, such as Bridgman crystal growth in which the furnace has an upper zone at temperature above the melting point and a lower zone at temperature below the melting point, have been developed (Kou 1996). The Czochralski method has dominated the production of single crystals for microelectronics and has been the subject of considerable research interest (Ostrach 1983; Prasad et al. 1997). The process involves buoyancy-driven convection due to temperature and concentration gradients, forced and mixed convection because of moving surfaces and materials, thermocapillary flows because of surface tension gradients, phase change, and thermal and mass transport processes.

Though silicon crystals have been of particular interest in the fabrication of electronic devices, there has been growing interest in GaAs, In*P*, and other such compounds because of their use in various electro-optic applications. An encapsulant layer of a very viscous melt such as boric oxide is placed over the melt to curb escape of volatiles. Figure 4.24a shows a schematic of the high-pressure

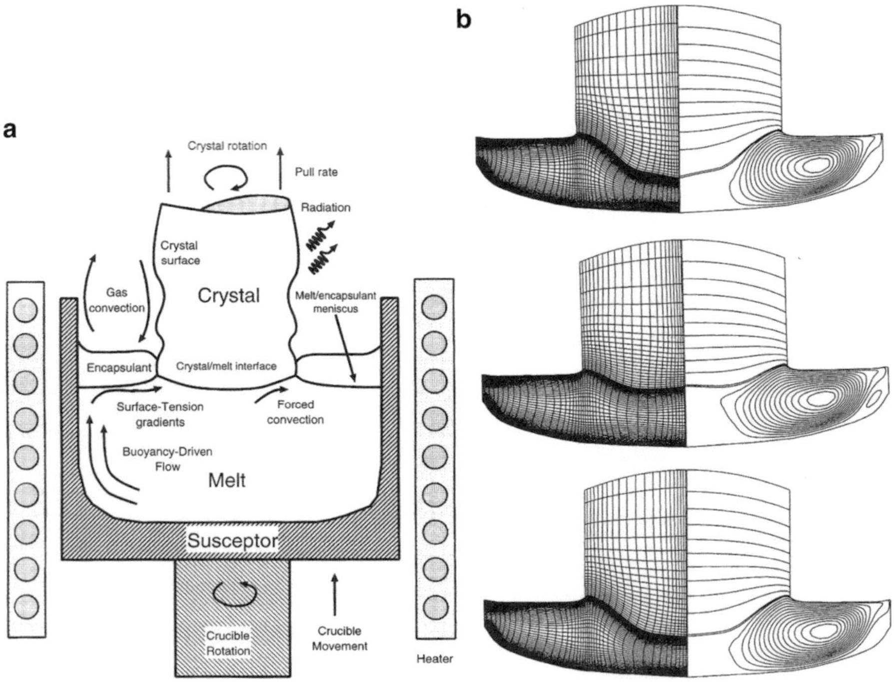

Fig. 4.24 (**a**) Schematic of the high-pressure liquid-encapsulated Czochralski crystal growing system; (**b**) grid distribution, flow field, and melt-crystal interface at three instants of time, showing strong oscillatory behavior which damps out at large time (Prasad et al. 1997)

liquid-encapsulated Czochralski process, indicating various mechanisms that arise (Brown 1988). The flow in the melt under the combined effects of buoyancy, surface tension, and rotation is shown in Fig. 4.24b, indicating oscillatory behavior of the flow and of the melt-crystal interface. Oscillations damp out at large time. However, at higher rotational speeds, the oscillatory behavior increases, and the process is difficult to simulate due to thin boundary layers. An applied magnetic field has been found to suppress oscillations and result in a flat interface (Prasad et al. 1997). Various other aspects, such as three-dimensional effects, continuous growth system, different thermal boundary conditions, and convection in high-pressure liquid-encapsulated Czochralski crystal growth, have been investigated.

4.8 Other Processes and Materials

There are several other manufacturing processes that involve phase change. These include processes like welding, soldering, and injection molding of plastics. In many polymer processing applications, including extrusion, the melting and solidification of the material are an important consideration (Jansson and Gedde 1992). In injection molding, for instance, the molten polymer is injected under pressure into a mold, and, after the mold is filled, the material is allowed to cool and thus solidify. Solidification in extrusion dies and in channels leading to a mold is not desirable since it affects the flow and the pressure due to the resulting blockage. The basic configuration for channel flow is shown in Fig. 4.25a, indicating a solidified layer near the boundaries and flow in the central core. Extensive experimental and numerical work has been done on this problem. For example, the enthalpy approach discussed earlier may be used to model the flow and thermal transport. The interface between the solid and melt regions and the velocity and temperature distributions may thus be computed (Richardson 1979; Lin and Jaluria 1997; Hassan et al. 2009). Some typical results are also shown in Fig. 4.25. It is important to note that as the temperature at the boundary is decreased, the thickness of the solidified layer increases, resulting in greater blockage to the flow. This, in turn, causes increased viscous dissipation, which heats up the fluid flowing in the central region. Thus, lowering the wall temperature ends up increasing the fluid temperature over the parametric ranges considered here. Complete blockage is not found to occur because of increasing viscous dissipation effects with greater blockage that heat up the fluid. Many such studies have been carried out and are discussed in a later chapter.

4.9 Summary

This chapter considers manufacturing processes that involve phase change of materials to obtain the desired product. Phase change arises in a fairly wide range of materials processing applications, such as casting, crystal growing, welding,

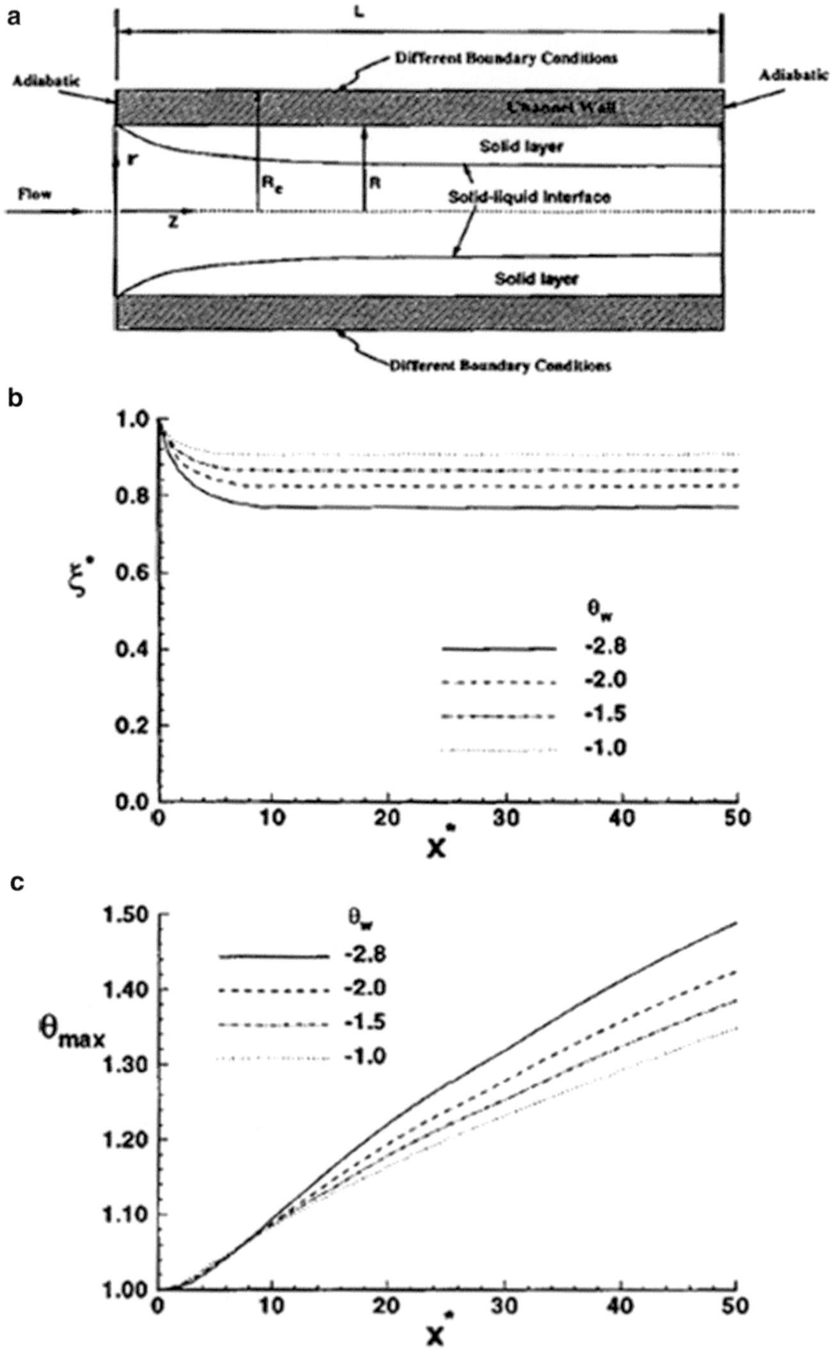

Fig. 4.25 (**a**) Schematic of polymer solidification in a channel; (**b**) dimensionless solid-liquid interface ξ*; (**c**) maximum temperature θ_{max} in the melt, with different outer wall temperatures θ_w

soldering, thermal sprays, injection molding, and additive manufacturing. As a result, extensive work has been done on melting and solidification processes to determine the basic characteristics of the process, time taken for solidification, quality of the solidified material, flow and temperature fields, thermal stresses, and the nature of impurities and defects in the product. Though simple conduction models have been traditionally used to determine solidification times and voids due to material shrinkage on solidification, the buoyancy-driven flow in the melt is critical to the behavior of the solid-liquid interface and the rate of solidification. The basic analytical and numerical approaches that may be employed are discussed, considering interface tracking and single-domain methods. The latter methods are based on enthalpy and are particularly suited to mixtures, polymers, and alloys that do not have a fixed melting point, and solidification occurs over a range of temperatures. For interface tracking, grid transformations are often used to obtain simpler domains. Some simple examples are taken from mold casting to illustrate the basic ideas. Essentially stationary interfaces arise in processes such as continuous casting and crystal growing in which the material moves as phase change occurs across a stationary interface whose stability, shape, and other characteristics are important in determining the quality of the product, as well as the rate of production. These aspects are discussed in terms of typical problems and systems.

Considerable interest and research activity has arisen in the recent years on the solidification of metal alloys due to their importance in various industrial applications and the stringent demands placed on the quality of the casting obtained. Similar considerations arise in other materials such as polymers and semiconductor materials. The basic approaches that have been employed are discussed here, along with some characteristic results. Some experimental results are presented in terms of benchmark problems for the validation of the numerical models.

Overall, materials processing based on phase change is of considerable importance, as seen from the extensive work done in this area. From classical processes such as mold casting to more recent processes such as crystal growing and injection molding, phase change is an important technique to obtain a given material in the form desired for a particular application. Though extensive literature is available on solidification, only the main aspects have been presented here. The various references given here may be consulted for further details.

References

J. Banaszek, Y. Jaluria, T.A. Kowalewski, M. Rebow, Semi-implicit FEM analysis of natural convection in freezing water. Numer. Heat Transf. **36**, 449–472 (1999)

A.G. Bathelt, R. Viskanta, W. Leidenfrost, An experimental investigation of natural convection in the melted region around a heated horizontal cylinder. J. Fluid Mech. **90**, 227–240 (1979)

C. Beckermann, C.Y. Wang, Multi-phase/-scale modeling of alloy solidification. Annu. Rev. Heat Transf. **6**, 115–198 (1995)

W.D. Bennon, F.P. Incropera, Developing laminar mixed convection with solidification in a vertical channel. J. Heat Transf. **110**, 410–415 (1988)

A.D. Brent, V.R. Voller, K.J. Reid, Enthalpy-porosity technique for modeling convection-diffusion phase change: Application to melting of a pure metal. Numer. Heat Transf. **13**, 297–318 (1988)
R.A. Brown, Theory of transport processes in single crystal growth from the melt. AICHE J. **43**, 881–911 (1988)
J.A. Dantzig, M. Rappaz, *Solidification*, 2nd edn. (EPFL Press, Lausanne, Switzerland, 2017)
G. de Vahl Davis, Natural convection of air in a square cavity: A bench mark solution. Int. J. Numer. Methods Fluids **3**, 249–264 (1983)
M.C. Flemings, *Solidification Processing* (McGraw-Hill, New York, 1974)
C. Gau, R. Viskanta, Melting and solidification of a pure metal on a vertical wall. J. Heat Transf. **108**, 174–181 (1986)
A. Ghosh, A.K. Mallik, *Manufacturing Science* (Ellis Horwood, Chichester, 1986)
H. Hassan, N. Regnier, C. Lebot, C. Pujos, G. Defaye, Effect of cooling system on the polymer temperature and solidification during injection molding. Appl. Therm. Eng. **29**, 1786–1791 (2009)
T. J. Hughel, G. F. Bolling (eds.), *Solidification* (American Society for Metals, Metals Park, 1971)
Y. Jaluria, *Design and Optimization of Thermal Systems*, 2nd edn. (CRC Press, Boca Raton, 2008)
S. Jana, S. Ray, F. Durst, A numerical method to compute solidification and melting processes. Appl. Math. Model. **31**, 93–119 (2007)
J. F. Jansson, U. F. Gedde (eds.), *Solidification Processes in Polymers* (Springer, Lausanne, Switzerland, 1992)
L.M. Jiji, *Heat Conduction*, 3rd edn. (Springer, Heidelberg, 2009)
B.H. Kang, Y. Jaluria, Thermal modeling of the continuous casting process. AIAA J. Thermophys. Heat Transf. **7**, 139–147 (1993)
S. Kou, *Transport Phenomena and Materials Processing* (Wiley, New York, 1996)
W. Kurtz, D.J. Fisher, *Fundamentals of Solidification*, 4th edn. (CRC Press, Boca Raton, 1998)
P. Lin, Y. Jaluria, Heat transfer and solidification of polymer melt flow in a channel. Polym. Eng. Sci. **37**, 1247–1258 (1997)
S. Ostrach, Fluid mechanics in crystal growth – The 1982 Freeman Scholar lecture. J. Fluids Eng. **105**, 5–20 (1983)
D.R. Poirier, Permeability for flow of interdendritic liquid in columnar-dendritic alloys. Metall. Trans. B **18**, 245–256 (1986)
V. Prasad, H. Zhang, A.P. Anselmo, Transport phenomena in Czochralski crystal growth processes. Adv. Heat Transf. **30**, 313–435 (1997)
P.J. Prescott, F.P. Incropera, Convection heat and mass transfer in alloy solidification. Adv. Heat Transf. **28**, 231–338 (1996)
N. Ramachandran, Y. Jaluria, J.P. Gupta, Thermal and fluid flow characteristics in one-dimensional solidification. Int. Commun. Heat Mass Transf. **8**, 69–77 (1981a)
N. Ramachandran, J.P. Gupta, Y. Jaluria, Two-dimensional solidification with natural convection in the melt and convective and radiative boundary conditions. Numer. Heat Transf. **4**, 469–484 (1981b)
N. Ramachandran, J.P. Gupta, Y. Jaluria, Thermal and fluid flow effects during solidification in a rectangular enclosure. Int. J. Heat Mass Transf. **25**, 187–194 (1982)
M. Rappaz, C. Beckermann, R. Trivedi (eds.), *Solidification Processes and Microstructures: A Symposium in Honor of Wilfried Kurz* (wiley, New York, 2004)
S.M. Richardson, Extended Leveque solutions for flows of power law fluids in pipes and channels. Int. J. Heat Mass Transf. **22**, 1417–1423 (1979)
F. Rosenberger, Fluid dynamics in crystal growth from vapors. Physicochem. Hydrodyn. **1**, 3–26 (1980)
W. Shyy, H.S. Udaykumar, M.M. Rao, R.W. Smith, *Computational Fluid Dynamics with Moving Boundaries* (Taylor & Francis, London, 1996)
R. Siegel, Shape of two-dimensional solidification interface during directional solidification by continuous casting. J. Heat Transf. **100**, 3–10 (1978)

R. Siegel, Two-region analysis of interface shape in continuous casting with superheated liquid. J. Heat Transf. **106**, 506–511 (1984)

E.M. Sparrow, S. Ramadhyani, S.V. Patankar, Analysis of melting in the presence of natural convection in the melt region. J. Heat Transf. **99**, 520–526 (1977)

E.M. Sparrow, S.V. Patankar, S. Ramadhyani, Effect of subcooling on cylindrical melting. J. Heat Transf. **100**, 395–402 (1978)

J. Szekely, V. Stanek, Natural convection transients and their effects in unidirectional solidification. Metall. Trans. **1**, 2243–2251 (1970)

M.E. Thompson, J. Szekely, Mathematical and physical modeling of double-diffusive convection of aqueous solutions crystallizing at a vertical wall. J. Fluid Mech. **187**, 409–433 (1988)

R. Viskanta, Heat transfer during melting and solidification of metals. J. Heat Transf. **110**, 1205–1219 (1988)

R. Viswanath, Y. Jaluria, A comparison of different solution methodologies for melting and solidification problems in enclosures. Numer. Heat Transf. **24**, 77–105 (1993)

R. Viswanath, Y. Jaluria, Numerical study of conjugate transient solidification in an enclosed region. Numer. Heat Transf. **27**, 519–536 (1995)

V.R. Voller, in *An Overview of Numerical Methods for Solving Phase Change Problems, Advances in Numerical Heat Transfer*, ed. by W. J. Minkowycz, E. M. Sparrow (Eds), vol. 1, (Taylor & Francis, Philadelphia, 1997), pp. 341–380

V.R. Voller, A model of microsegregation during binary alloy solidification. Int. J. Heat Mass Transf. **43**, 2047–2052 (2000)

V.R. Voller, An enthalpy method for modeling dendritic growth in a binary alloy. Int. J. Heat Mass Transf. **51**, 823–834 (2008)

V.R. Voller, Numerical methods for phase-change problems, Chapter 19, in *Handbook of Numerical Heat Transfer*, 2nd, W.J. Minkowycz, E.M. Sparrow and J. Murthy, Wiley, New York (2006)

V.R. Voller, Frequency analysis of macrosegregation measurements and simulations. Int. J. Heat Mass Transf. **79**, 468–471 (2014)

V.R. Voller, C. Prakash, A fixed grid numerical modeling methodology for convection diffusion mushy region phase change problems. Int. J. Heat Mass Transf. **30**, 1709–1719 (1987)

C.Y. Wang, C. Beckermann, A unified solute diffusion model for columnar and equiaxed dendritic alloy solidification. Mater. Sci. Eng. **A171**, 199–211 (1993)

F. Wolff, R. Viskanta, Melting of a pure metal from a vertical wall. Exp. Heat Transf. **1**, 17–30 (1987)

F. Wolff, R. Viskanta, Solidification of a pure metal at a vertical wall in the presence of liquid superheat. Int. J. Heat Mass Transf. **31**, 1735–1744 (1988)

Chapter 5
Continuous Materials Processing

In many manufacturing processes, the material undergoing processing is continuously moving. An example of this circumstance was seen in the last chapter as continuous casting. The moving material is subjected to heat and mass transfer in order to transform it into a desired configuration or to obtain certain characteristics. Processes such as hot and cold rolling, wire and fiber drawing, extrusion, and metal forming generally involve continuously moving materials (Arridge and Prior 1964; Fisher 1976; Altan et al. 1983; Tadmor and Gogos 1979). The speed at which the material moves can be as low as a few centimeters per hour, as in the case of Czochralski crystal growth process (Jackson 1974), or as high as a few meters per second, as in the case of hot rolling or wire drawing in cable manufacture (Hollander 1970). Figure 5.1 shows sketches of a few manufacturing processes in which transport from continuously moving materials is of interest.

Continuous processing is attractive because of higher production rates, as compared to batch processes, where the system has to be stopped after each batch of raw material and then restarted for the next one. Efforts are often made to convert traditional batch operations to continuous ones, for instance, in the manufacture of solar cells and flexible electronic equipment. Because of the growing interest in and importance of continuous processing, substantial work has been done on this problem for many diverse applications, which involve different geometries, boundary and operating conditions, materials, requirements, and constraints. For instance, Glicksman (1968, 1974), Matovich and Pearson (1969), and Gifford (1982) studied the formation and cooling of glass and metal fibers, Tseng (1984a, b) considered continuous metal forming, and Chopra and Niessen (1975) and Kuiken (1977) studied solidification of a moving sheet. The cooling of flat sheets and cylinders, as related to hot rolling and metal forming, has been studied extensively, such as by Yangai (1976) and many others. Many of these are traditional manufacturing processes, but demands on higher quality and the emergence of many new materials have provided a strong impetus to further research on these processes, as well as design and optimization of relevant systems. Later chapters discuss optical fiber drawing and polymer extrusion, which involve continuous processing, in detail.

Y. Jaluria, *Advanced Materials Processing and Manufacturing*, Mechanical Engineering Series, https://doi.org/10.1007/978-3-319-76983-7_5

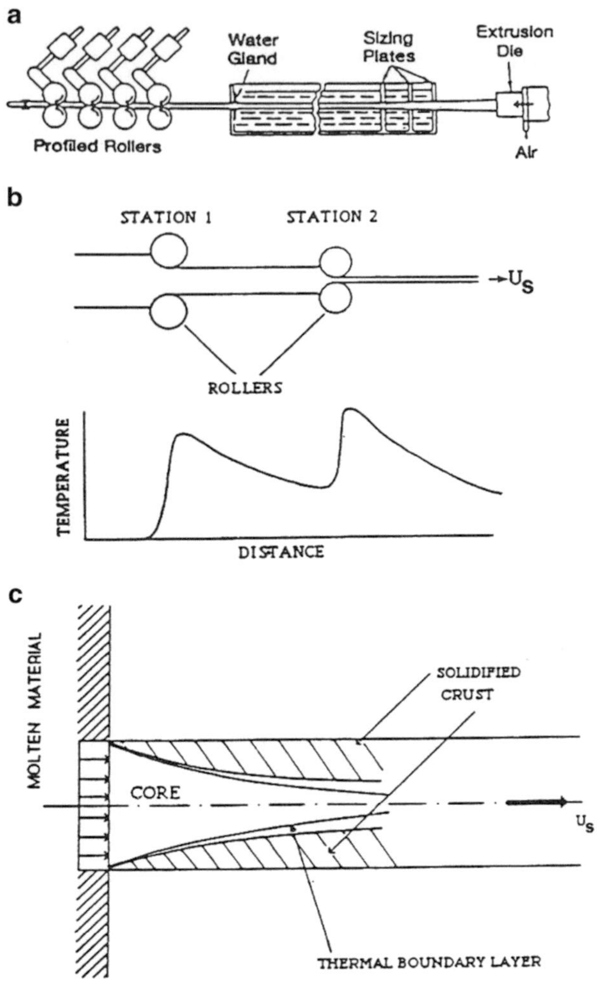

Fig. 5.1 Examples of a few manufacturing systems that involve transport in a continuously moving material. (**a**) Wire drawing; (**b**) hot rolling; (**c**) continuous casting

5.1 Basic Process

The basic process involves a continuously moving material, subjected to heat and mass transfer at the surface. Typically, convection and radiation, as well as possibly phase change due to evaporation cooling or condensation heating, are considered at the surface. The transport in the moving material is coupled with the convection outside, arising due to the motion of the material, buoyancy, and forced flow. Therefore, a conjugate problem is typically of interest, with temperature variation in the moving material and in the ambient fluid, as sketched in Fig. 5.2. Also, the process goes through a transient at the start of the process and, in most cases, reaches steady-state conditions at large time. Figure 5.3 shows a sketch of an extrusion process and the transient phenomenon that arises as the material first emerges from

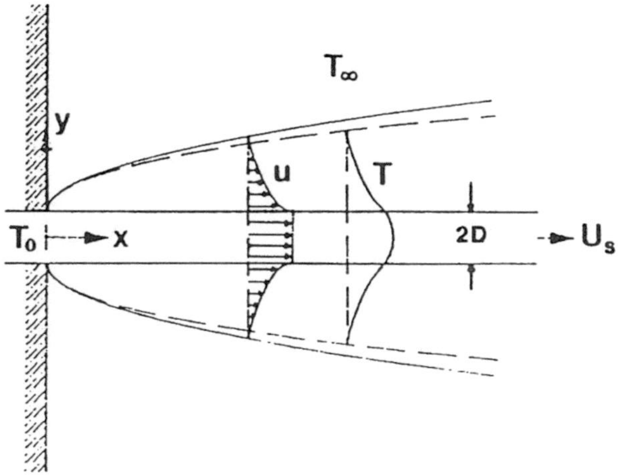

Fig. 5.2 Schematic of the velocity and temperature profiles in the conjugate problem for a continuously moving material

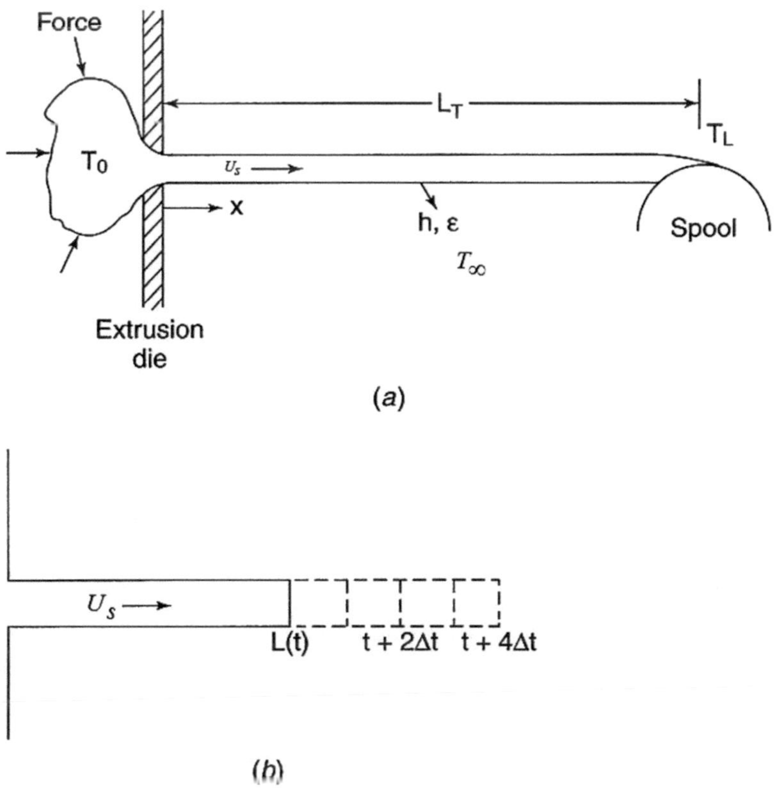

Fig. 5.3 (**a**) Sketch of a metal extrusion process. (**b**) The transient process, showing the edge of the moving material at different time intervals with a time step of Δt

the extrusion die. As the edge of the moving material moves away from the die, a steady-state situation is finally attained. The figure also shows the position of the edge of the material at different time intervals, t, $t + \Delta t$, $t + 2\Delta t$, etc., where Δt is a time step.

Therefore, the general problem to be solved is a time-dependent, three-dimensional, conjugate transport problem with a moving boundary. However, depending on the materials, geometry, and application, several simplified circumstances may be considered to obtain the desired information on the process. These include:

1. Transport within the moving material for a specified thermal condition at the surface
2. Convective flow due to moving material and transport at the surface
3. Conjugate transport
4. Transient effects

In the first case, convective heat transfer coefficients, based on empirical correlations derived from experimental results for similar configurations, are employed to impose the boundary conditions at the surface. The second case focuses on flow induced in the ambient fluid due to material movement, along with forced flow and buoyancy effects, to determine the resulting thermal field and associated transport to a moving rod or plate at given temperature. The third case couples the transport in the moving material with the convection and radiation at the surface. Finally, the last case focuses on transient effects, such as those at the start of the process. The dimensions and boundary conditions dictate whether a one-, two-, or three-dimensional problem is to be solved and whether time dependence must be included. In addition, sprays may be used to cool the material, resulting in phase change, and other heating/cooling approaches may be used, depending on the application. A review of the different regimes and solution methods for several different cases is given by Jaluria (1992). Analytical, numerical, and experimental approaches have been adopted in different circumstances, and the corresponding results are available in the literature, as briefly presented here. The results shown are for common materials and typical operating conditions for a range of manufacturing processes.

5.2 Heat Transfer Within the Moving Material

The transport processes within the moving material are of particular importance in materials processing, since the resulting temperature distribution affects the material microstructure, thermal stresses, and transformations within the material. It also determines the distance needed for achieving a desired temperature level for a particular speed, the energy transferred to and from the material, and several other aspects that are related to product quality and production rate, depending on the process under consideration. The boundary conditions at the surface are given in terms of the convective heat transfer coefficient, which is based on available correlations or experimental data. Radiation may be included by the use of simple enclosure models, for given temperatures of the surrounding surfaces, or as specified heat flux input from a high-temperature source such as radiant heating panels.

5.2.1 Analysis

Consider a cylindrical rod of diameter D moving at constant speed U_s, as shown in Fig. 5.3. Then, for constant material properties and axisymmetry, the time-dependent temperature T in the material is governed by the equation

$$\frac{\partial T}{\partial t'} + U_s \frac{\partial T}{\partial x} = \alpha \left(\frac{\partial^2 T}{\partial r^2} + \frac{1}{r} \frac{\partial T}{\partial r} + \frac{\partial^2 T}{\partial x^2} \right) \tag{5.1}$$

where r is the radial coordinate distance from the axis, x is the axial coordinate distance from the point of emergence of the solid material, t' is physical time, and α is the material thermal diffusivity. The equation may be nondimensionalized with D and U_s as the characteristic quantities to yield

$$\frac{\partial \theta}{\partial t} + Pe \frac{\partial \theta}{\partial X} = \frac{\partial^2 \theta}{\partial R^2} + \frac{1}{R} \frac{\partial \theta}{\partial R} + \frac{\partial^2 \theta}{\partial X^2} \tag{5.2}$$

where $R = r/D$, $X = x/D$, $\theta = (T - T_\infty)/(T_0 - T_\infty)$, $t = \alpha t'/D^2$, and $Pe = U_s D/\alpha$. Here, the ambient temperature is taken as T_∞. Thus, the Peclet number Pe arises an important parameter. Similarly, for two-dimensional transport in a wide plate of thickness H, the equation is

$$\frac{\partial \theta}{\partial t} + Pe \frac{\partial \theta}{\partial X} = \frac{\partial^2 \theta}{\partial X^2} + \frac{\partial^2 \theta}{\partial Y^2} \tag{5.3}$$

where H replaces D in the preceding cylindrical case.

The boundary and initial conditions for these equations may be written as

$$\begin{aligned} &t = 0 : L(t) = 0 \\ &t > 0 : \text{at } X = 0, \theta = 1.0 \quad \text{for} \quad 0 \le R \le 0.5 \\ &\text{at } X = L(t), \frac{\partial \theta}{\partial X} = -Bi_L \theta \quad \text{for} \quad 0 \le R \le 0.5 \\ &\text{at } R = 0, \frac{\partial \theta}{\partial R} = 0 \quad \text{for} \quad 0 \le X \le L(t) \\ &\text{at } R = 0.5, \frac{\partial \theta}{\partial X} = -Bi\theta \quad \text{for} \quad 0 \le X \le L(t) \end{aligned} \tag{5.4}$$

Here, $L = L'/D$, *where* $L' = U_s t'$ *and is* the physical length, h the convective heat transfer coefficient, k the material thermal conductivity, the Biot number $Bi = hD/k$ and Bi_L is for convection at the end surface of the rod. As L increases, it is expected that the steady-state circumstance will be attained eventually, so that $L \to \infty$, $\frac{\partial \theta}{\partial X} \to 0$, implying that the end surface heat loss is negligible for a long plate or rod and the temperature attains the ambient temperature, or $\theta \to 0$.

In many practical problems, such as hot rolling of steel sheets, the thickness is small, and a simpler one-dimensional model, with uniform temperature across any cross-section, may be employed to obtain θ (X, t). This model applies for small values of Bi. The governing equation for the moving plate becomes

$$\rho C_{\mathrm{p}}\left(\frac{\partial T}{\partial t}+U_{\mathrm{s}}\frac{\partial T}{\partial x}\right)=k\frac{\partial^2 T}{\partial x^2}-\frac{hP}{A}(T-T_{\infty}) \tag{5.5}$$

with the initial and boundary conditions

$$t=0: L(t)=0 \quad t>0: \text{at } x=0, T=T_0; \text{at } x=L(t), -k\frac{\partial T}{\partial x}=h_L(T-T_{\infty}) \tag{5.6}$$

where h_L is the heat transfer coefficient at the end surface of the moving rod, P the perimeter, and A the cross-sectional area. These can again be written in dimensionless terms, with H as the characteristic dimension, as

$$\frac{\partial \theta}{\partial t}+Pe\frac{\partial \theta}{\partial X}=\frac{\partial^2 \theta}{\partial X^2}-\frac{PH}{A}Bi\theta \tag{5.7}$$

The initial conditions are the same as given earlier in Eq. (5.4), and the boundary conditions are

$$t>0: \text{at } X=0, \theta=1.0 \quad \text{and} \quad \text{at } X=L(t), \frac{\partial \theta}{\partial X}=-Bi_L\theta \tag{5.8}$$

Therefore, the Peclet number Pe and the Biot number Bi arise as the main governing parameters. If the end convective heat transfer coefficient is different from h, Bi_L is an additional parameter. The geometry is reflected in the boundary conditions or in the value of P/A in the 1D model. For a wide plate, the parameter PH/A simply becomes 2, and for an infinite cylinder, it becomes 4, with D replacing H as the characteristic dimension. If radiative heat loss at the surface is also included, the Biot number may be defined in terms of the overall heat transfer coefficient, including a radiation heat transfer coefficient for small temperature differences from the surroundings. Otherwise, a radiation term must be included in the analysis. However, inclusion of radiation would lead to nonlinearity due to the dependence of transport on temperature as T^4, unless a simplification such as the radiation heat transfer coefficient is made, and the problem has to be solved numerically in most cases (Jaluria and Singh 1983).

The temperature θ at the point of emergence of the material from a die, rollers, or a preceding processing system, $X = 0$, is taken as 1.0 in the above formulation. However, a more realistic boundary condition would be one that takes into account the upstream thermal condition in, say, a furnace, oven, or extruder. Also, the temperature θ may be imposed at 1.0 far upstream, that is, $X \to -\infty$, indicating the temperature at which the material is maintained before moving toward the point of emergence. This is an important consideration, since the temperature at $X = 0$ is a

Fig. 5.4 Sketch of a continuously moving material whose surface is subjected to convective heat transfer over a region of length L', with no surface heat transfer in the other regions, and which has a specified temperature T_0 far upstream

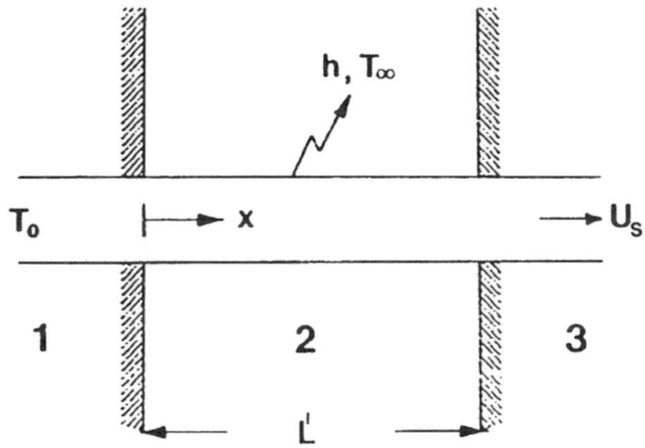

consequence of material motion and the downstream heat transfer mechanisms. This aspect is discussed below.

Consider the circumstance shown in Fig. 5.4, where regions 1 and 3 have no surface heat transfer and surface convective transport exists in region 2. The temperatures T_1, T_2, and T_3 in the three regions are governed by the following equations, under steady-state 1D conditions:

$$\frac{d^2T_1}{dx^2} - \frac{U_s}{\alpha}\frac{dT_1}{dx} = 0 \tag{5.9}$$

$$\frac{d^2T_2}{dx^2} - \frac{U_s}{\alpha}\frac{dT_2}{dx} - \frac{hP}{kA}(T_2 - T_\infty) = 0 \tag{5.10}$$

$$\frac{d^2T_3}{dx^2} - \frac{U_s}{\alpha}\frac{dT_3}{dx} = 0 \tag{5.11}$$

The temperature is assumed to approach T_0 as $x \to -\infty$ and to remain finite as $x \to \infty$. Continuity of temperature and heat flux at $x = 0$ and $x = L'$ provide the remaining boundary conditions. Thus, this problem brings up the question of temperature at $X = 0$ and the upstream penetration of thermal effects due to conduction. Also, this is a typical circumstance in many materials processing operations, such as metal forming and heat treatment, where the material is subjected to heat transfer over only a limited region, while the regions before and after the heated section are unheated.

5.2.2 Analytical and Numerical Results

The steady-state one-dimensional problems can generally be solved analytically. But transient problems, as well as two- or three-dimensional problems and those with radiation, generally have to be solved numerically. For the numerical modeling of the transient problem, if the location of the end or edge of the moving surface is known, as shown in Fig. 5.3, the continuous movement of the material may be replaced by finite steps, so that the length L is kept constant over a time increment Δt and the transient conduction problem is solved over this interval. The length L is then taken at the increased value for the next time interval, with the additional finite region adjacent to the base taken at temperature T_0, and the computation is carried out for this interval. This numerical procedure is carried out until results are obtained over a given time interval or until the steady-state circumstance is obtained for large L (Jaluria 1992). The appropriate numerical method for all these circumstances may be chosen from the discussion in Chap. 3. Some of the typical results obtained by analysis or computation are given here.

Figure 5.5a shows the temperature variation with downstream distance X for the 1D steady-state case at different values of the dimensionless distance L_T between the inlet and the location beyond which the adiabatic boundary condition is imposed. Clearly, the temperature at the end depends on the distance L_T. As L_T increases to large values, being 7.0 for the conditions employed here, the end temperature attains the ambient value, i.e., $\theta = 0$. This result is important in choosing L_T for a physical, practical, process. Similar trends are obtained for the two-dimensional problem (Jaluria and Singh 1983). Figure 5.5b shows the transient results, indicating the approach to steady-state conditions at large time t. The length $L(t)$ increases with time, ultimately becoming large enough for the end temperature to attain the ambient value. Similar behavior is observed for a flat plate and 2D and axisymmetric cases. The isotherms for some of these cases are shown in Fig. 5.6. The time-dependent isotherms vary gradually as time elapses until the steady-state situation is reached and the 1D model is seen to be appropriate for small values of Bi.

Similarly, the circumstance shown in Fig. 5.4 is considered, and Eqs. (5.9)–(5.11) are solved analytically. The results on the temperatures in the three regions are obtained as closed-form algebraic expressions in terms of the Peclet number $\hat{Pe} = U_s\gamma/\alpha$ and the Biot number $\hat{Bi} = 4h\gamma/k$, where $\gamma = A/P$, as given by Karwe and Jaluria (1986). Figure 5.7 shows the results graphically for typical values of the governing parameters $\hat{Pe}$ and the aspect ratio parameter δ, which is the ratio of the characteristic dimension to the length L'. The downstream temperature decay becomes more gradual as the Peclet is increased, due to shorter time for heat loss up to a given distance, and this results in an increase in temperature at a given location. At larger Peclet numbers, the diffusion effects upstream are smaller due to decreasing conduction effects, as compared to convection due to the moving material. A larger Biot number implies more rapid cooling downstream and the upstream penetration of thermal effects are also found to increase with increasing Biot number. A decrease in δ implies an increase in L' for a given cross-section. This results in lower temperature in region 3, since a larger distance is available for heat loss. For very small δ, being around 0.2 in this figure, the temperature at the end of region 2 is already close to the ambient temperature, and a further decrease in δ will not affect the temperature distribution. All these observations and results are useful in the design of the system to achieve the desired control on material temperature for processing. Further work on this problem could involve 3D and transient effects, inclusion of material property variations, dependence of material microstructure on

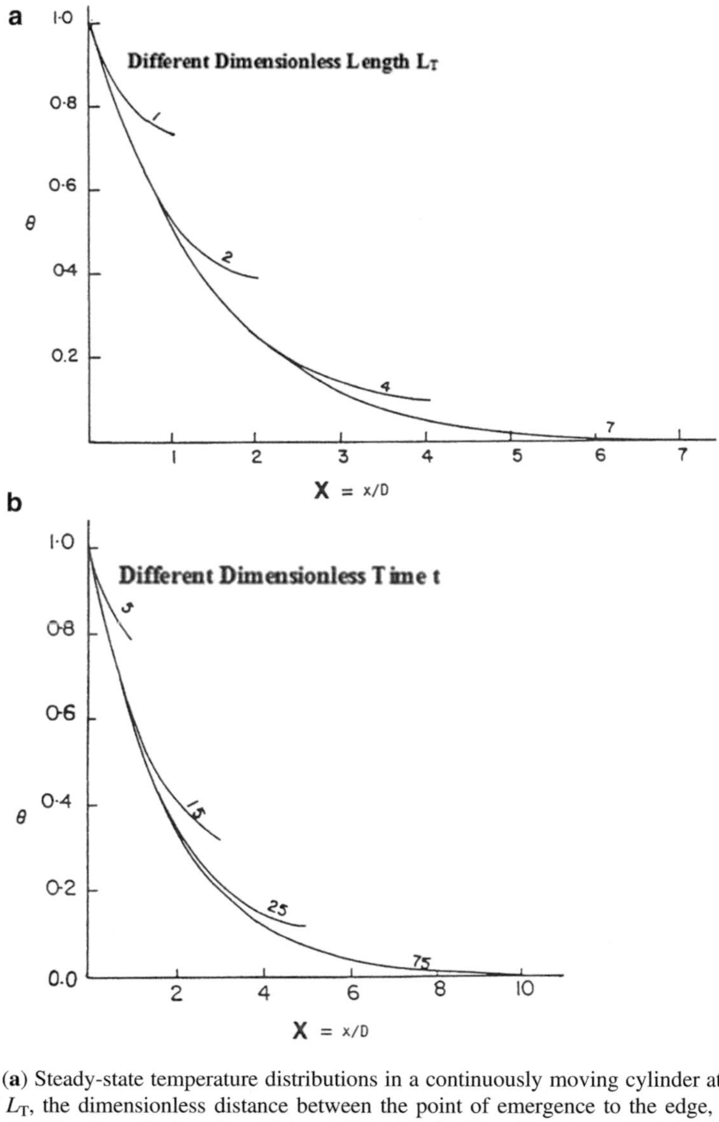

Fig. 5.5 (**a**) Steady-state temperature distributions in a continuously moving cylinder at different values of L_T, the dimensionless distance between the point of emergence to the edge, where an adiabatic condition is applied, at $Pe = 0.4$ and $Bi = 0.2$. (**b**) Transient temperature distributions at various dimensionless time t after the start of the process for $Pe = 0.2$ and $Bi = 0.1$

the thermal field, different thermal conditions at the surface, other geometries, and inclusion of other modes of heat transfer. Several of these aspects are brought up again later in optical fiber drawing and other continuous processes (see Jaluria (2013, 2016)).

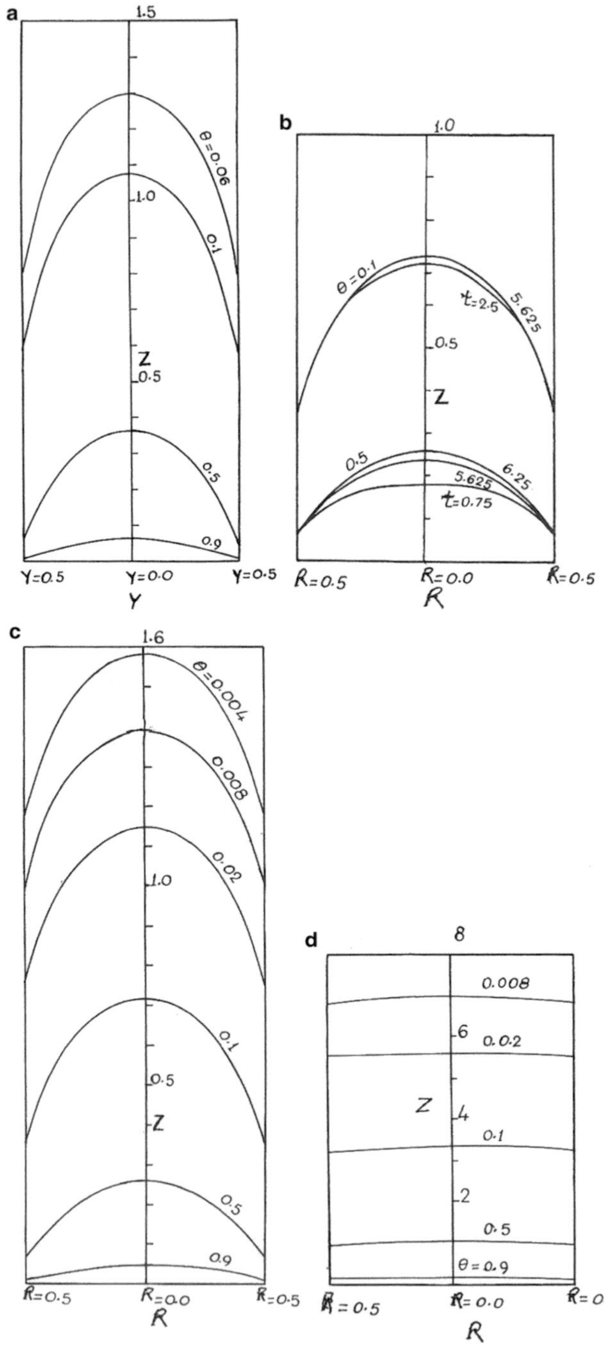

Fig. 5.6 (**a**) Isotherms in a moving 2D strip at $Pe = 0.4$ and $Bi = 10.0$. (**b**) Time-dependent isotherms for the axisymmetric problem at $Pe = 0.4$ and $Bi = 10.0$. (**c**) Steady-state isotherms in a long moving cylindrical rod at $Pe = 0.4$ and $Bi = 10.0$. (**d**) Isotherms in a long moving cylindrical rod at $Pe = 0.4$ and $Bi = 0.2$

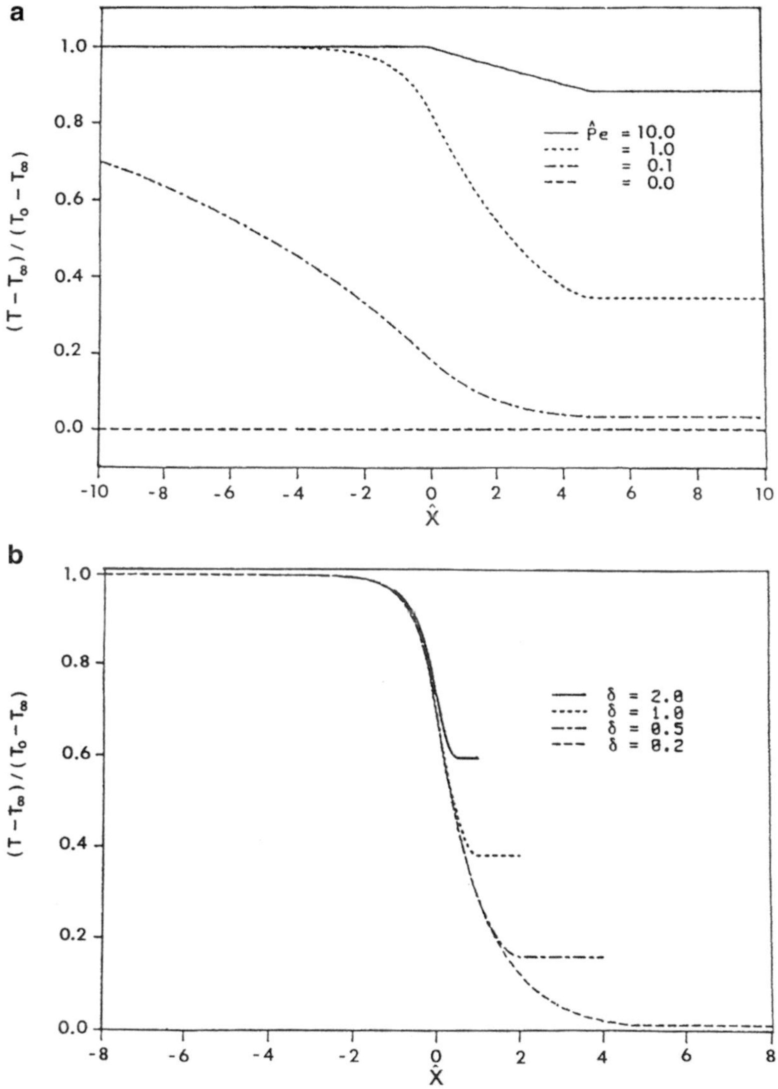

Fig. 5.7 (**a**) Effect of varying the Peclet number on the temperature distribution for the problem shown in Fig. 5.4 at a Biot number of 1.0 and $\delta = 0.2$. (**b**) Effect of varying the parameter δ on the temperature distribution at a Peclet number of 2.0 and Biot number of 10.0

5.3 Convective Transport Due to a Moving Surface

The heat transfer coefficient h at the surface is not known but arises from the fluid flow adjacent to the moving surface. Therefore, it is important to consider the flow induced by a moving surface in a quiescent ambient medium to determine the heat

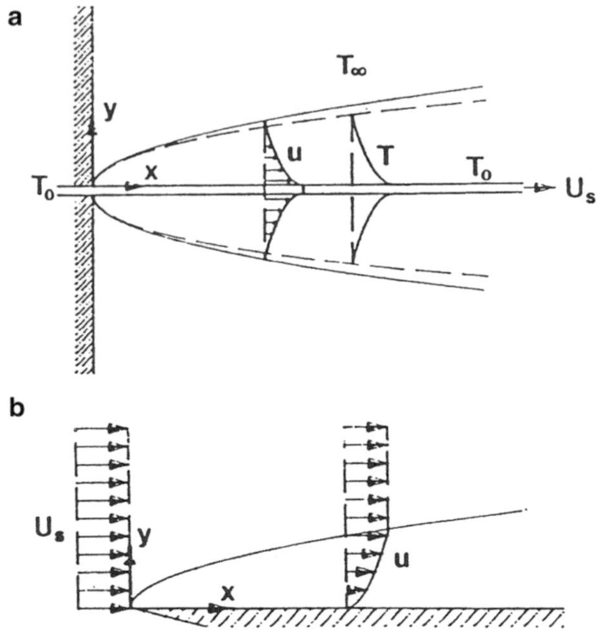

Fig. 5.8 (**a**) Sketch of the velocity and temperature profiles due to a moving isothermal surface at temperature T_0. (**b**) Sketch of the velocity profiles for a uniform parallel flow over a stationary surface

transfer due to this flow. Of course, in many cases, an additional forced flow, as well as significant buoyancy effects, may be present, and the resulting transport is due to the combined effect of these three mechanisms (Roy Choudhury and Jaluria 1995). Figure 5.8 shows the velocity and temperature boundary layers generated by a continuously moving, isothermal, flat surface. This problem was first studied by Sakiadis (1961a, b), who employed a similarity transformation to convert the governing partial differential equations for a semi-infinite surface to ordinary differential equations. The boundary-layer thickness grows in the direction of surface motion, and the velocity decreases from U_s at the surface to zero in the ambient medium. This velocity variation is quite different from the boundary layer on a stationary flat surface in uniform parallel flow. Similarly, other analytical and experimental studies have been carried out by Koldenhof (1963), Erikson et al. (1966), Tsou et al. (1967), Griffin and Thorne (1967), Chen and Strobel (1980), and Moutsoglou and Chen (1980), considering different geometries, turbulence, mass transfer, and other additional effects. Recent years have seen many studies on continuously moving surfaces, or stretching sheets, considering effects such as magnetic fields, nanofluids, viscous dissipation, non-Newtonian fluids, porous media, and combined heat transfer mechanisms, (see, e.g., the papers by Khan and Pop (2010), Jafar et al. (2012), and Mastroberardino and Mahabaleshwar (2013)). Many of these papers are largely mathematical in nature and may not relate to actual physical processes of interest in manufacturing or other practical systems.

For a 2D laminar boundary-layer flow, the governing equations for constant fluid properties, with negligible viscous dissipation and pressure work effects, are

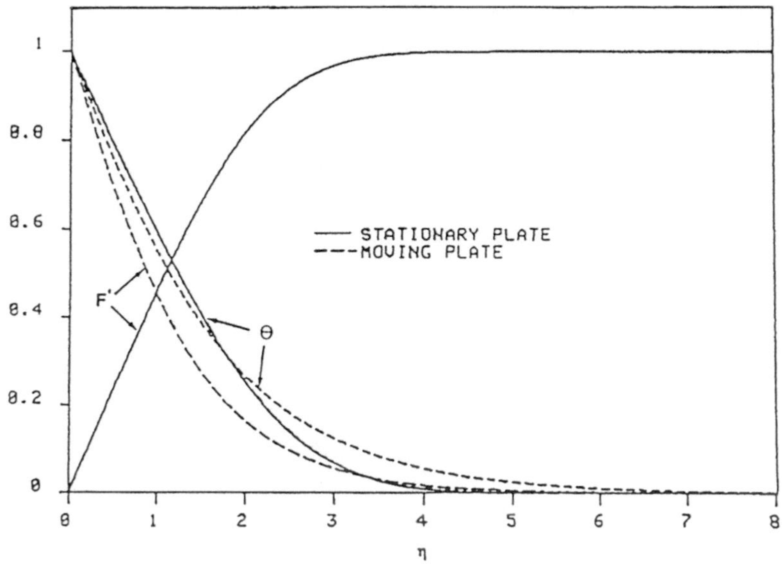

Fig. 5.9 Variation of dimensionless velocity component along the surface, F', and the dimensionless temperature θ with the similarity variable η for a surface moving in a quiescent medium, as well for a stationary surface with a uniform parallel flow over it

$$\frac{\partial u}{\partial x} + \frac{\partial v}{\partial y} = 0 \tag{5.12}$$

$$u\frac{\partial u}{\partial x} + v\frac{\partial u}{\partial y} = \upsilon\frac{\partial^2 u}{\partial y^2} \tag{5.13}$$

$$u\frac{\partial T}{\partial x} + v\frac{\partial T}{\partial y} = \alpha_f\frac{\partial^2 T}{\partial y^2} \tag{5.14}$$

where u and v are the velocity components in the x and y directions, respectively, υ is the kinematic viscosity, and α_f is the fluid thermal diffusivity. The no-slip conditions result in $u = U_s$ and $v = 0$ at the moving surface. These equations may be solved by the similarity method, which converts these equations into ordinary differential equations (Karwe and Jaluria 1986) and solved using methods such as the Runge-Kutta scheme. The finite difference method may also be used, with marching in the x-direction for these boundary-layer equations (Jaluria and Torrance 2003; Minkowycz et al. 2006).

Figure 5.9 shows the results from the similarity analysis for Prandtl number, $Pr = \nu/\alpha_f$, of 0.7, which applies for air. The figure shows the calculated distributions of F', which represents the dimensionless velocity component in the x-direction, u/U_s, and dimensionless temperature θ (η), where the similarity variable $\eta = \frac{y}{x}\sqrt{\frac{U_s x}{2\upsilon}}$. The corresponding distributions for uniform flow over a stationary plate are also shown for comparison. The dimensionless temperature gradient at the surface $\theta'(0)$ is obtained as

−0.4939 and −1.9457 for $Pr = 0.7$ and 7.0, respectively. These values can be used to obtain the convection coefficient $h(x)$ and the heat transfer rate from an isothermal moving surface. These were found to be larger than those for a stationary surface, since entrainment brings the colder ambient fluid toward the moving surface. The heat transfer in the moving material is not considered here, and only the convective boundary-layer flow is studied. The conjugate problem, which couples the conduction in the moving material and the convective flow, is obviously of much greater interest in materials processing and is discussed next.

5.4 Conjugate Transport

The conjugate problem involves solving for coupled transport within the material and the moving material. Thus, the heat transfer coefficient at the surface of the material is not assumed but is obtained from a solution of the governing equations for the fluid flow. In a sense, this combines the two approaches outlined in the preceding two sections and presents a more realistic model for heat transfer to or from a moving material undergoing thermal processing. The flow is generated by the moving surface, along with buoyancy effects, if significant, and any forced flow employed in the process.

5.4.1 *Boundary-Layer Formulation*

A few studies have considered the conjugate problem for a continuous, moving material (Chida and Katto 1976; Kuiken 1974, 1975; Karwe and Jaluria 1988). Different materials, geometries, and operating conditions have been considered. In some cases, the boundary-layer assumptions may be made to simplify the problem. Otherwise, the full, elliptic equations need to be solved. Boundary-layer assumptions may be made for external flow on the surface, with negligible upstream conduction effects in the material. Equations (5.12)–(5.14) apply for the fluid. The heat transfer within the moving material, for a flat plate, is governed by the equation

$$U_s \frac{\partial T}{\partial x} = \alpha \frac{\partial^2 T}{\partial y^2} \tag{5.15}$$

This circumstance is shown in Fig. 5.2, along with the coordinate system. The diffusion in the direction of motion is neglected in the governing equations in order to obtain a parabolic boundary-layer formulation that can be solved by marching in the x-direction (Jaluria and Torrance 2003; Pletcher et al. 2016).

Some typical results from the preceding boundary-layer solution are shown in Figs. 5.10 and 5.11. The governing parameters are $\widetilde{Pe}$, Pr, and R, which is the ratio

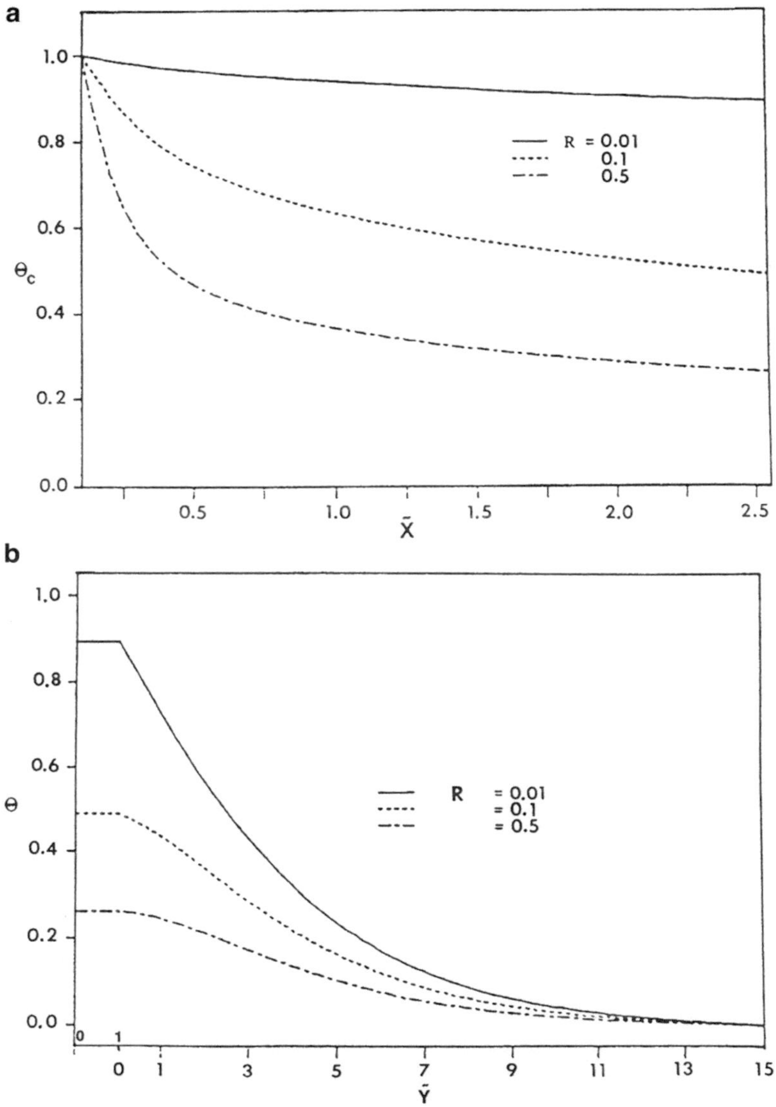

Fig. 5.10 (**a**) Downstream variation of the centerline temperature θ_c for the conjugate boundary-layer formulation at $Pr = 0.7$ and $\widetilde{Pe} = 0.01$. (**b**) Temperature variation along the *y*-direction within the plate and the fluid at $\widetilde{X} = 2.4$

$(k\rho C_p)_f/(k\rho C_p)_s$ of the fluid to solid properties, where ρ is the density. Here, $\widetilde{Pe} = U_s H^2/\alpha l$ for a plate of thickness H, where l is an arbitrary length scale in the x-direction and is chosen as the characteristic dimension in keeping with the usual

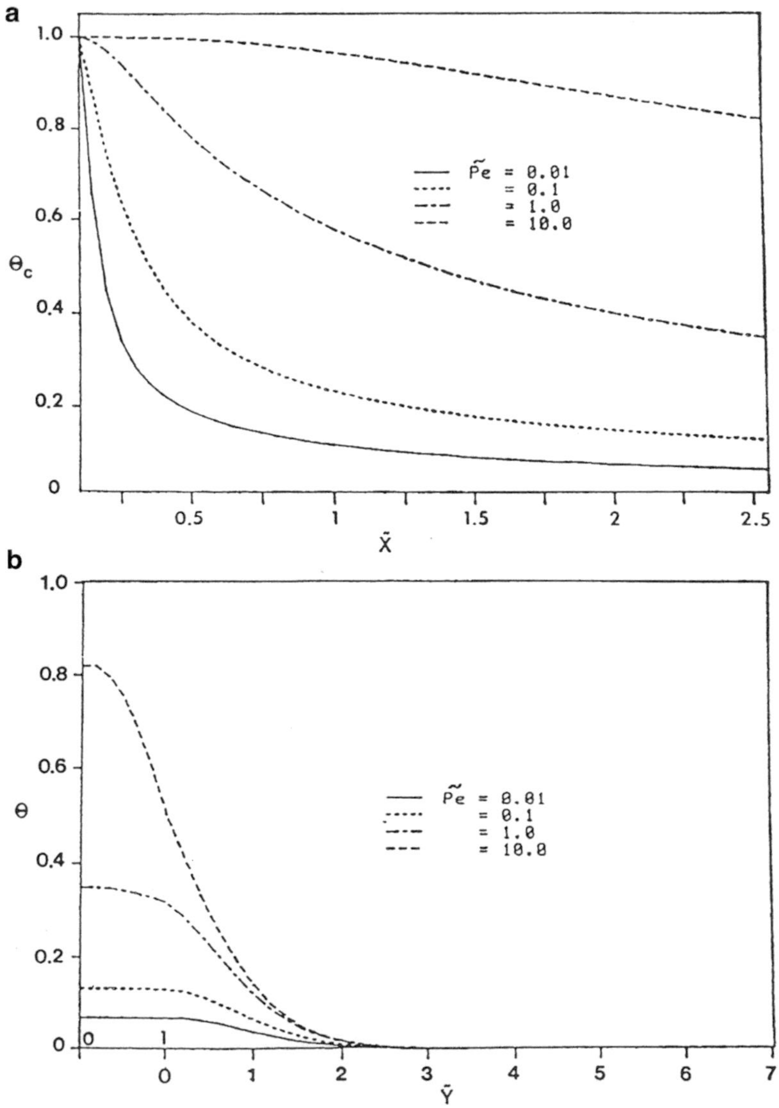

Fig. 5.11 (**a**) Downstream variation of the centerline temperature θ_c for a heated glass plate moving in water, assuming a 2D boundary-layer flow and conjugate transport at $Pr = 7.0$ and $R = 1.092$. (**b**) Temperature distributions in the y-direction at $\widetilde{X} = 2.5$

formulation for boundary-layer flow over a semi-infinite surface. Thus, $\widetilde{X} = x/l$ and $\widetilde{Y} = \frac{y}{1}\sqrt{Re}$, where the Reynolds number $Re = \frac{U_s l}{\upsilon}$. It is seen that for very small values of R, which would arise for, say, a highly conducting plate material, the centerline temperature of the plate drops very gradually, and the downstream penetration of thermal effects is large, as expected. Such results can, therefore, be used to determine the distance needed for the material to cool to a desired level. For the parametric values employed, the temperature distribution within the material is fairly uniform. A larger temperature variation obviously arises for a material of lower thermal conductivity. The temperature gradient at the surface is larger for smaller values of R, indicating higher heat transfer coefficients for a given fluid if the quantity $(k\rho C_p)_s$ of the moving material is increased. Again, this trend reflects the higher temperature levels downstream for smaller R. The downstream temperature decay decreases as the Peclet number increases due to shorter time to a given location, as seen earlier. The temperature distribution within the material at a given downstream location is more uniform for smaller Peclet number, indicating the result of a longer exposure of the plate to the ambient at the corresponding smaller velocity U_s.

Thus, to achieve a desired cooling rate or temperature uniformity within the material, the speed of the moving material U_s may be varied. Temperature uniformity across the material is particularly important in the thermal processing of glass, ceramics, and other such brittle materials. The preceding formulation can easily be extended to axisymmetric cases, such as moving cylindrical rods.

5.4.2 *Solution of the Full Equations (Elliptic Formulation)*

Even though the boundary-layer formulation provides a significant simplification, the transport processes in regions close to the point of emergence from, say, a die, slot, or rollers, are important for a realistic modeling of practical manufacturing systems. The boundary-layer formulation does not capture these processes accurately. Also, these regions often involve large axial gradients and thus require the solution of the full equations, which are elliptic in nature. Far downstream, the axial diffusion terms are small, and a parabolic, marching, scheme may be adopted to reduce the computational time.

The conjugate problem based on the full, elliptic, equations is modeled by the following equations:

For the fluid

$$\nabla . \bar{V} = 0 \tag{5.16}$$

$$\frac{\partial \bar{V}}{\partial t'} + \bar{V} . \nabla \bar{V} = -\frac{1}{\rho_f} \nabla p + \upsilon \nabla^2 \bar{V} \tag{5.17}$$

$$\frac{\partial T}{\partial t'} + \bar{V} . \nabla T = \alpha_f \nabla^2 T \tag{5.18}$$

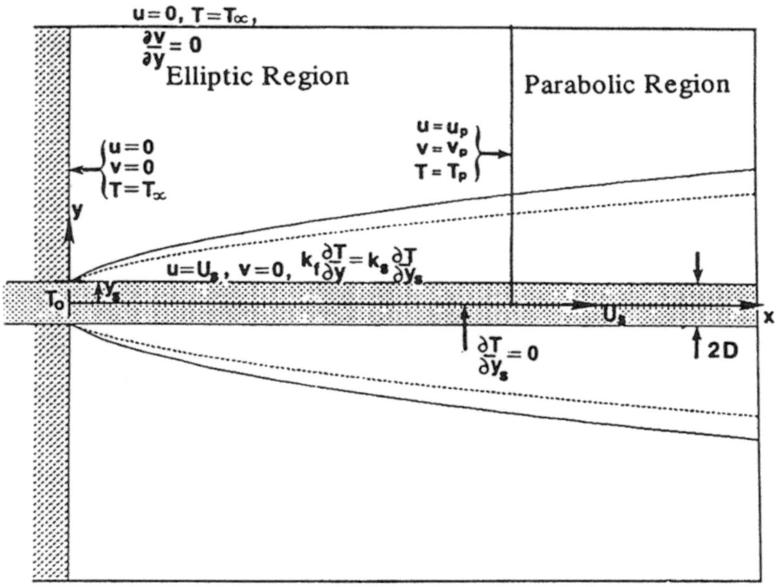

Fig. 5.12 The computational domain for the solution of the full equations, indicating elliptic and parabolic regions

For the moving solid material

$$\frac{\partial T}{\partial t'} + U_s \frac{\partial T}{\partial x} = \alpha \nabla_s^2 T \tag{5.19}$$

where V is the velocity vector in the flow and p the local pressure; subscripts f and s refer to the fluid and solid, respectively; and the half-thickness D is taken as the characteristic dimension for the coordinates, *Re* and *Pe*. Figure 5.12 shows the relevant computational domain, indicating the elliptic and parabolic regions, the coordinate system, and several important boundary conditions. The results from the full equations and the boundary-layer formulation are matched at the interface, whose location is varied to ensure that the overall results are independent of the chosen location. The problem may be solved by employing the vorticity stream-function formulation for 2D problems or by using primitive variables for both 2D and 3D cases, as outlined in Chap. 3. For further details, see Karwe and Jaluria (1988) and Jaluria (2006).

Figure 5.13a shows the typical flow generated by a plate moving in a quiescent medium. The ambient fluid is drawn toward the moving surface, and the pressure gradients directed toward the origin give rise to a small reverse flow in this region. Farther downstream, this effect dies down, and the flow approaches the characteristics of a boundary-layer flow, with its thickness growing in the direction of motion. However, this flow is quite different from the usual boundary-layer flow over a stationary surface. The corresponding thermal field is shown in Fig. 5.13b, indicating the downstream temperature decay in the solid and the fairly uniform temperature across the solid thickness due to its high thermal conductivity. Figure 5.14

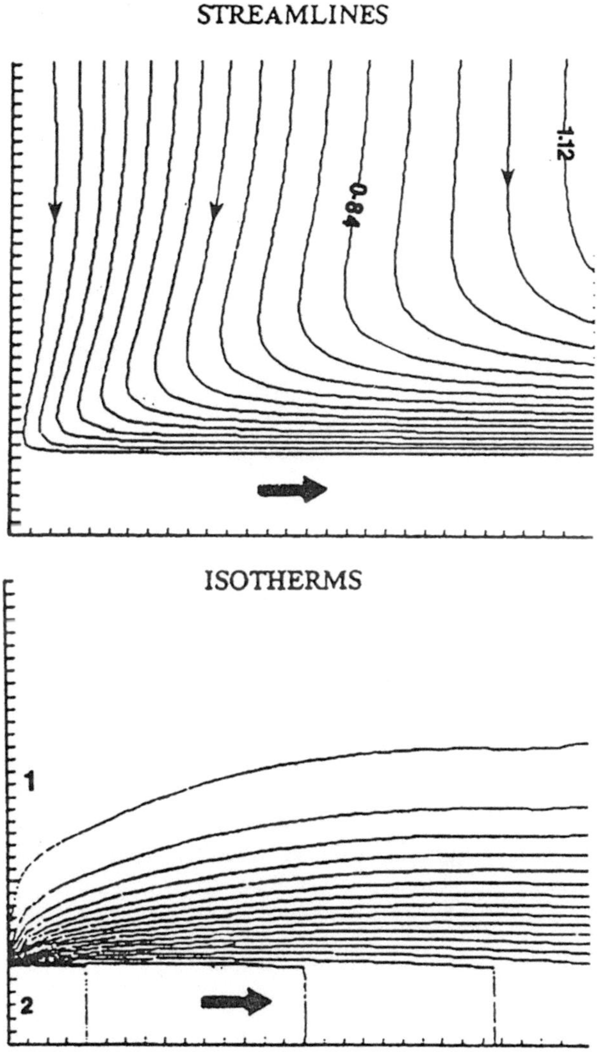

Fig. 5.13 Calculated streamlines and isotherms for a heated aluminum plate moving in air, as obtained from the solution of the full equations at $Pe = 0.05$, $Pr = 0.7, k/k_f = 1.27 \times 10^{-4}$, and $Re = 20.0$

shows two possibilities for the upstream boundary conditions at the point of emergence, $x = 0$. The upstream penetration of thermal effects and the influence of the upstream conditions on the temperature distributions downstream may be determined.

Figure 5.15 shows these upstream effects that are often critically important in continuous materials processing. The upstream boundary condition is applied at the location $X = -X_b$. As X_b increases to large values, the results become independent of a further increase, giving the behavior for $X_b \rightarrow \infty$, which is representative of the situation where the material emerges from a long furnace, oven, or extruder. In most practical cases, X_b is finite, representing, for instance, a heated region for the thermal

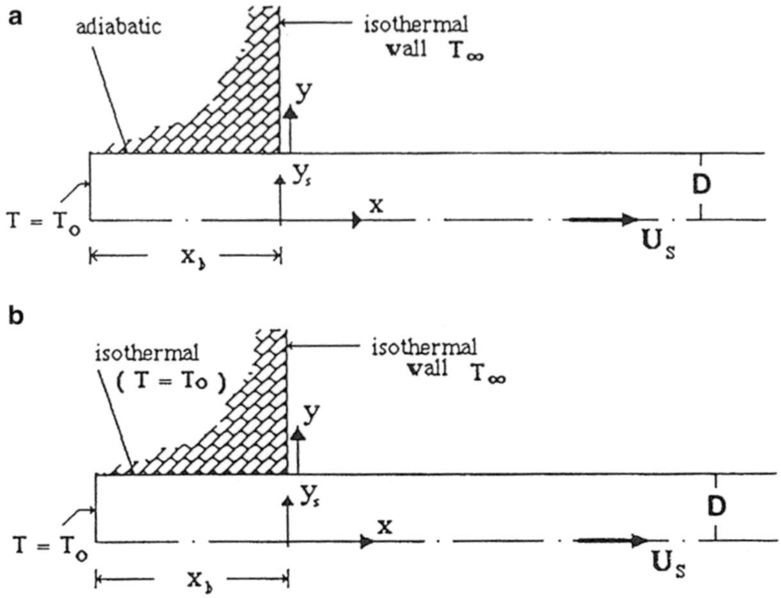

Fig. 5.14 Two circumstances in the simulation of the conditions upstream of the point of emergence of the moving material. (**a**) Adiabatic surface boundary condition; (**b**) isothermal surface boundary condition

processing of the material. The effects of X_b and the upstream boundary conditions are shown in this figure. The centerline temperature is higher for the isothermal case, implying that the back-conduction effects are smaller in this case. Several variations in these conditions, such as radiative heating in a furnace before the material emerges, are obviously possible, depending on the application being considered. Similarly, different geometries, materials, heat transfer mechanisms at the surface, and other operating conditions may be considered.

5.5 Buoyancy and Forced Flow Effects

The inclusion of buoyancy effects due to the temperature differences is important in many applications, particularly at low values of U_s and high temperature T_0 at the inlet. Figure 5.16 shows the sketches for a plate moving vertically up or down, along with the coordinate system. If the Boussinesq approximations, which neglect density change in the continuity equation and take density difference to vary linearly with temperature difference, are employed, an additional term $-\bar{g}\beta(T - T_\infty)$ arises on the right-hand side of Eq. (5.17). Here, $\bar{g}$ is the gravitational acceleration and β the coefficient of volumetric thermal expansion, as given earlier in Chap. 2. Therefore, if buoyancy effects are important in the flow, the orientation will influence the flow

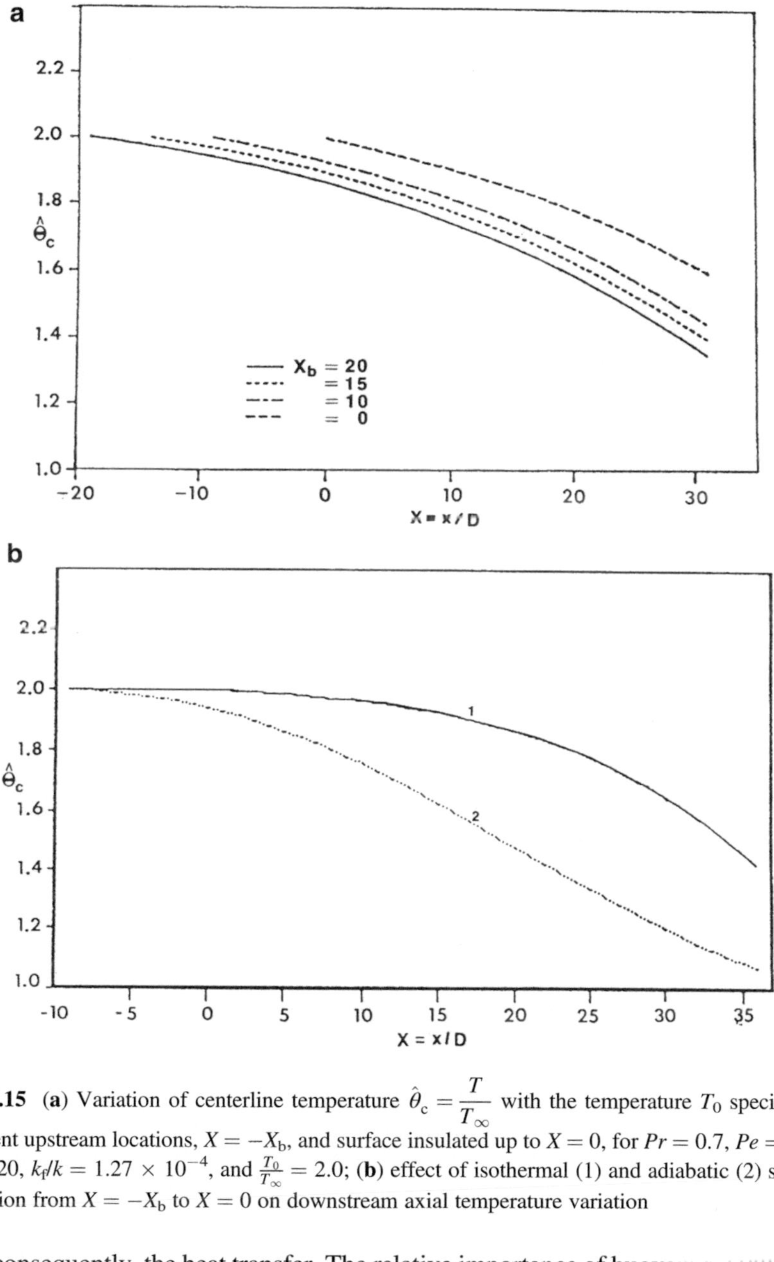

Fig. 5.15 (**a**) Variation of centerline temperature $\hat{\theta}_c = \dfrac{T}{T_\infty}$ with the temperature T_0 specified at different upstream locations, $X = -X_b$, and surface insulated up to $X = 0$, for $Pr = 0.7$, $Pe = 0.05$, $Re = 20$, $k_f/k = 1.27 \times 10^{-4}$, and $\frac{T_0}{T_\infty} = 2.0$; (**b**) effect of isothermal (1) and adiabatic (2) surface condition from $X = -X_b$ to $X = 0$ on downstream axial temperature variation

and, consequently, the heat transfer. The relative importance of buoyancy, compared to flow due to plate motion, is given by the parameter Gr/Re^2, where the Grashof number Gr and the Reynolds number Re are defined as

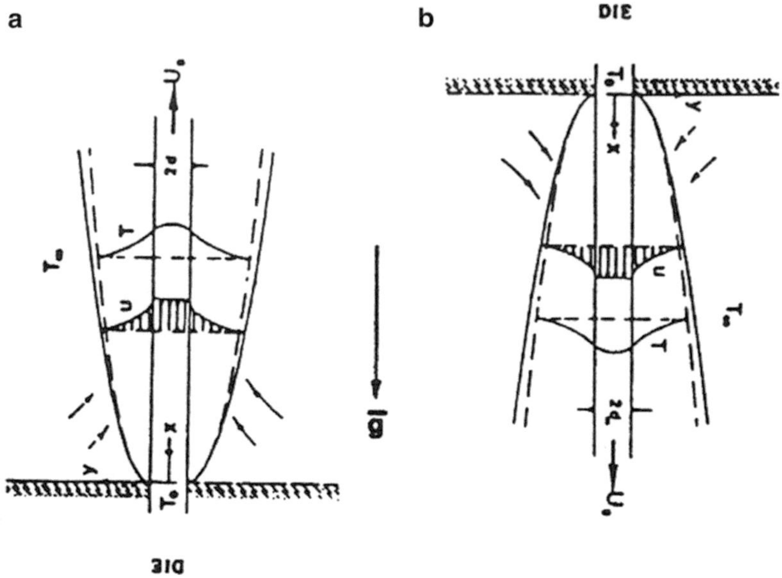

Fig. 5.16 Flow geometry near the die showing the plate, slot, and the walls for the plate moving vertically (**a**) upward and (**b**) downward

$$Gr = \frac{g\beta(T_0 - T_\infty)D^3}{\upsilon^2} \quad \text{and} \quad Re = \frac{U_s D}{\upsilon} \tag{5.20}$$

with D being the characteristic dimension such as diameter or thickness and g the magnitude of gravitational acceleration (Gebhart et al. 1988). For a heated plate moving upward, the buoyancy force is aligned with the plate motion and opposes the flow due to plate motion if it moves downward, yielding $\pm g\beta(T - T_\infty)$ as the additional term in the x component of the momentum equation for the two cases, respectively (Chen and Strobel 1980; Moutsoglou and Chen 1980).

Figure 5.17 shows the computed results for a heated plate moving vertically upward at different values of the mixed convection parameter Gr/Re^2. The maximum velocity in the flow is much larger than the plate speed U_s at large values of Gr/Re^2, as shown in the figure. This flow increases the heat transfer from the plate, which is reflected in lower temperature levels downstream. These effects are smaller for a horizontal plate, where the buoyancy affects the pressure which in turn affects the flow. For the vertical downward motion, buoyancy acts against the plate motion and the velocity and heat transfer decrease, for a dominant shear flow. Similarly, other orientations and geometries may be considered, with a horizontally moving cylindrical rod requiring a 3D solution because of non-axisymmetric flow.

In many practical processes, the flow due to the motion of the material and buoyancy does not provide adequate heating/cooling rates and an additional, externally induced, forced flow is employed to increase the heat transfer rate (Ramachandran et al. 1987; Kyriacou et al. 1990; Kang and Jaluria 1992). Other

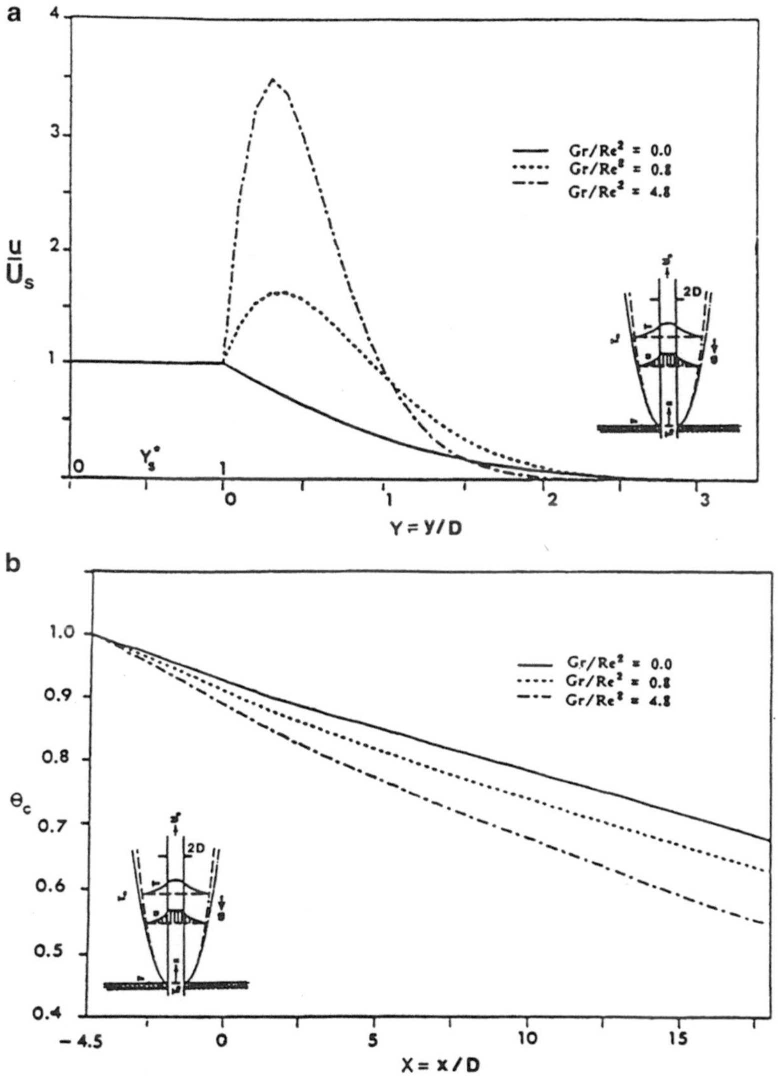

Fig. 5.17 (**a**) Velocity profiles at $X = 15.0$, for $Pe = 0.07$, $Pr = 7.0$, $Re = 25.0$, and $k_f/k = 0.0128$ at various values of Gr/Re^2; (**b**) downstream temperature variation

heat transfer mechanisms, such as radiation and evaporation with the use of sprays, are also used for enhancing the heat transfer (Zumbrunnen et al. 1989). The forced flow may be in an extensive ambient medium, as is the case with rolled sheets and cooling of extruded material in a large water trough in cable manufacture. It may also be in a channel or duct surrounding the moving material, which is a common circumstance in the cooling of glass fibers where a forced flow of inert gases may be blown over the fiber in an enclosed region (Roy Choudhury and Jaluria 1994).

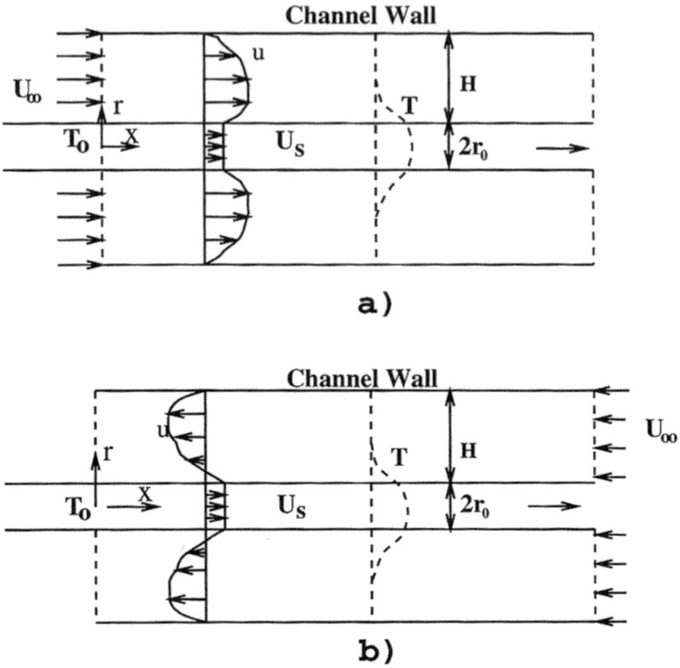

Fig. 5.18 Schematic of a moving axisymmetric solid in forced flow in a channel. (**a**) Aiding flow; (**b**) opposing flow

Figure 5.18 shows schematics of an axisymmetric solid moving in aiding or opposing forced flow in a channel. The basic governing equations are Eqs. (5.16)–(5.19), including buoyancy effects if considered, but the boundary conditions are changed to account for the forced flow in the extensive medium or in the channel.

Figure 5.19 shows the computed flow and thermal field for the vertical orientation, in terms of velocity vectors and isotherms. A uniform flow at velocity U_∞ is taken at $x = 0$, and the diffusion terms are dropped far downstream. The flow develops rapidly downstream to yield velocity levels higher than both U_s and U_∞. The thermal field similarly develops downstream, reaching a fully developed condition downstream for a long channel and a boundary layer for an extensive environment. Figure 5.20 shows the effect of *Gr* on the flow for a plate moving vertically in a water channel. The maximum velocity increases with increasing buoyancy effects, given by *Gr*. A fairly uniform velocity is seen across much of the channel, and it then drops to zero at the far wall of the channel, as expected from the no-slip conditions there. The temperature level is highest near the surface, resulting in the strongest buoyancy effect there.

Figure 5.21 shows the results for three different materials, Teflon, glass, and aluminum, in terms of the downstream temperature variation. The upstream penetration of thermal effects is small for Teflon, due to its low thermal conductivity. The surface temperature cools rapidly because of small diffusion in the radial and axial

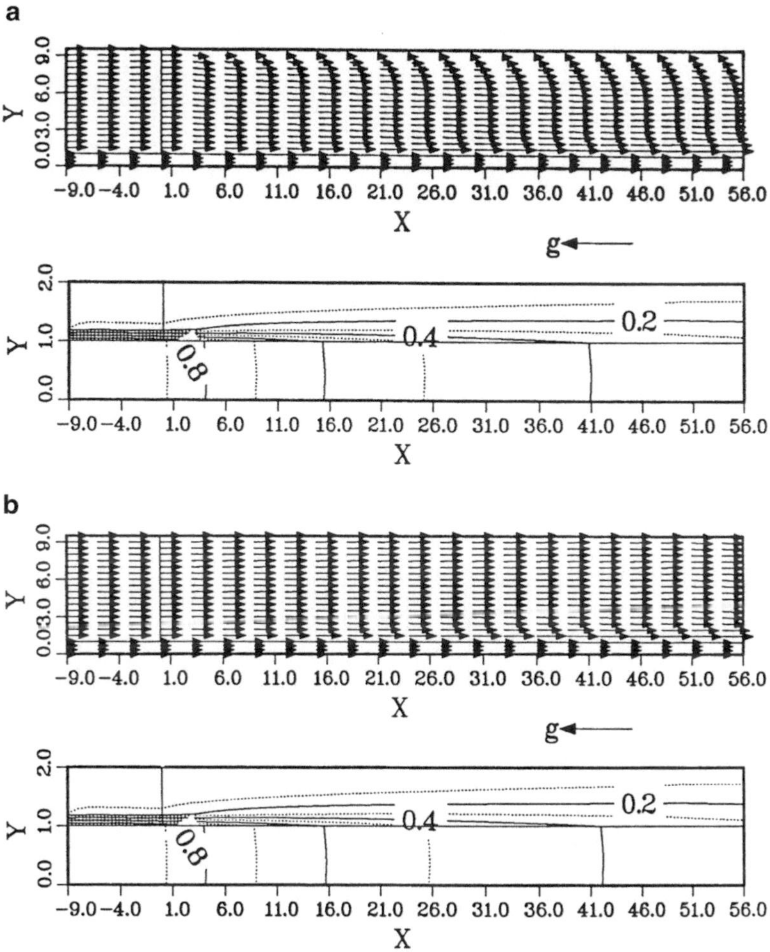

Fig. 5.19 Calculated velocity vectors (upper) and isotherms (lower) for typical circumstances, $Pr = 7.0$, $Pe = 0.29$, $Re = 25$, $Gr = 1500$, $\frac{U_\infty}{U_s} = 2.0$, and $k_f/k = 0.0029$. (**a**) Aluminum plate moving in a water channel flow; (**b**) aluminum plate moving in a free stream of water

directions. This gives rise to the largest difference between the surface and midplane temperatures, among these three materials. These observations are important for materials processing applications with respect to, say, thermal stresses that may damage brittle materials or the resulting microstructure. Similar trends were observed for forced flow in an extensive environment. Opposing forced flow streams are also used in various manufacturing processes, such as cooling of wires and optical fiber to remove contaminants from the system. Though laminar flow provides better control on the process, turbulent flows are often used to enhance the transport. Similarly, different geometries, 3D transport, different operating conditions,

Fig. 5.20 Computed velocity profiles for a flat plate moving vertically upward in a channel with the forced flow in the direction of motion at $Re = 25.0$, $Pr = 7.0$, $Pe = 0.3$, $\frac{U_\infty}{U_s} = 2.0$, and $k_f/k = 0.0029$

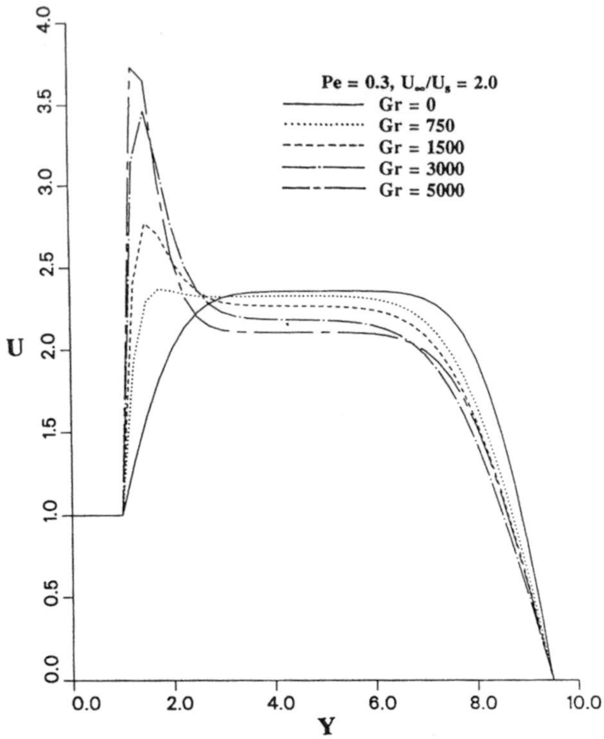

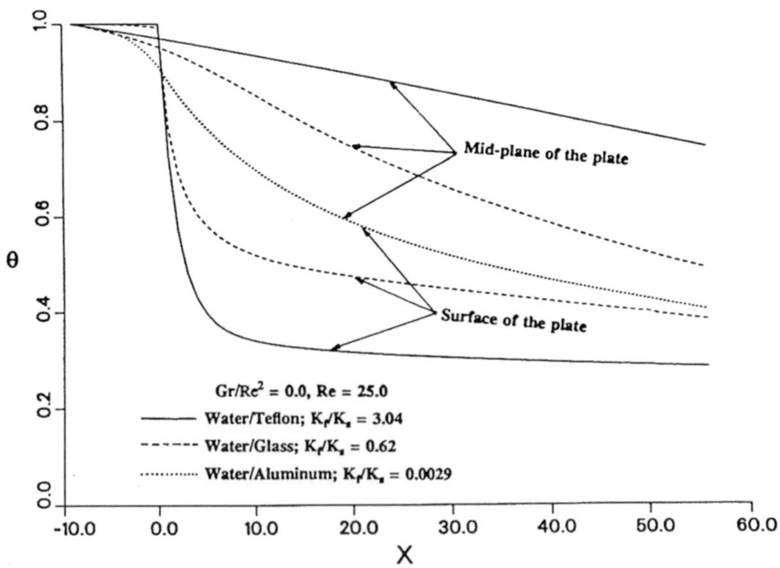

Fig. 5.21 Downstream variation of the dimensionless temperature θ at $Re = 25.0$, $Pr = 7.0$, and $Pe = 0.3$. $Gr = 1500.0$ and $k_f/k = 0.0029$ for vertical aiding forced flow in a channel

cross-flow, and other configurations are of interest and have been considered in a few studies.

5.6 Transient Effects

The transient behavior, particularly at the initial stages of the process, is also important with respect to the product quality, as outlined earlier for the relatively simple problem of assumed heat transfer coefficient at the surface. The conjugate problem can be similarly considered, extending the preceding formulation to include the transient terms (Kang et al. 1991). Figure 5.22 shows numerical results for a heated aluminum plate moving vertically upward in water. A long plate is assumed to start moving at time $t = 0$, when the surface temperature is also raised to T_0, which is higher than the ambient temperature T_∞. It is seen that the flow starts near the moving surface due to the no-slip conditions. A recirculating flow appears and gradually moves downstream, finally being swept away by the main flow. The boundary-layer thickness grows along the direction of motion, as expected. The minimum heat transfer coefficient from the plate occurs at the recirculation region. Thus, local hot spots can arise in this region. Buoyancy effects increase with time as the fluid temperature rises. Therefore, the transient flow and the heat transfer rates can be significantly different from steady-state conditions which are eventually reached.

Figure 5.23a shows the temperature level rising with time from the no-flow, unheated situation, at $X = 30$, to steady state, with the surface temperature held at T_0. The corresponding transient variation of the local Nusselt number, $Nu\ (X) = h(x)D/k_f$, is shown in Fig. 5.23b. The Nusselt number is calculated only up to the downstream location where the surface temperature $\theta_s = 0.05$, in order to obtain accurate results. Beyond this point, the temperature and the temperature gradient are too small to yield accurate values. It is seen that the Nusselt number decreases downstream, reaches a minimum value, and then increases again. The minimum arises in the recirculating region, as seen in Fig. 5.22 and which reduces the heat transfer from the plate due to low velocities encountered in this region. All these trends are important in practical situations. Of particular interest are the time for which the transient process lasts and the temperature distributions in the material. Methods to shorten the transient for better process control and enhanced temperature uniformity in the material can be obtained from such results.

Not much work has been done on the transient problem, particularly when the material length increases from zero to large values, as discussed earlier, even though these can significantly affect the product quality and productivity. Also, the transport processes over the transient regime may lead to material characteristics that are unacceptable, and the corresponding product may have to be discarded. One particular time-dependent circumstance is outlined here. Similarly, other scenarios and boundary conditions, such as the uniform heat flux condition, may be considered, depending on the application.

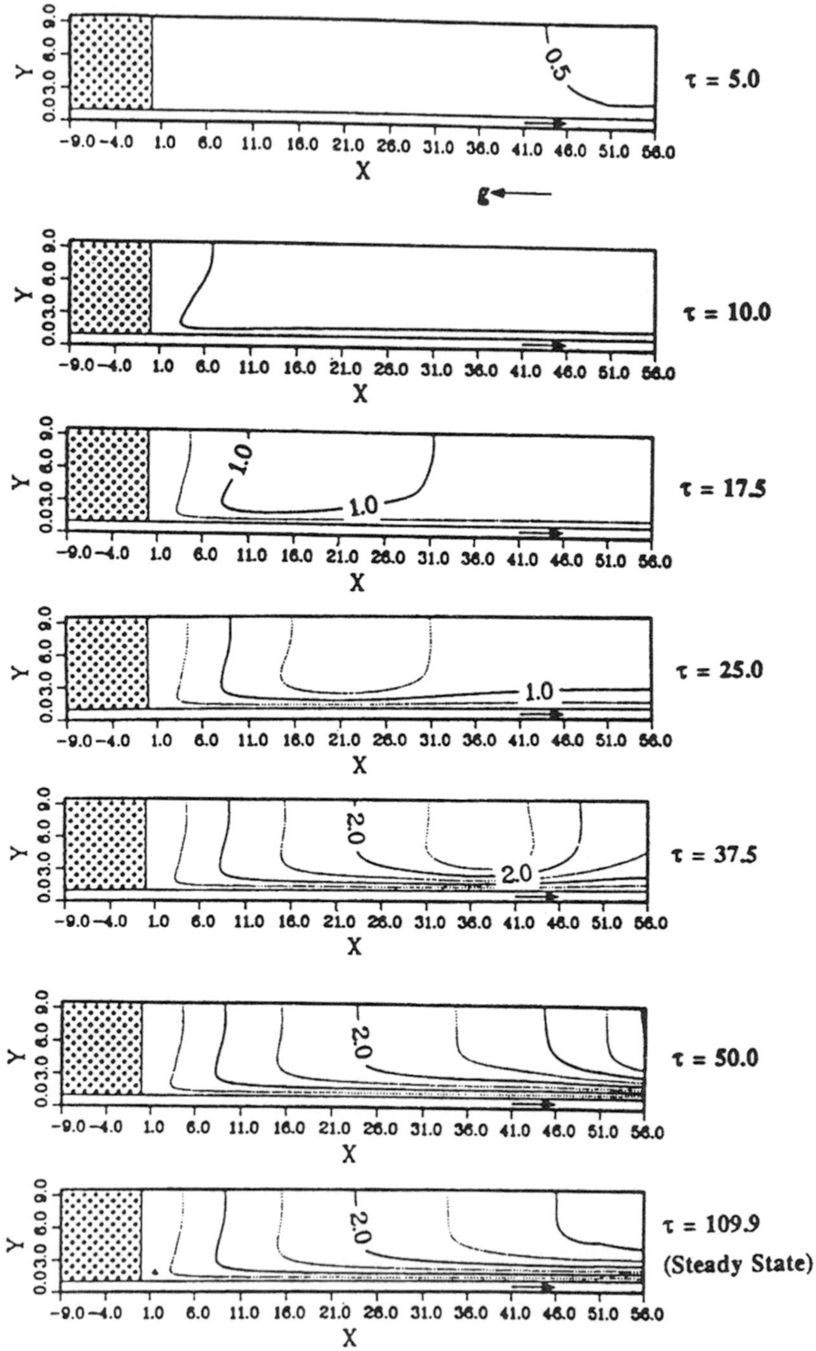

Fig. 5.22 Time-dependent streamlines for an aluminum plate moving vertically upward in water at $Pr = 7.0$, $Re = 25.0$, $Pe = 0.29$, $Gr = 1000.0$, and $k_f/k = 0.0029$

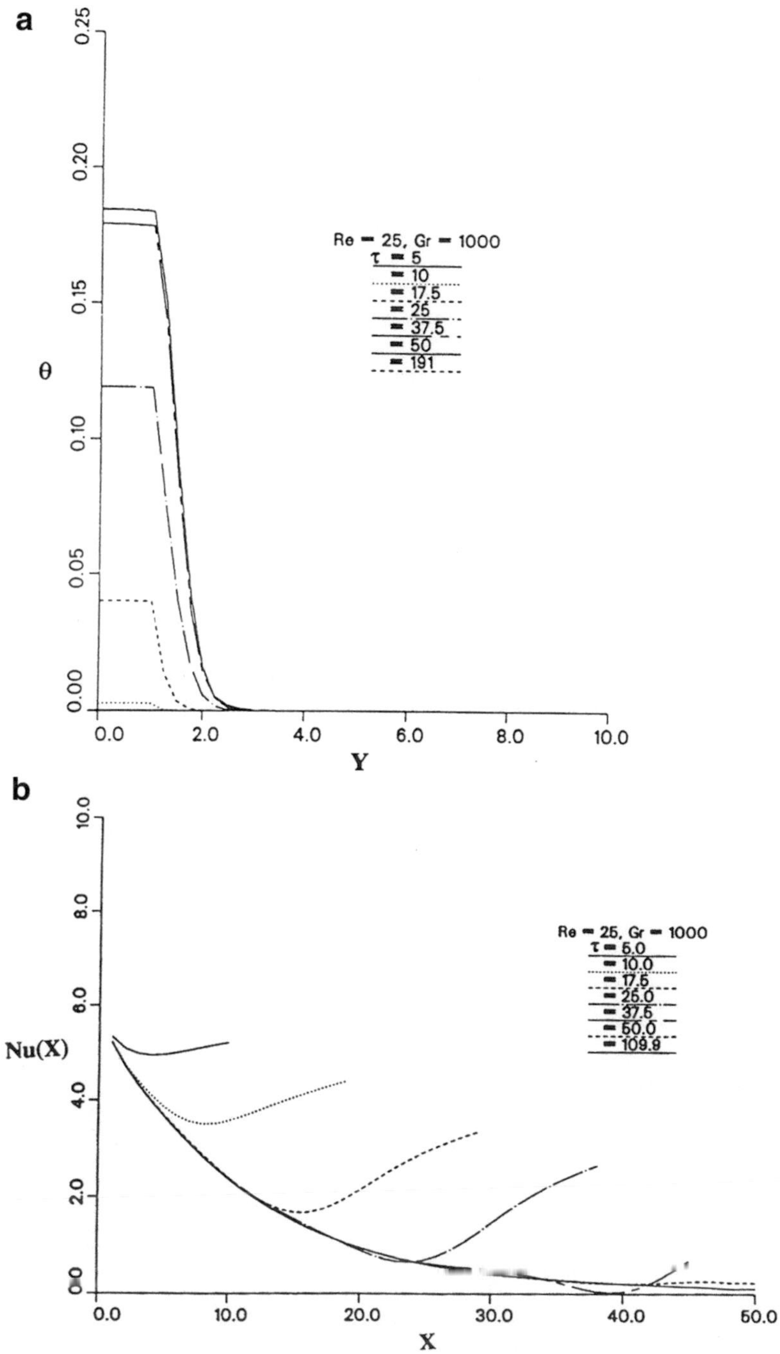

Fig. 5.23 Transient results for the conditions of Fig. 5.21. (**a**) Transient variation of the temperature distributions at $X = 30.0$. (**b**) Transient downstream variation of the local Nusselt number

5.7 Experimental Work

Several experimental investigations have been carried out on various aspects of continuous processing and provide results that can be used for the validation of the mathematical/numerical models. Tsou et al. (1967) and Chida and Katto (1976) presented several comparisons between their numerical and experimental results on the flow and temperature distributions. Generally, good agreement was obtained for a range of materials and speed U_s. Due to the practical relevance of this manufacturing process, different fluids, materials, and flow conditions, including channel flows and turbulence, have been considered for a wide variety of applications in materials processing. Detailed experimental work has been carried out by Karwe and Jaluria (1992) and Kang et al. (1994) on flat plates and cylinders moving in air or water.

Figure 5.24 shows some typical comparisons between experimental data and numerical predictions. The numerical results in Fig. 5.24a, b were obtained by Roy Choudhury and Jaluria (1994) and those in Fig. 5.24c, d by Kang (1990). It is seen that the agreement, in terms of the temperatures in the material, is fairly good at relatively small buoyancy effects and low speeds of the moving material. At large buoyancy levels, the aiding circumstance was modeled quite well. But the opposing case was more complicated due to separation and recirculation resulting from the two opposing effects. Good agreement was also obtained on the flow and thermal field in the fluid for various circumstances. However, the occurrence of instability, transition, and turbulence at large speeds or at large temperature differences has been found to increase the difference between numerical and experimental results, indicating the need for better modeling of these flows.

Figure 5.25 shows a sequence of shadowgraph photographs of the flow near the surface of an aluminum plate moving vertically downward. As time increases, a large disturbance leading to flow separation is seen to arise. The flow under consideration is an opposing buoyancy circumstance, since the plate is hot and the motion is downward. This could lead to flow separation downstream. The disturbance was found to remain essentially at one location and not move downstream, indicating that the observed phenomenon is due to opposing buoyancy and not transient effects. The effect is governed by the local mixed convection parameter $Gr_x/\mathrm{Re}_x^2 = g\beta\Delta Tx/U_s^2$, where ΔT is the temperature difference between surface and ambient temperatures. The effect was found to be larger at larger values of this parameter. Such disturbances in the flow affect the local transport at the surface and thus the local characteristics of the product. It is important to understand and control these effects for better consistency in the processed material.

Much of the experimental work has been directed at practical problems such as fiber cooling, rolling, and extrusion processes. Work has also been done on continuous processes such as heat treatment and deposition, particularly chemical vapor deposition that may be used to make continuous photovoltaic panels. Clearly, detailed, controlled experiments are needed for improving the existing models to make them more realistic and dependable.

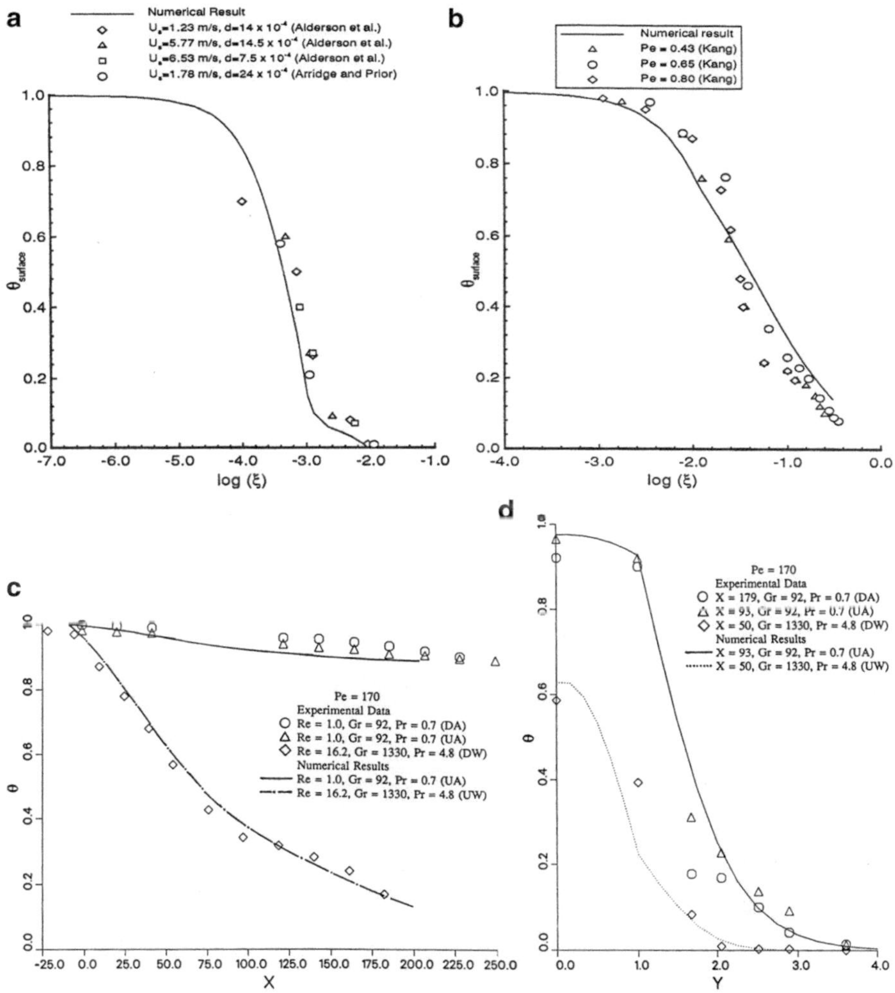

Fig. 5.24 Comparisons between numerical and experimental results for continuously moving materials in the vertical direction. (**a**) Glass rod moving in air; (**b**) aluminum rod moving in water, ξ being a dimensionless coordinate distance in the direction of material motion defined by Chida and Katto (1976); (**c**) midplane temperature in plate moving in air; (**d**) temperature distribution at $X = 50$ for plate moving in air. Here, D refers to downward motion, U to upward, A to air, and W to water

5.8 Additional Aspects

Continuous materials processing is of interest in a wide range of manufacturing processes. Figure 5.26 shows schematics of some of the typical systems employed in processes, which are of particular interest in metal forming, such as hot rolling, wire drawing, extrusion, and roll bending. Many books, such as those by Blazynski

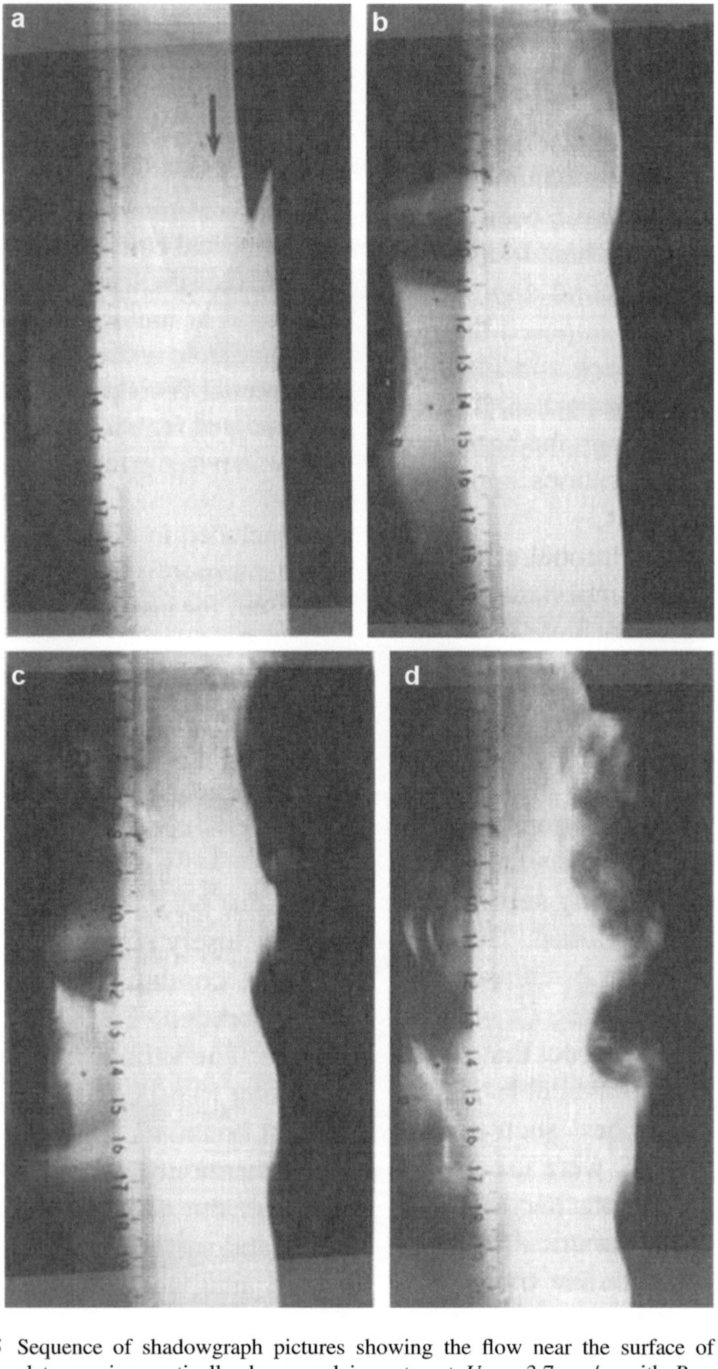

Fig. 5.25 Sequence of shadowgraph pictures showing the flow near the surface of a heated aluminum plate moving vertically downward in water at $U_s = 3.7$ cm/s, with $Re = 140.36$, $Pe = 1.05$, $Pr = 4.31$, and $Gr/Re^2 = 0.45$. Time t increases from (**a**)–(**d**) (Karwe and Jaluria 1992)

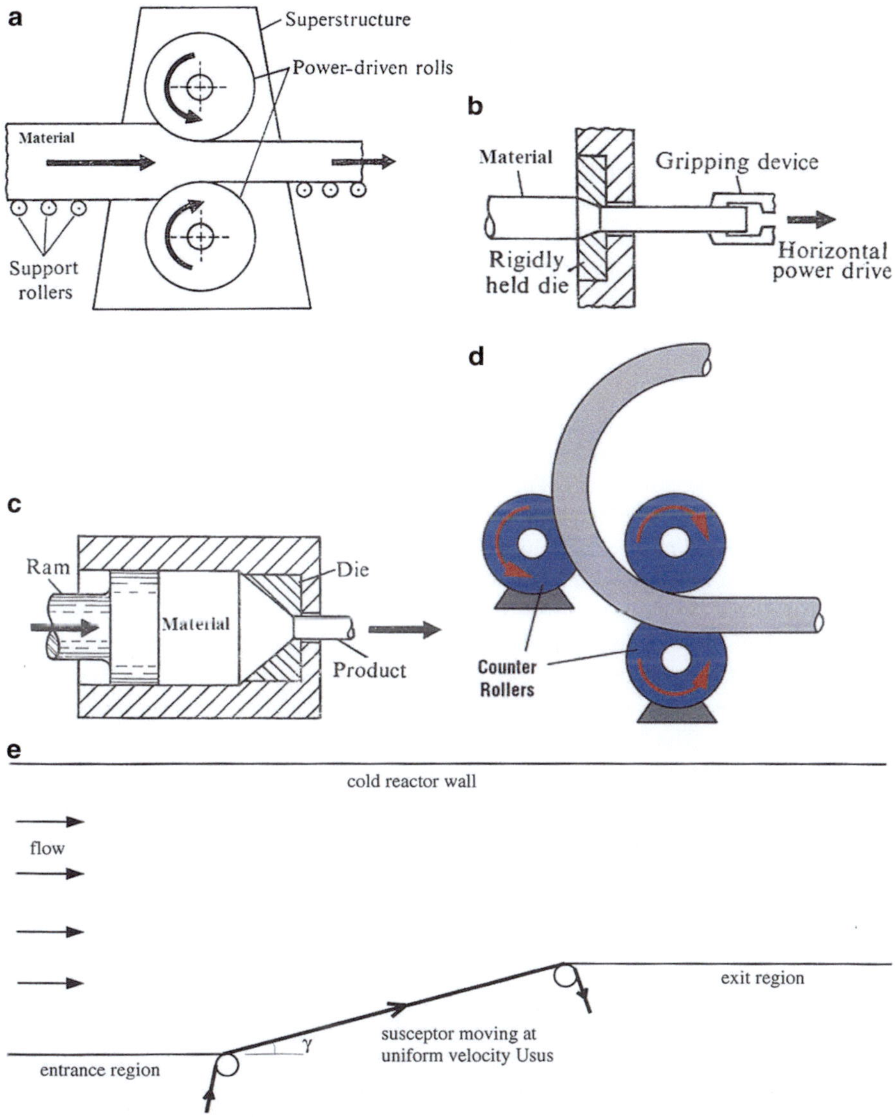

Fig. 5.26 (**a**) Rolling; (**b**) drawing; (**c**) extrusion; (**d**) roll bending; (**e**) continuous chemical vapor deposition (Ghosh and Mallik 1986)

(1976), Altan et al. (1983), Ghosh and Mallik (1986), Schey (1999), and Kalpakjian and Schmid (2010), discuss the basic aspects of these processes and present different system designs to achieve the desired product characteristics. For instance, besides the system shown for metal extrusion, we can employ hydrostatic extrusion, impact extrusion, and inverse extrusion. Similarly, different materials such as polymers and brittle materials may be considered.

Hot rolling is carried out at temperatures higher than the recrystallization temperature of the material and cold rolling at temperatures below the recrystallization temperature, which varies with the material, being typically between 400 and 700 °C for steels. Similar considerations apply for other forming operations, such as hot extrusion. Higher temperatures reduce the stresses in the material and the force needed for the force. But it does not lead to significant work hardening, for which cold working is preferable. The material flow in an extruder, a rolling station, or a drawing die can be modeled and simulated like fluid flow, discussed in earlier sections, using the material characteristics and properties related to stress and strain or strain rate. The strain and strain rate dependence may be of the form

$$\sigma = a + b(\varepsilon)^n \text{ and } \sigma = c + d(\dot{\varepsilon})^m \tag{5.21}$$

where σ, ε and $\dot{\varepsilon}$ are stress, strain, and strain rate, respectively, and a, b, c, d, m, and n are temperature-dependent constants that are generally determined by curve fitting of experimental measurements. The basic principles of mass and energy conservation and force-momentum balance apply and lead to the governing equations. Many of the analytical and numerical solution strategies discussed earlier for fluid flow and heat transfer can be employed here as well.

Thus, the results presented in this chapter can be used to determine the temperature variation in the moving material and include heating or cooling stations at appropriate distances to maintain the temperatures at desired levels (see Fig. 5.1b). The heat generated due to friction and the forming operation must be included in these calculations. The forces acting on the material and the system are determined to obtain the force to be exerted in extrusion or wire drawing. The analysis and results on these processes are beyond the scope of this book. The preceding references may be consulted for details on metal forming and other similar processes for different materials. The forming processes can then be coupled with the downstream transport processes in the moving material.

There are several other manufacturing processes that involve continuous materials processing. We will discuss polymer extrusion and optical fiber drawing in later chapters. Heat treatment of continuous sheets and rods, drying of paper sheets and rolls, baking of sheets and rods of food materials, cooling of polymer jacketing in the manufacture of cables, and drying of paper pulp in paper-insulated wires are some of the examples that may be mentioned. Figure 5.26e shows deposition on a continuously moving surface. Such a process may be used for chemical vapor deposition on continuous sheets, serving as susceptors or base surfaces, to fabricate photovoltaics or solar cells (Chiu and Jaluria 2000). Thus, many traditional batch manufacturing processes may be converted into continuous ones to reduce costs by using the results and discussions presented here.

5.9 Summary

This chapter discusses continuous materials processing, as contrasted with batch processing. There is growing interest in continuous processes because of greater production rates and lower costs. There are a wide range of processes that can be considered for continuous processing. These include hot and cold rolling, extrusion, wire and fiber drawing, heat treatment of moving materials, and deposition on continuously moving surfaces. The problem is time-dependent at the start of the process but generally approaches steady-state conditions as time elapses. Both situations are of practical interest. Fluctuations and instabilities can also lead to time dependence. Of particular interest is the thermal field in the moving material, since the resulting thermal stress and microstructure are determined by the temperature distribution. The temperature rise or decay with distance is also critical in designing the system to maintain temperatures above certain values, such as the recrystallization temperature, for hot rolling. Though a convective heat transfer coefficient, obtained from empirical data and correlations, may be employed at the surface, the problem is actually a conjugate one, and both the flow and thermal field in the solid have to be considered. Though boundary-layer assumptions are valid in some cases, solution of the full elliptic equations is generally needed for a realistic simulation. Buoyancy effects are important in many circumstances, and an additional forced flow, in an extensive environment or in a channel, may be employed to enhance the heat transfer. Experiments are used to validate the analytical and numerical models, to provide physical insight into the basic mechanisms, and to provide realistic and practical operating conditions. The basic formulation, solution strategies, and typical results are presented on this materials processing technique. The relevance of such results in the operation, control, design, and optimization of practical continuous manufacturing processes is discussed.

References

J.V. Alderson, J.B. Caress, R.L. Sager, The cooling rate of glass fiber in the continuous filament process, Laboratory Report No. L.R. 235 of Pilkington Bros., Lathom, Lancashire (1968)

T. Altan, S. Oh, H. Gegel, *Metal Forming Fundamentals and Applications* (American Society for Metals, Metals Park, 1983)

R.G.C. Arridge, K. Prior, Cooling time of silica fibers. Nature **203**, 386–387 (1964)

T.Z. Blazynski, *Metal Forming: Tool Profiles and Flow* (Wiley, New York, 1976)

T.S. Chen, F.A. Strobel, Buoyancy effects in boundary layer adjacent to a continuous, moving horizontal flat plate. J. Heat Transf. **102**, 170–172 (1980)

K. Chida, Y. Katto, Conjugate heat transfer of continuously moving surface. Int. J. Heat Mass Transf. **19**, 461–470 (1976.) and pp. 453-460

W.K.S. Chiu, Y. Jaluria, Continuous chemical vapor deposition processing with a moving finite thickness susceptor. J. Mater. Res. **15**, 317–328 (2000)

O. Chopra, P. Niessen, Solification of lead abd lead alloys in continuous drum sheet casting. Z. Metallk. **66**, 10–16 (1975)

L.E. Erikson, L.T. Fan, V.G. Fox, Heat and mass transfer on a moving continuous flat plate with suction or injection. Ind. Eng. Chem. Fund. **5**, 19–25 (1966)
E.G. Fisher, *Extrusion of Plastics* (Wiley, New York, 1976)
B. Gebhart, Y. Jaluria, R.M. Mahajan, B. Sammakia, *Buoyancy-Induced Flows and Transport* (CRC Press, Boca Raton, 1988)
A. Ghosh, A.K. Mallik, *Manufacturing Science* (Ellis Horwood, Chichester, 1986)
W.A. Gifford, A finite element analysis of isothermal fiber formation. Phys. Fluids **25**, 219–225 (1982)
L.R. Glicksman, The cooling of glass fibers. Glass Technol. **9**, 131–138 (1968)
L.R. Glicksman, A prediction of upper temperature limit for glass fiber spinning. Glass Technol. **15**, 16–20 (1974)
J.F. Griffin, J.L. Thorne, On thermal boundary layer growth on continuous moving belts. AICHE J. **13**, 1210–1211 (1967)
F. Hollander, A model to calculate the complete temperature distribution in steel during hot rolling. Iron & Steel Inst **23**, 46–74 (1970)
U.A. Jackson, The present state of the theory of crystal growth from the melt. J. Crystal Growth **24**, 130–136 (1974)
K. Jafar, R. Nazar, A. Ishak, I. Pop, MHD flow and heat transfer over stretching/shrinking sheets with external magnetic field, viscous dissipation and Joule effects. Can. J. Chem. Eng. **90**, 1336–1346 (2012)
Y. Jaluria, Transport from continuously moving materials undergoing thermal processing. Ann. Rev. Heat Transf. **4**, 187–245 (1992)
Y. Jaluria, Numerical modeling of manufacturing processes, in *Handbook of Numerical Heat Transfer*, ed. by W. J. Minkowycz, E. M. Sparrow, J. Murthy, (Wiley, New York, 2006), pp. 729–784
Y. Jaluria, Thermal issues in materials processing. J. Heat Transf. **135**, 061701-1 to 14 (2013)
Y. Jaluria, Heat and mass transfer in materials processing and manufacturing. Adv. Heat Transf. **48**, 1–94 (2016)
Y. Jaluria, A.P. Singh, Temperature distribution in a moving material subjected to surface energy transfer. Meth. Appl. Mech. Eng. **41**, 145–157 (1983)
Y. Jaluria, K.E. Torrance, *Computational Heat Transfer*, 2nd edn. (Taylor & Francis, New York, 2003)
S. Kalpakjian, S. Schmid, *Manufacturing Processes for Engineering Materials*, 6th edn. (Pearson, Boston, 2016)
B.H. Kang, Conjugate heat transfer from a continuously moving material and from an isolated heat source, Ph.D. Thesis, Rutgers University, New Brunswick, 1990
B.H. Kang, Y. Jaluria, A numerical study of the fluid flow and heat transfer due to a heated plate moving in uniform forced flow. Numer. Heat Transf. **22**, 143–165 (1992)
B.H. Kang, Y. Jaluria, M.V. Karwe, Numerical simulation of conjugate transport from a continuous moving plate in materials processing. Numer. Heat Transf. **19**, 151–176 (1991)
B.H. Kang, J. Yoo, Y. Jaluria, Experimental study of the convective cooling of a heated continuously moving material. J. Heat Transf. **116**, 199–208 (1994)
M.V. Karwe, Y. Jaluria, Thermal transport from a heated moving surface. J. Heat Transf. **108**, 728–733 (1986)
M.V. Karwe, Y. Jaluria, Fluid flow and mixed convection transport from a moving plate in rolling and extrusion processes. J. Heat Transf. **110**, 655–661 (1988)
M.V. Karwe, Y. Jaluria, Experimental investigation of thermal transport from a heated moving plate. Int. J. Heat Mass Transf. **35**, 493–511 (1992)
W.A. Khan, I. Pop, Boundary-layer flow of a nanofluid past a stretching sheet. Int. J. Heat Mass Transf. **53**, 2477–2483 (2010)
E.A. Koldenhof, Laminar boundary layer on continuous flat and cylindrical surfaces. AICHE J. **9**, 411–418 (1963)

H.K. Kuiken, The cooling of a low-heat-resistance sheet moving through a fluid. Proc. Roy. Soc. London **A341**, 233–252 (1974)
H.K. Kuiken, The cooling of a low-heat-resistance cylinder moving through a fluid. Proc. Roy. Soc. London **A346**, 23–35 (1975)
H.K. Kuiken, Solidification of a liquid on a moving sheet. Int. J. Heat Mass Transf. **20**, 309–314 (1977)
S. Kyriacou, C.E. Polymeropoulos, V. Sernas, Accelerated cooling of optical fiber. Mat. Res. Soc. Proc. **172**, 49–54 (1990)
A. Mastroberardino, U.S. Mahabaleshwar, Mixed convection in viscoelastic flow due to a stretching sheet in a porous medium. J. Porous Media **16**, 483–500 (2013)
M.A. Matovich, J.R.A. Pearson, Spinning a molten threadline. Ind. Eng. Chem. Fund. **8**, 512–520 (1969)
W. J. Minkowycz, E. M. Sparrow, J. Y. Murthy (eds.), *Handbook of Numerical Heat Transfer*, 2nd edn. (Wiley, New York, 2006)
A. Moutsoglou, T.S. Chen, Buoyancy effects in boundary layers on inclined, continuous, moving sheets. J. Heat Transf. **102**, 371–372 (1980)
R.H. Pletcher, J.C. Tannehill, D. Anderson, *Computational Fluid Mechanics and Heat Transfer*, 3rd edn. (CRC Press, Boca Raton, 2016)
N. Ramachandran, T.S. Chen, B.F. Armaly, Correlations for laminar mixed convection in boundary layers adjacent to horizontal, continuous moving sheets. J. Heat Transf. **109**, 1036–1039 (1987)
S. Roy Choudhury, Y. Jaluria, Forced convective heat transfer from a continuously moving heated cylindrical rod in materials processing. J. Heat Transf. **116**, 724–734 (1994)
S. Roy Choudhury, Y. Jaluria, Cylinder moving in pressure- and buoyancy-induced channel flow: A numerical study of transport due to three aiding/opposing mechanisms. Numer. Heat Transf. **27**, 373–393 (1995)
B.C. Sakiadis, Boundary layer behavior on continuous solid surfaces: II. Boundary layer equations for two-dimensional and axisymmetric flows. AICHE J. **7**, 26–28 (1961a)
B.C. Sakiadis, Boundary layer behavior on continuous solid surfaces: I. The boundary layer on continuous flat surface. AICHE J. **7**, 221–225 (1961b)
J.A. Schey, *Introduction to Manufacturing Processes*, 3rd edn. (McGraw-Hill, New York, 1999)
Z. Tadmor, C. Gogos, *Principles of Polymer Processing*, 3rd edn. (Wiley, New York, 1979)
A.A. Tseng, Finite-difference solutions for heat transfer in a roll rotating at high speed. Numer. Heat Transf. **7**, 113–125 (1984a)
A.A. Tseng, A generalized finite-difference scheme for convection-dominated metal forming problems. Numer. Meth. Eng. **20**, 1885–1900 (1984b)
F.K. Tsou, E.M. Sparrow, R.J. Goldstein, Flow and heat transfer in the boundary layer on a continuously moving surface. Int. J. Heat Mass Transf. **10**, 219–235 (1967)
K.T. Yangai, Prediction of strip temperature for hot strip mills. Trans. Iron Steel Inst. Jpn. **26**, 11–19 (1976)
D.A. Zumbrunnen, R. Viskanta, F.P. Incropera, The effect of surface motion on forced convection film boiling heat transfer. J. Heat Transf. **111**, 760–766 (1989)

Chapter 6
Polymer Processing

Polymers consist of long-chain molecules composed of many repeated smaller molecules and include materials such as plastics, rubber, resins, silicones, elastomers, acrylics, and biopolymers. The processing of polymers is of considerable interest in a wide variety of applications such as manufacture of consumer products, biological systems, composite materials, pharmaceutical products, and processed food items. Among the most widely used polymers are thermoplastics, which become pliable or moldable beyond a specific temperature. These include common materials like polyvinylchloride (PVC), low-density polyethylene (LDPE), polystyrene (PS), and nylon. Since heating of these materials is needed for flow, forming, phase change, chemical conversion, and other changes associated with processing, most of the relevant manufacturing processes are thermally based, i.e., heat and mass transfer, thermodynamics, and fluid flow are the dominant mechanisms that govern the process. Similarly, biopolymers include proteins, DNA, sugar, and starch. Polymer matrix composites are similarly important in many different applications. Thermoset polymers, which become permanently hard when heat is applied and do not soften on further heating, include epoxies, acrylic resins, and vulcanized rubber. Additional new polymeric materials continue to be developed for new and emerging technological applications.

The focus in this chapter is on thermoplastic polymers. Because of the tremendous importance of these materials in our lives and in a wide diversity of commercial applications, a lot of work has been done on the development of processing techniques and the design of the relevant systems. Detailed information is also available in the literature on the processing of different materials and development of new products. This chapter focuses on the important manufacturing processes involved in polymer processing and discusses the underlying transport mechanisms and relevant practical issues. The processes considered here include:

Extrusion
Injection molding
Thermoforming

Y. Jaluria, *Advanced Materials Processing and Manufacturing*, Mechanical Engineering Series, https://doi.org/10.1007/978-3-319-76983-7_6

Several other processes are based on combinations of these basic processes or other similar processes. The transport phenomena involved in these processes are discussed, along with solution techniques and typical results that may be used to predict, control, and design the process. The extrusion process is discussed in detail to bring out the major considerations, followed by a discussion of other techniques.

6.1 Material Properties

The most important concern in polymer processing, as is true for most materials processing applications, is the material itself. The temperature and shear it undergoes during processing change its chemical and physical structure, resulting in changes in properties during the process and of the final product. Material properties thus vary due to temperature and pressure changes, chemical reactions, phase changes, and conversion or structural changes in the material. Since the transport processes are strongly influenced by the material properties, it is important to employ accurate property data and appropriate property characterization in the analysis and in the experiments. A brief discussion of the material characteristics and commonly used constitutive equations is given here. For further details, the books on polymers and on polymer processing, such as those by Han (1976), Tadmor and Gogos (1979), Pearson and Richardson (1983), Crochet et al. (1984), and Bird et al. (1987), may be consulted.

As mentioned earlier in Chap. 2, these materials are generally non-Newtonian in behavior, with strong property dependence on temperature and composition as well. These are fluids whose viscosity is dependent on the shear rate and, thus, on the flow. Figure 2.4 sketched the typical behavior of non-Newtonian fluids in terms of the shear stress versus shear rate curves. The flow dependence of the viscosity significantly complicates the solution for the flow and heat transfer. To differentiate between the flow behavior of a Newtonian fluid and that of a non-Newtonian fluid, let us consider the flow between two parallel surfaces with one surface stationary and the other moving at constant velocity. If the fluid is Newtonian, the velocity distribution is linear and the shear rate du/dy, where u is the velocity parallel to the surfaces and y the distance from the stationary plate, is constant across the fluid layer. The shear stress τ_{yx} is the force applied to the moving plate divided by its area. The slope of the shear stress variation with the shear rate is the coefficient of viscosity μ of the fluid. For a Newtonian fluid, μ is independent of the shear rate and depends only on the temperature and pressure. For non-Newtonian fluids, the plot of τ_{yx} versus du/dy is not a straight line through the origin. The apparent viscosity which is the ratio of the shear stress to the shear rate may increase or decrease with du/dy, as shown in Fig. 2.4. If there is no recovery of deformation on removal of applied stress, i.e., there is no elastic response, the fluid is termed viscoinelastic or purely viscous. If it does have an elastic response, it is known as viscoelastic. The former type of fluids are shown in Fig. 2.4. In time-independent fluids, the duration of the shear has no effect on the viscosity. Various types of polymeric materials are shown. If the viscosity decreases with increasing shear rate, it is known as shear thinning or pseudoplastic. Most common plastics can be characterized as shear thinning materials. If the viscosity increases with increasing shear rate, the fluid is known as dilatant, or shear thickening, and this applies to some suspensions.

As discussed earlier, the fluid is often treated as a generalized Newtonian fluid, with the viscosity given in terms of the shear rate $\dot{\gamma}$ and temperature T. The equation for a power-law variation was given earlier as

$$\mu = \mu_o \left[\frac{\dot{\gamma}}{\dot{\gamma}_o}\right]^{n-1} \exp[-b(T - T_o)] \tag{2.12}$$

where the subscript o refers to a reference value, n is the power-law index, and b is a constant. The shear rate is related to the flow through the velocity gradients. Many fluids of practical interest can be approximated by the power-law relationship. Also, the exponential dependence on temperature may be replaced by the Arrhenius dependence, exp (b/T), for many fluids as

$$\mu = \mu_o \left[\frac{\dot{\gamma}}{\dot{\gamma}_o}\right]^{n-1} \exp\left[\frac{b}{T}\right] \tag{6.1}$$

Changes in the material characteristics and structure may also be due to chemical reactions, as seen later. If mass transfer is included, the viscosity will also depend on the species concentration c. Then, the viscosity expression includes a term such as exp $[-b_m(c - c_o)]$ or exp (b_m/c), where b_m is a constant that multiplies the temperature dependence term. Therefore, the constitutive equation may be written, for instance, as the following for such fluids (Lai and Kokini 1990):

$$\mu = \mu_o \left[\frac{\dot{\gamma}}{\dot{\gamma}_o}\right]^{n-1} \exp[-b(T - T_o)]\exp[-b_m(c - c_o)] \tag{6.2}$$

Several other viscosity expressions have been obtained for different fluids, such as the Bingham fluid, shown in Fig. 2.4. These are viscoplastic materials that behave as rigid body at low stresses but flow as viscous fluid at large stress, with toothpaste as a common example. In addition, viscoelastic behavior, in which the material exhibits both viscous and elastic characteristics during deformation, must also be modeled in some cases, particularly for flow emerging from a die, where it gives rise to die swell.

Various other approaches have been developed to model different types of non-Newtonian materials. Some fluids may be represented by the Carreau model (Kokini et al. 1992). A modified version of the Carreau model is used for a variety of materials, particularly dimethyl silicone fluid (Viscasil-300M) which is used for some experiments and numerical simulations (Esseghir and Sernas 1991). The model is given as

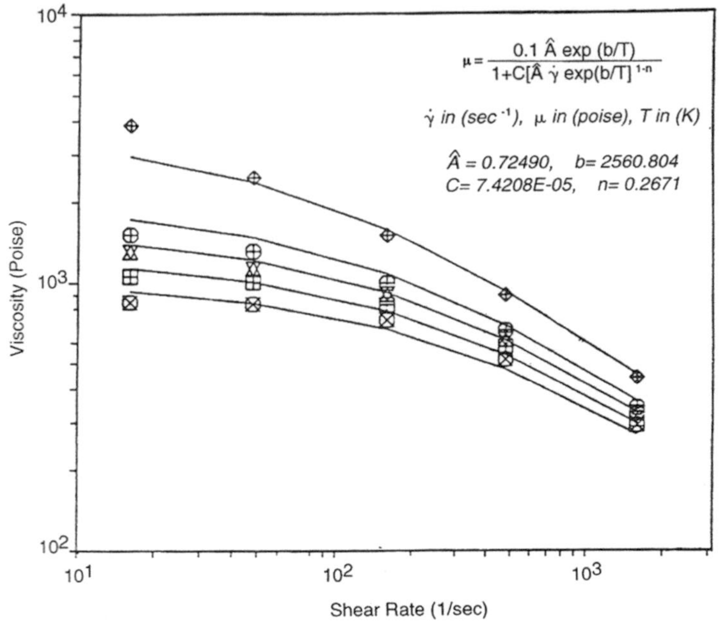

Fig. 6.1 Viscosity measurements for dimethyl silicone fluid (Viscasil-300M)

$$\mu = \frac{0.1\hat{A}\exp(b/T)}{1 + C\,[\hat{A}\,\dot{\gamma}\exp(b/T)]^{1-n}} \tag{6.3}$$

where $\hat{A}$, C, b, and n are constants that are determined by curve fitting the experimental results; see Fig. 6.1. The approach to Newtonian behavior at small shear rates and a power-law dependence at larger values is clearly seen for these fluids. For carboxymethyl cellulose (CMC) solutions, n is found to be 0.445 and 0.81 for pure and degraded fluids, respectively, whereas corn syrup is Newtonian ($n = 1$). The values of the constants for Viscasil-300M were determined by Esseghir and Sernas (1991), as given in Fig. 6.1. Values for other fluids may be obtained from the references given.

However, despite the critical importance of accurate material property data in polymer processing, usually only limited data are available in the literature, and the dependence on temperature, concentration, shear rate, etc. is not known accurately. Also, many of the properties are available at room temperature, whereas the processing is at much higher temperatures. This aspect remains a major concern in the modeling of manufacturing processes involved with polymeric materials.

6.2 Single Screw Extrusion

Extrusion of polymers, such as plastics and food, is an important process and has been employed for a wide range of materials and applications. This process is particularly useful when thermal and/or mechanical means are to be employed to obtain a uniformly processed product in a continuous operation. The screw extrusion process for polymeric materials usually includes the following stages: feeding, conveying, plasticizing, homogenizing, and pressurizing (Tadmor and Gogos 1979). These processes take place within a special reactor consisting of one or more screws rotating inside a barrel, usually referred to as single, twin, or multiscrew extruders; see Fig. 6.2. The pressure and temperature of the material rise as they flow down the extruder channel toward the die. The cross-sectional shape of the product is obtained by pushing the molten material through an appropriately designed die. In order to obtain desired product quality and characteristics, a given set of operating parameters has to be maintained at certain values, for a particular design of the extruder. The main operating parameters are the screw rotational speed (rpm), mass flow rate, and the barrel temperature, while the extruder design parameters are screw

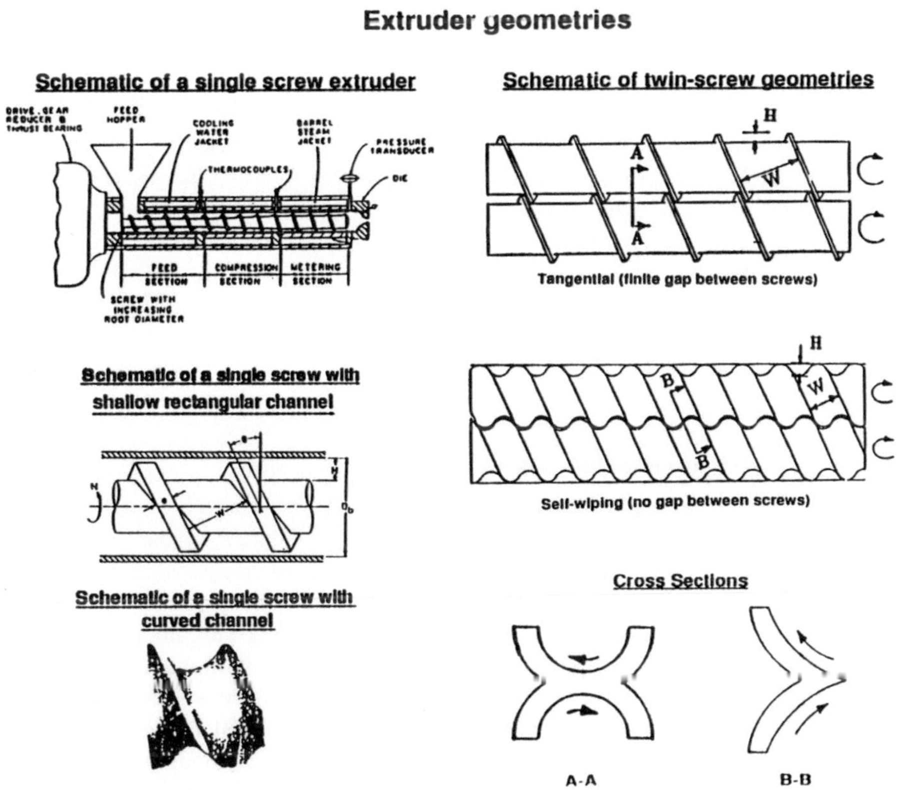

Fig. 6.2 Schematics of single and twin screw polymer extruders

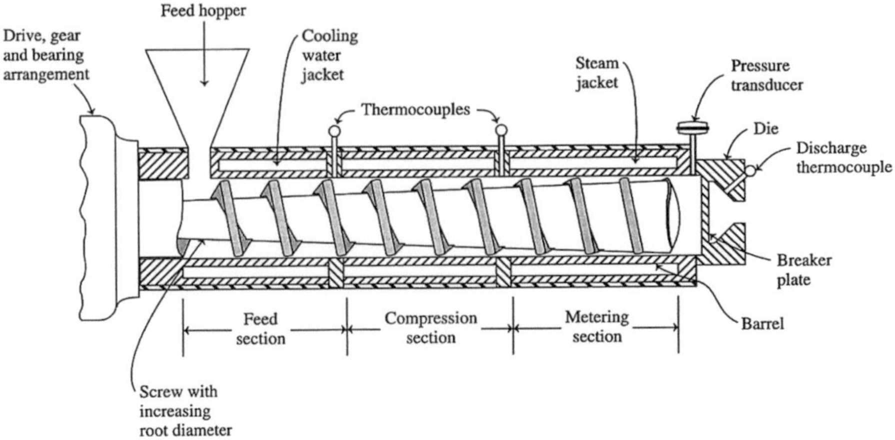

Fig. 6.3 Schematic of a single screw extruder

diameter, screw profile, barrel diameter, extruder length, other dimensions, and materials used. The desired temperature level at the barrel is maintained by electrical heating, by using a hot fluid, and/or by cooling of the barrel.

A single screw extruder, as shown schematically in Fig. 6.3, is an important manufacturing system and was briefly discussed in the preceding chapters. It is used for the fabrication of a variety of polymer-based products and is also used to feed molten material into injection molding systems (Fenner 1979; Tadmor and Gogos 1979; Jaluria 1996). Though twin screw extruders have recently become important in the industry because of better mixing and control characteristics, single screw extruders continue to be used extensively for a range of products and applications. Therefore, we start with a consideration of the transport processes involved in single screw extrusion.

6.2.1 Two-Dimensional Model

The geometry of a single screw extruder is very complex, and the presence of screw flights with a rotating screw makes the problem even more complicated. Even if the boundary conditions, such as barrel temperature, flow rate, and screw speed, are kept constant with time, the flow at a given point in the flow varies with time. Therefore, a time-dependent, three-dimensional problem arises with the additional complexity of non-Newtonian fluid with temperature-dependent properties. A solution of the complete problem is, therefore, very involved, and simplified models have been developed to understand the basic phenomena and provide inputs that may be used for the design of practical extruders. For a simplified two-dimensional model, the coordinate system is fixed to the screw root, and curvature effects are neglected. This allows the screw channel to be "opened out," resulting in the rectangular channel shown earlier in Fig. 2.7, with the barrel moving in a direction opposite to the screw rotation at the velocity V_b at the screw helix angle ϕ. This transformation is commonly employed in the literature (Fenner 1979; Karwe and Jaluria 1990). Karwe et al. (1995) have shown analytically that the results are independent of the coordinate system for a Newtonian fluid. The analysis applies to the metering section of the screw extruder. Curvature effects are important in sections with deep channels, such as the solid conveying zone. However, for the shallow metering section considered here, curvature effects are neglected (Gopalakrishna et al. 1992).

For steady, developing, two-dimensional flow of a homogeneous fluid in a single screw extruder with shallow and long channels, i.e., for $H << B$, in Fig. 2.7, after neglecting the inertia terms (creeping flow approximation), the equations for the conservation of momentum become

$$\frac{\partial p}{\partial x} = \frac{\partial(\tau_{yx})}{\partial y} \qquad \frac{\partial p}{\partial y} = 0 \qquad \frac{\partial p}{\partial z} = \frac{\partial(\tau_{yz})}{\partial y} \tag{6.4}$$

where p is the local pressure and τ is the shear stress. The creeping flow approximation is valid because the Reynolds number Re based on channel height H is much smaller than unity, typically of order 10^{-3}, making the inertia terms much smaller than the viscous terms. The clearance between the screw flights and the barrel is assumed to be small enough to neglect the leakage across the flights from one screw channel to the neighboring one. However, for a more realistic simulation of the process, the effects of leakage must be included, particularly the viscous dissipation generated in the clearance region.

In the presence of strong viscous dissipation effects and/or heat addition from the barrel, the thermal convection along the z-direction (along the down-channel direction) is significant. Therefore, the temperature field develops along the z-direction. The velocities will also change with the downstream position as a result of this change in temperature, if the fluid viscosity is dependent upon its temperature. It is assumed that diffusion in the z-direction is negligible in comparison with convection. This assumption can be shown to be valid for the extruder length $L >> B$, using a simple scaling analysis, and is borne out by experimental data. The temperature gradient in the y-direction is expected to be much greater than that in the down-channel direction. The diffusion term in the y-direction is, therefore, retained. Thus, the energy equation becomes

$$\rho\, C_p\, w\left(\frac{\partial T}{\partial z}\right) = \frac{\partial}{\partial y}\left(k\frac{\partial T}{\partial y}\right) + \tau_{yx}\left(\frac{\partial u}{\partial y}\right) + \tau_{yz}\left(\frac{\partial w}{\partial y}\right) \tag{6.5}$$

Here, T is the local temperature, ρ the density, C_p the specific heat at constant pressure, and k the thermal conductivity of the fluid. The first term on the right-hand side is due to thermal diffusion in the y-direction, and the last two terms are due to viscous dissipation. The shear stress τ_{yx} and τ_{yz} are given for this two-dimensional flow by

$$\tau_{yx} = \mu\left(\frac{\partial u}{\partial y}\right) \qquad \tau_{yz} = \mu\left(\frac{\partial w}{\partial y}\right) \tag{6.6}$$

The shear rate for this two-dimensional circumstance is given by the expression

$$\dot{\gamma} = \left[\left(\frac{\partial u}{\partial y}\right)^2 + \left(\frac{\partial w}{\partial y}\right)^2\right]^{1/2} \tag{6.7}$$

with the viscosity μ given by one of the various models discussed in the preceding section, whichever is appropriate for a given fluid. If the power-law model is used, the singularity at zero shear rate must be avoided, often by simply putting $n = 1$ as shear rate approaches zero, i.e., Newtonian behavior at small shear rates.

The above energy equation is parabolic in the z-direction, and marching may be used to obtain the solution. A restriction to the flow is imposed by the presence of a die at the end of the extruder. The flow in the extruder is strongly coupled with that inside the die. For very narrow dies and at large extruder speeds, a backflow may arise in the extruder channel in terms of the coordinate system described. This makes the problem elliptic, requiring a different approach for the solution. The boundary conditions are also shown in Fig. 2.7. The temperature distribution at the barrel is specified as T_b (z), which, in many cases, is a constant or has different values in different sections. The screw is taken as adiabatic. Since the energy equation is parabolic in z, boundary conditions are necessary only at $z = 0$ to allow marching in the z-direction and, thus, obtain the solution in the entire domain. The boundary conditions at $z = 0$ are provided in terms of the developed velocity profiles, denoted by subscript *dev*, at the inlet temperature $T = T_i$. These are obtained by solving the momentum equations, keeping the temperature constant at T_i, by means of an implicit finite-difference scheme (Karwe and Jaluria 1990).

The governing equations and boundary conditions may be nondimensionalized with channel height H, barrel velocity component in the z-direction V_{bz}, barrel and inlet temperatures, T_b and T_i, and μ_o as characteristic quantities. The continuity equation for the conservation of mass is

$$\frac{\partial u^*}{\partial x^*} + \frac{\partial w^*}{\partial z^*} = 0 \tag{6.8}$$

However, the constraints on the flow are generally written in integral form, given in dimensionless terms as

$$\int_0^1 u^* dy^* = 0 \qquad \int_0^1 w^* dy^* = q_v = \frac{\dot{Q}/B}{HV_{bz}} \tag{6.9}$$

where the first condition ensures that the net flow across the channel is zero, if the leakage flow across the screw flights is negligible, and the second condition gives the down-channel flow rate. Therefore, the parameter q_v is the dimensionless volumetric flow rate, generally called the throughput, emerging from the extruder. If the screw channel is not rectangular, q_v is defined as Q/AV_{bz}, where A is the channel cross section. The integral form of the continuity equation is generally used since the throughput can be specified as an operating condition. Dimensionless parameters Peclet number Pe, temperature difference β, and Griffith number G arise, where

$$Pe = \frac{V_{bz}H}{\alpha} \quad \beta = b(T_b - T_i) \quad G = \frac{\mu V_{bz}^2}{k(T_b - T_i)} \tag{6.10}$$

Here, the shear rate is taken as V_{bz}/H in the viscosity expression, α is the fluid thermal diffusivity, T_b is the barrel temperature, and T_i is the inlet temperature (Jaluria 1996).

The governing equations are conveniently solved by means of finite-difference techniques for simple geometries. The computational domain is the rectangular channel shown. The iterative Newton-Raphson method (Jaluria and Torrance 2003) may be used to satisfy the conditions on the total flow rates. The iteration is terminated when the pressure gradients satisfy a chosen convergence criterion. Using the boundary conditions in terms of u, w, and T at any upstream z location, the energy equation is solved to obtain the temperature distribution at the next downstream z location. The same approach is extended to solve the mass transfer problem. The numerical scheme is validated by comparisons with experimental data on actual, full-size extruders, using both Newtonian and non-Newtonian fluids, as discussed later.

6.2.2 *Fully Developed Flow*

The simplest solution is that of the fully developed circumstance for which the temperature and the velocity fields are assumed to remain unchanged downstream. Though analytical solutions can be obtained for certain channel flows driven by pressure or shear, the present circumstance, with combined pressure and shear effects, screw helix, and non-Newtonian fluids, requires a numerical solution of the governing equations. Figure 6.4 shows the calculated w^* velocity profiles for different values of q_v. The characteristic curves in terms of the throughput and the dimensionless pressure gradient are shown for $n = 0.5$ and different Griffith numbers in Fig. 6.4b.

When the pressure gradient is zero, the flow is only due to the viscous effect of the moving barrel and is termed as drag flow. For Newtonian flow, the velocity profile is linear for this circumstance and $q_v = 0.5$. This situation, which was discussed earlier in Chap. 2, is similar to the Couette flow between two parallel plates in the absence of a pressure gradient, with one plate stationary and the other moving at constant speed. For a favorable pressure gradient, the throughput exceeds 0.5, and for an adverse pressure gradient, it is less; see Fig. 3.1. The velocity profile bulges outward, with the velocity exceeding the linear variation, for the favorable case. The opposite

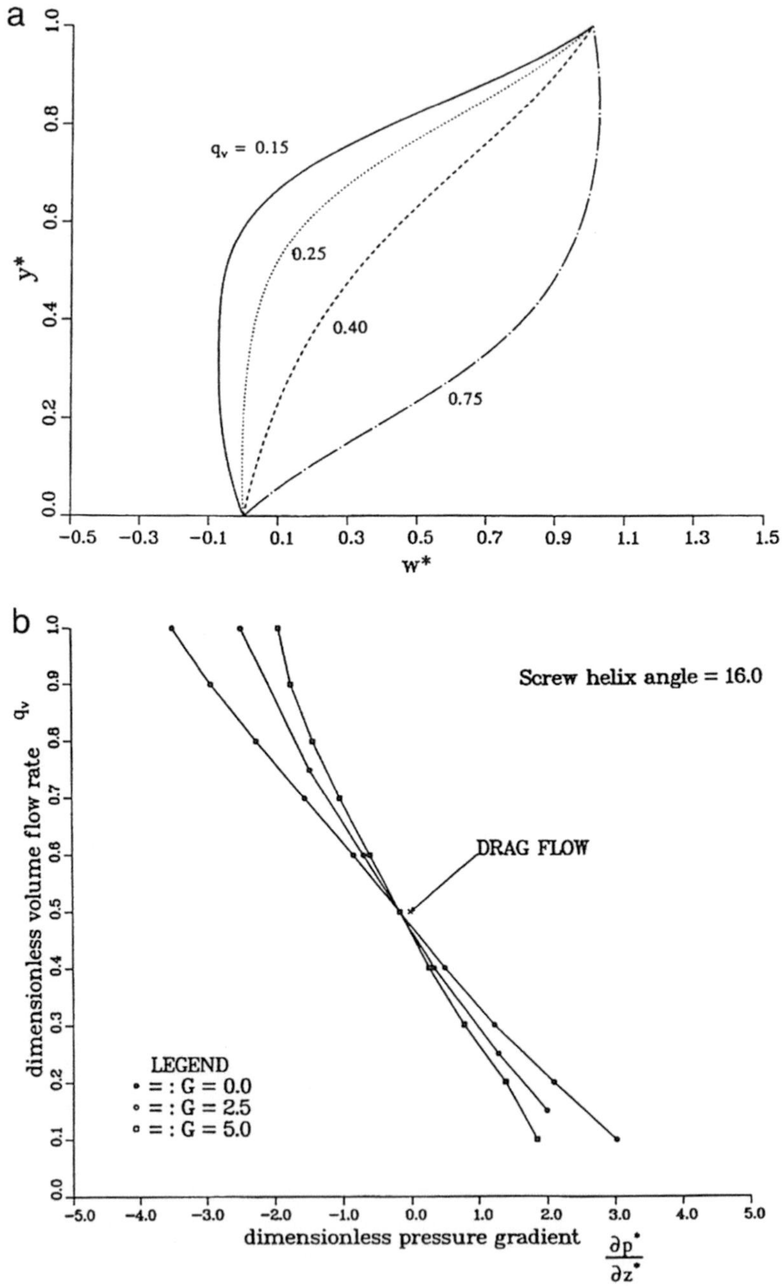

Fig. 6.4 Results for fully developed flow of a non-Newtonian fluid in a channel at $n = 0.5$. (**a**) Profiles of dimensionless down-channel velocity component w^* at Griffith number $G = 2.5$ and different throughput q_v; (**b**) variation of dimensionless pressure gradient with throughput at different values of G

occurs for the adverse pressure case. Similarly, for non-Newtonian flow, the profiles are seen to be strongly dependent on the throughput, though drag flow does not arise at $q_v = 0.5$ but at a value which depends on the fluid, temperature, and other conditions. The screw channel is assumed to be completely filled with the non-Newtonian fluid. Therefore, a decrease in the throughput at a given screw speed implies a smaller diameter die, i.e., greater obstruction to the flow. This results in a greater pressure rise downstream and an increased adverse pressure gradient, which is reflected in decreased velocity levels. Reverse flow may also arise at very small throughputs in this coordinate system. Larger throughputs are obtained with a favorable pressure gradient, i.e., pressure decreasing downstream. In an extruder, the obstruction provided by the die and by a tapered screw in many cases increases the pressure, resulting in an opposing pressure circumstance, as seen for small values of q_v. A higher Griffith number implies greater viscous heating. This results in higher temperatures and lower viscosity. This gives rise to a smaller pressure gradient at a given throughput in the positive pressure gradient range, which is of interest in extrusion.

6.2.3 Developing Flow

The computations are then extended to developing flow. For the results presented here, the ratio of axial screw length L to channel height H is taken as about 70, corresponding to practical extruders. This results in a dimensionless down-channel distance of around 200. Once the numerical results for the velocity and temperature fields are obtained, various other quantities of interest, such as the heat input to/from the barrel, local Nusselt number Nu_H, bulk temperature, shear stress, and pressure at various downstream locations, including that at the die, may be calculated. A few typical results are presented in Fig. 6.5 in terms of isotherms and constant velocity lines along the screw channel. The temperature and velocity profiles at four locations are also shown. The temperature of the fluid far downstream is seen to increase above the barrel temperature. Therefore, beyond a certain downstream location, heat transfer occurs from the flow to the barrel if the barrel is maintained at a fixed temperature. This implies initial heat transfer to the fluid by the barrel followed by heat removal from the fluid further downstream. This effect is due to the viscous heating of the fluid and varies strongly with the Peclet and Griffith numbers. Due to larger viscous dissipation, the fluid temperature may be much higher than the barrel temperature, as seen in Fig. 6.6.

It is seen from these results that the constant velocity lines are almost parallel to the barrel, indicating very little mixing in the fluid, which is largely stretched as it goes through the extruder. To enhance mixing, reverse screw elements and breaks in the screws are often employed in practical single screw extruders. Mixing is also substantially increased in twin screw extruders, as discussed later. It is also seen that large temperature differences exist in the fluid, from the barrel to the screw. This is largely due to the low thermal conductivity of typical polymeric materials. At

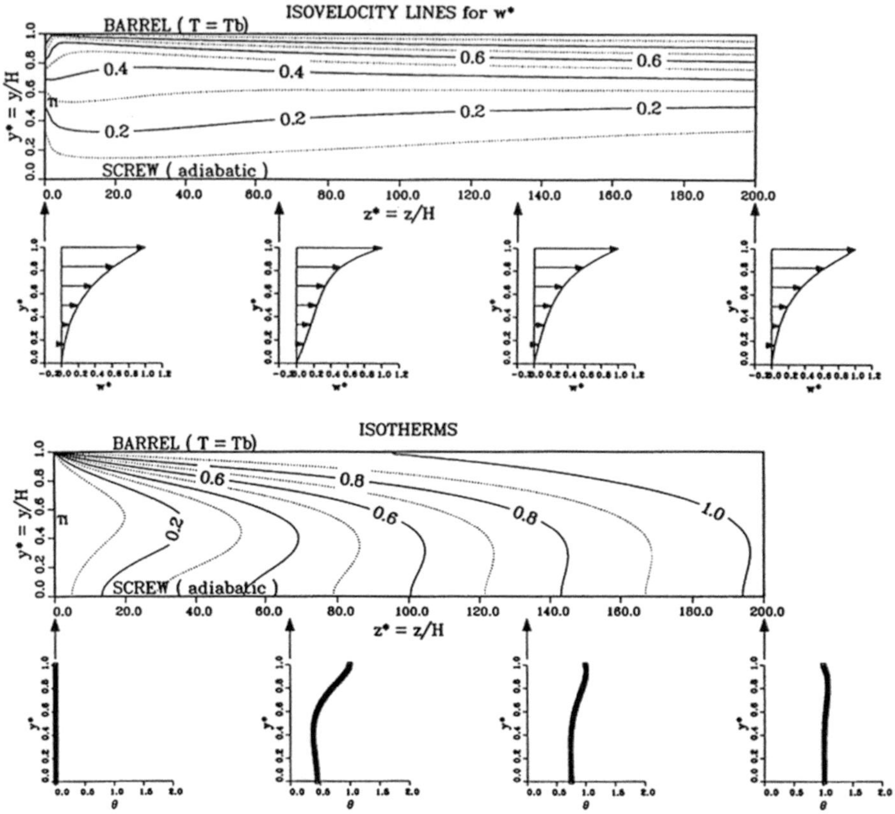

Fig. 6.5 Calculated velocity and temperature fields in the extruder at $n = 0.5$, $G = 10.0$, $Pe = 3427$, $\beta = 1.61$, and $q_v = 0.3$

relatively high values of the dimensionless throughput q_v, the backflow is small, and much of the fluid near the screw root remains unheated. Viscous dissipation is important and affects the thermal field substantially.

Figure 6.7 shows the variation of the dimensionless pressure p^* and the pressure gradient $\partial p^*/\partial z^*$ along the screw channel. The pressure rises toward the outlet, as expected, and the pressure gradient varies gradually downstream. The bulk temperature and the heat transfer rate to the fluid q_{in}, as given by the local Nusselt number $Nu = q_{in}H/[k(T_b - T_i)]$, at the barrel were also calculated. The downstream variations and the values attained vary strongly with q_v, as mentioned earlier. The bulk temperature increases, as expected, and the heat transfer rate was found to decrease, approaching zero far downstream, indicating the small amount of energy transfer needed to maintain the barrel temperature at a given value. These trends are physically similar to those observed in practical extruders.

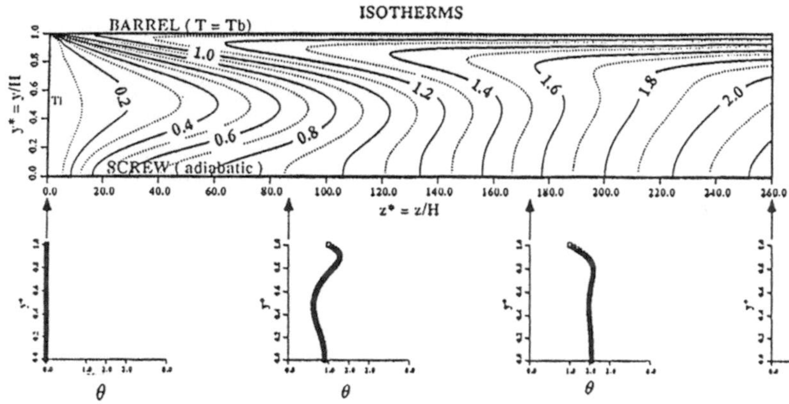

Fig. 6.6 Temperature profiles in single screw extruder at $n = 0.3$ and $Pe = 7050$

6.2.4 Three-Dimensional Transport

The basic transport processes in the extruder channel are three-dimensional, though two-dimensional models, as outlined above, have been used extensively to model the flow and generate data needed for design. The main problem with such two-dimensional models is that the effect of the flights is brought in by mass conservation considerations only, approximating the flights as being very far apart. Thus the recirculating flow generated in a screw channel, between two flights separated by a finite distance, is not simulated. Though this is applicable for shallow channels, three-dimensional models are needed for deep channels and for a more realistic modeling of practical extruders. However, three-dimensional modeling is fairly involved, and a relatively simple model to simulate 3D flows has been developed by Sastrohartono et al. (1995) and is outlined here.

In the unwrapped extruder channel, it is reasonable to assume that the velocity vector $\bar{V}$ does not change significantly along the screw channel direction, i.e., z-coordinate direction. Then, $\partial \bar{V}/\partial z << \partial \bar{V}/\partial y$ or $\partial \bar{V}/\partial x$. Based on these assumptions and using the coordinate system sketched in Fig. 2.7, a three-dimensional flow analysis for the screw channel can be carried out in terms of in-plane and down-channel flows, using a three-dimensional finite element model, with marching in the z-direction. This approach is simpler than the full three-dimensional elliptic solution and requires much less computer storage and computational time. It can be used for typical extruder conditions, except when the throughput q_v is small, of order 0.1 or less.

The screw channel is assumed to be fully filled with the material. The computational domain is taken as the screw channel cross section and any profile can easily be modeled using the finite element method. The mesh discretization for a particular, self-wiping, screw profile was shown earlier in Fig. 3.9. This screw profile is used in twin screw extruders to allow one screw to wipe the surface of the other as it rotates, as discussed later. Figure 6.8 shows typical in-plane (u–v) and down-channel (w)

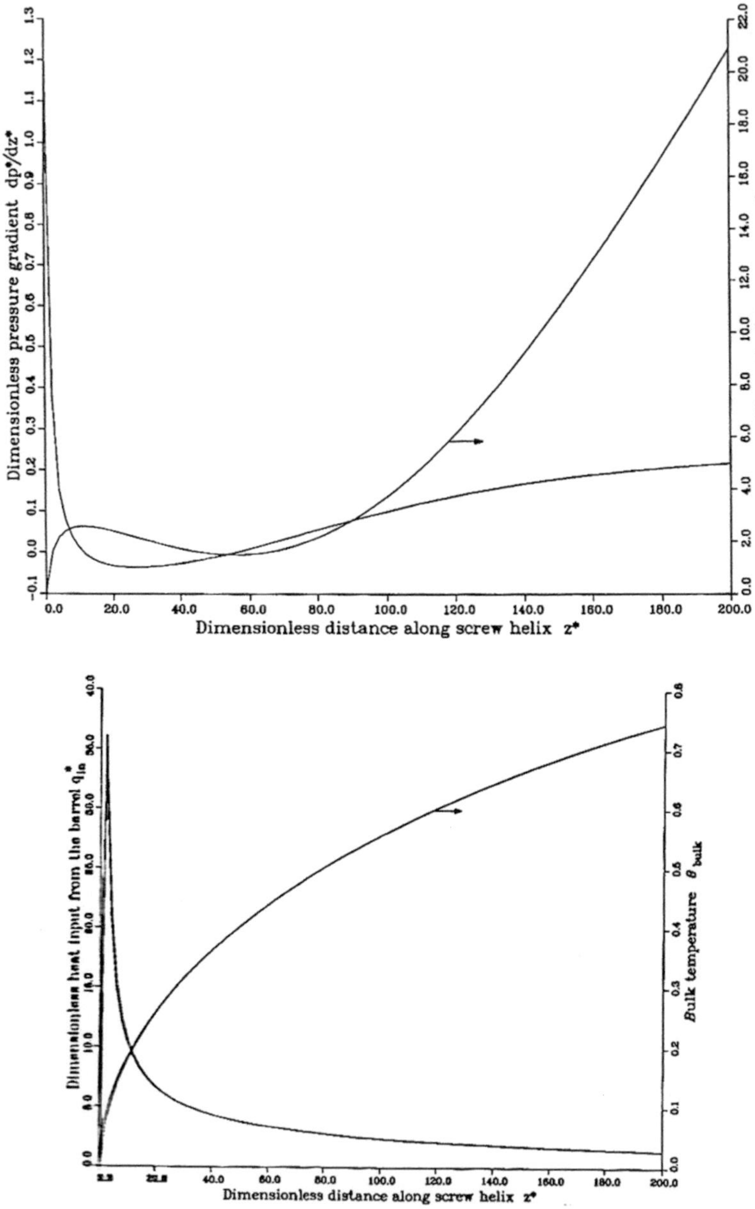

Fig. 6.7 Variation of pressure, pressure gradient, bulk temperature, and local Nusselt number in the down-channel direction for the conditions of Fig. 6.5

velocity fields and the temperature and pressure contours for pressure increasing in the down-channel direction. Velocity and temperature profiles are also shown. As can be seen from this figure, there is a recirculating flow that arises in the channel and

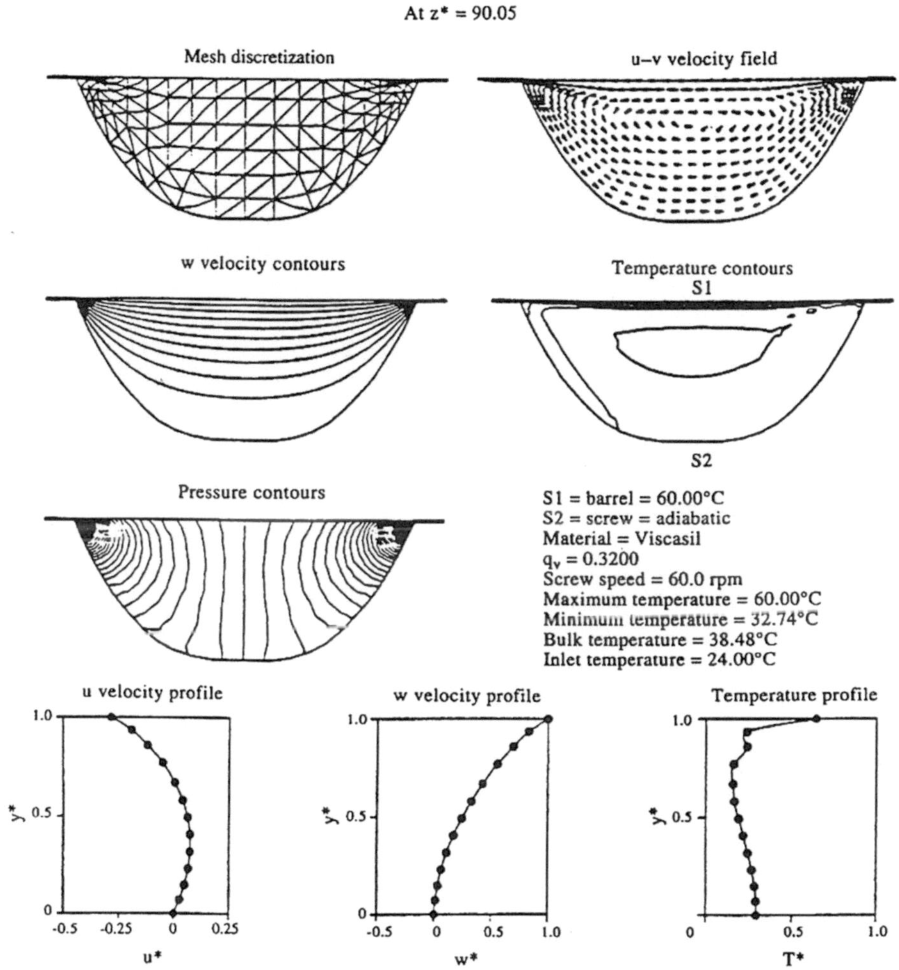

Fig. 6.8 Finite element grid, cross-sectional flow, pressure, and temperature field from 3D modeling for Viscasil-300M at $z^* = 90.05$, with $T_i = 24$ °C, $T_b = 40/60/80$ °C at a particular down-channel location, $q_v = 0.32$, and N (rpm) = 60. (Sastrohartono et al. 1995)

conveys fluid from near the barrel to the bottom of the channel. Therefore, this recirculation provides a mechanism for convective mixing of the fluid. Hot fluid near the barrel is conveyed downward in the channel, resulting in larger temperature near the bottom than that predicted by a two-dimensional model which is unable to capture this recirculating flow. Therefore, it enhances uniformity of fluid temperature across the cross section. As expected, the effect increases if the throughput is reduced, resulting in greater pressure rise and recirculation. This flow is also significantly affected by leakage at the gaps between the screw and the barrel. Additional results from this three-dimensional model are given by Jaluria (1996).

6.2.5 Axial Formulation

The marching procedure outlined above for two- and three-dimensional modeling is quite satisfactory for a wide range of operating conditions. However, at very small values of q_V, typically of order 0.1 or less, appreciable backflow due to opposing pressure arises in the extruder channel, in the moving barrel coordinate frame employed here for analysis. Then the numerical schemes used earlier do not work. This is largely due to flow and convection in the negative z-direction resulting from the reverse flow. As discussed by Chiruvella et al. (1996a), two different approaches may be employed to obtain a solution under these circumstances. The first one solves the elliptic problem instead of the parabolic problem, in which marching is used, by including diffusion of energy in the down-channel direction. However, this solution requires boundary conditions at the end of the screw channel, and the numerical computation is quite involved. The flow through a die may be solved and then coupled with the extruder to obtain the needed downstream boundary conditions, or developed conditions may be imposed at the end of an extended computational domain. The second approach uses a different coordinate frame, involving the axial direction parallel to the screw axis. At each point in the extruder channel, the axial and tangential directions are considered. This is known as the axial formulation which can be visualized as the flow along the screw axis as one looks at the extruder from the front. Since there is no backflow occurring in the direction parallel to the screw axis, one can choose to march along this direction from the hopper to the die for essentially any throughput. Both these approaches were found to yield results that are fairly close for typical extruders. However, the axial formulation is preferable due to the much smaller computational effort needed for the marching scheme as compared to the elliptic formulation. It can be shown analytically that, for an isothermal Newtonian fluid, the down-channel and axial formulations yield the same pressure and velocity distributions. The same concept is used for non-isothermal, non-Newtonian flow.

The variation of pressure in the axial direction is shown in Fig. 6.9, indicating expected trends. A comparison between the results for pressure at the die from the two formulations, along with experimental data from Esseghir and Sernas (1992) for Viscasil-300M, is also shown. Close agreement is found between all three, lending strong support to the two models (Chiruvella et al. 1996a). The axial formulation is particularly attractive because of the applicability of the marching procedure which has much small computer storage and computational time requirements than the elliptic code.

6.2.6 Tapered Screw

Single screw extruders are often tapered to increase the pressure generated in the extruder channel. The barrel is typically cylindrical with grooves on the inside

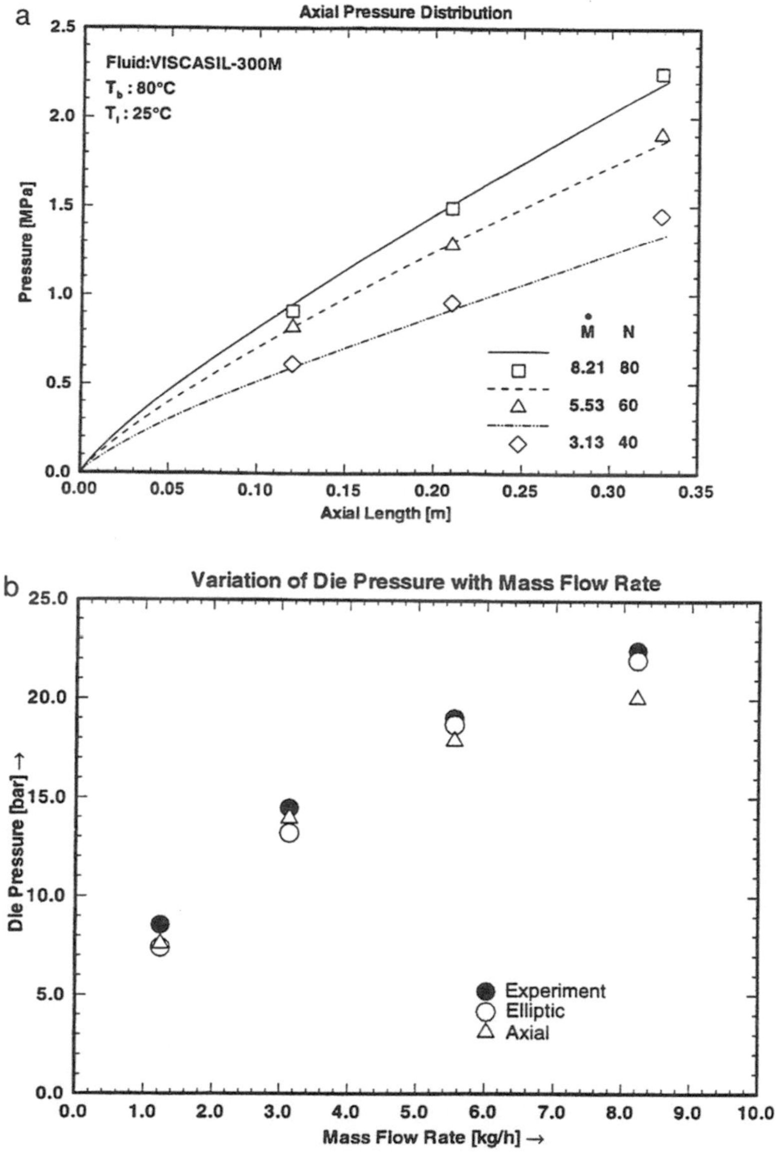

Fig. 6.9 Variation of pressure with axial distance and prediction of pressure at the die by the elliptic and axial formulations, along with experimental results (Chiruvella et al. 1996a)

surface to promote mixing, and the screw root is continuously tapered from the feed section to the die. This taper angle corresponds to a change in channel depth from H_1 at the inlet to H_2 at the outlet just before the die. The height H is then a function of z. An additional parameter H_2/H_1, or the taper angle $\psi = \arctan\,[(H_1 - H_2)/L]$, where L is the helical length of the screw channel, arises in the analysis. The results of the

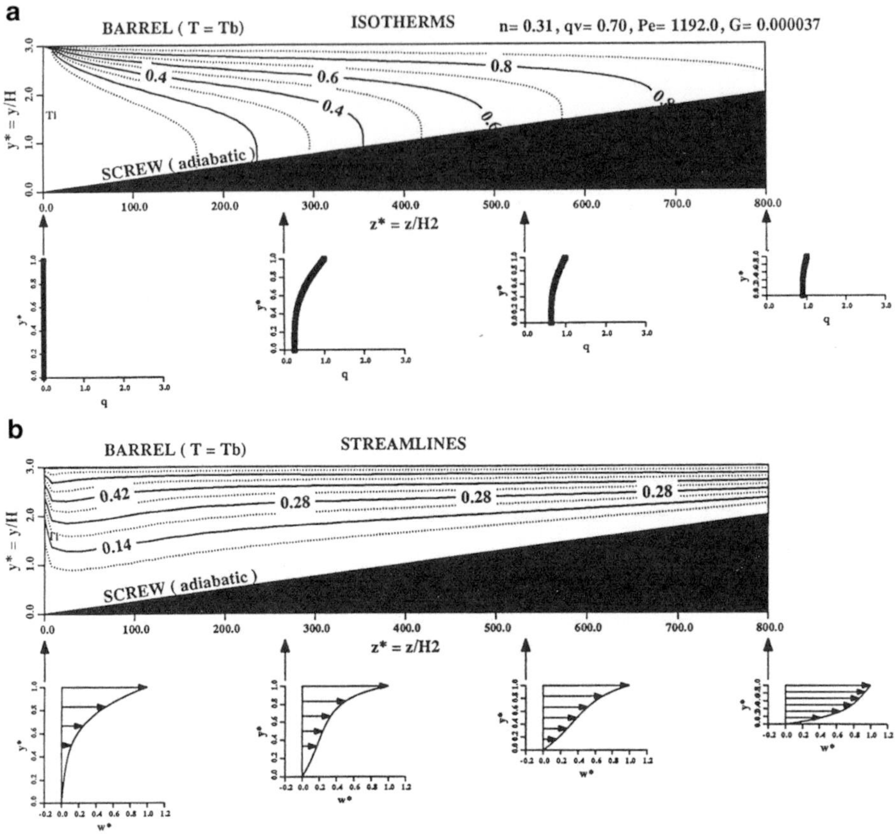

Fig. 6.10 Calculated results for a tapered extruder channel at $n = 0.31$, q_v (based on exit channel height) $= 0.7$, $Pe = 1192$, $G = 3.7 \times 10–5$, and taper angle $= 0.1432°$. (**a**) Isotherms and temperature profiles; (**b**) streamlines and velocity profiles

simulation for a typical taper angle of 0.1432° are shown in Fig. 6.10, in terms of the temperature and velocity distributions. In the figure, the taper section is shown enlarged for clarity. The temperature profile becomes almost completely uniform across the channel depth in the regions close to the die for a tapered channel. As expected, the velocity increases and the spacing of the streamlines is seen to become smaller as the fluid moves toward the die. The velocity profiles show a markedly different behavior for tapered channels as compared to the case with no taper. It is noted that the initially curved velocity profile becomes linear (corresponding to drag flow) as the section becomes shallower. A maximum in the pressure profile was also observed at a distance of about two-thirds of the total channel length. This arises because with a narrowing channel, the pressure must decrease to maintain the given flow rate. This results in a maximum in the pressure upstream of the die.

6.3 Residence Time

An important output from the numerical solution is the residence time distribution (RTD). The residence time refers to the time spent by a fluid particle in the extruder from the hopper to the die. Excessive residence time can lead to the material being over-processed, over-cooked, if it is food, or degraded. A relatively short residence time may lead to under-processing. Therefore, the characteristics of the product strongly depend on the residence time distribution, since structural changes due to thermal processing and chemical reactions are usually time-dependent. The residence time distribution is experimentally obtained by releasing a fixed amount of color dye or tracer in the material at the hopper and measuring the flow rate of the dye material as it emerges from the extruder at the die. The time it takes for the dye to first appear is the minimum residence time and relates to the fastest moving fluid. Similarly, an average residence time may be defined. Overall, the distribution, which is clearly a function of the flow field, indicates the spread of residence time as the material goes through the extruder.

The residence time may be numerically determined by following the approach used in experimentation (Karwe and Jaluria 1990). The flow of a finite amount of dye is simulated as it moves from the hopper to the die, as sketched in Fig. 6.11a. The velocity field, obtained from the solution of the governing equations, is employed to numerically track a fluid particle by integrating the velocity over time. The axial component of the velocity is used to trace the axial movement of the particles. As expected, the particles close to the barrel and the screw take a long time to come out, as compared to the particles near the middle portion of the screw channel. The amount of color dye emerging from the extruder as a function of time is obtained. This may then be used to obtain the minimum, average, and the spatial distribution of the residence time.

Figure 6.11b shows these results in terms of the dye flow rate, normalized by the total flow rate. In this particular case, most of the dye emerges over a fairly short time interval, with the extended regions representing material near the barrel and the screw root. The function $F(t)$, which indicates the cumulative fraction of the total amount of dye emerging up to time t, is defined as

$$F(t) = \int_0^t f(t)dt \tag{6.11}$$

Here, $f(t)\,dt$ is the amount of material that has a residence time between t and $t + dt$. The average residence time $\bar{t}$ is V_e/Q_e, where V_e is the total internal volume of the extruder and Q_e the volume flow rate. In Fig. 6.11c, $F(t)$ is plotted as a function of time, along with the distributions for a few common flows. The RTD is affected by the nature of the fluid and the flow and only slightly by the barrel temperature. It is mainly affected by the flow rate, or throughput, which substantially influences the flow field. Experimental and numerical results on RTD have been given in the

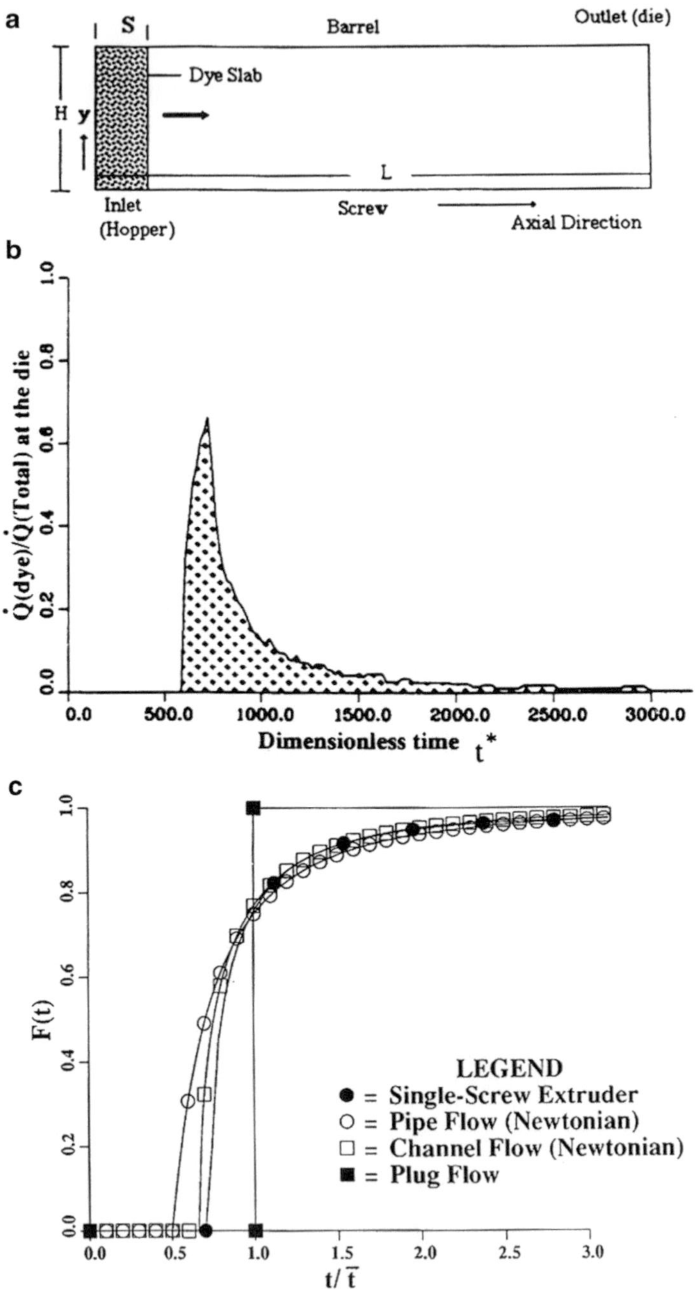

Fig. 6.11 Residence time distribution calculations. (**a**) Schematic diagram showing the input of a dye slab and the computational domain for RTD calculations; (**b**) variation of the dye flow rate, normalized by the total flow rate, with time for the conditions in Fig. 6.5 with $Pe = 5000$; (**c**) variation of the cumulative distribution function $F(t)$ for different flow configurations, with $\bar{t}$ as the average residence time

literature for different extruders and operating conditions. These may be used for selecting the appropriate conditions for a given material or desired thermal process.

6.4 Mixing

As mentioned earlier, mixing inside the extruder channel is an important consideration since it determines the uniformity in the product. The movement of the material particles as the flow proceeds downstream may be investigated for a better insight into the mixing process. A particular screw channel, obtained for a self-wiping screw profile mentioned earlier, is shown in Fig. 6.12. The distributive mixing inside the channel is investigated in terms of mixing between two different fluid regions, shown as white and black portions, with each initially occupying one half of the channel. These are followed in time, using the computed three-dimensional flow field in the screw channel. Clearly, the process is a slow one, and the materials gradually mix with each other as time elapses. A slow recirculating flow is evident from these results. Several other measures of mixing have been considered in the literature, such as movement of chosen particles through tracking. The effect of chaotic mixing due to changes in the geometry and boundary conditions has also been studied (Kwon et al. 1994; Xu et al. 2014). The development and increasing use of twin screw extruders are largely due to enhanced mixing achieved in these extruders.

6.5 Experimentation

Detailed experimentation on the flow and heat transfer in this problem is involved because of the complex domain, movement of the screw, and generally opaque nature of the typical materials. However, experimental data on the overall behavior of the extrusion process can be obtained more easily. Thus, extensive data are available in the literature on temperature and pressure at the die, residence time distribution, total heat transfer, total torque exerted on the screw, flow rate, and characteristics of the extrudate (Tadmor and Gogos 1979; Harper 1981; Rauwendal 2014). These results can be used for the validation of the numerical model, provide physical insight into the extrusion process, and provide inputs for the prediction, control, design, and optimization of the extruder. Figure 6.13 shows some typical results on axial pressure variation and temperature and pressure at the die. A close agreement with numerical predictions in terms of pressure and temperature measured at the die was obtained, lending strong support to the model. For a given die, the die pressure was found to increase with the flow rate which is increased by raising the screw speed. Also, as expected, the pressure increases as the fluid moves from the inlet toward the die, providing an adverse pressure gradient to the shear-driven flow arising from the screw rotation.

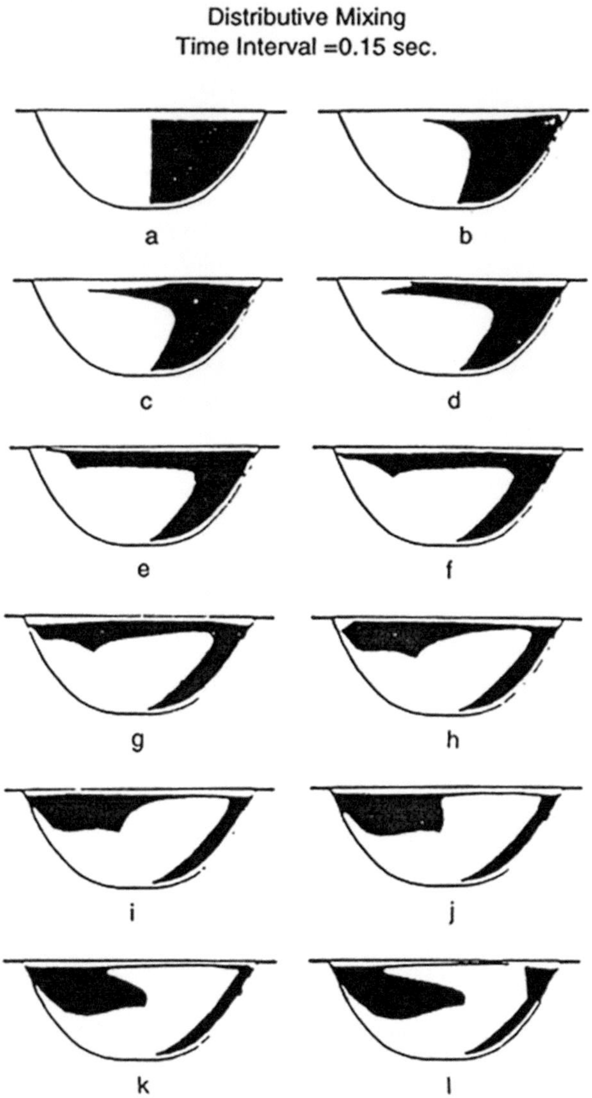

Fig. 6.12 Mixing characteristics in a single screw extruder channel shown in terms of time sequence of distributive mixing of two different fluid regions inside the screw channel. (Sastrohartono and Kwon 1990)

A few well-designed, accurate, controlled, and innovative experiments on single and twin screw extruders have also been carried out to obtain the temperature profile in the extruder channel (Esseghir and Sernas 1991, 1992). These results have been used for validating the numerical model presented here, as well as for providing a better understanding of the basic flow and heat transfer processes associated with extrusion. Some characteristic experimental results, along with numerical results from two-dimensional finite volume and three-dimensional finite element calculations,

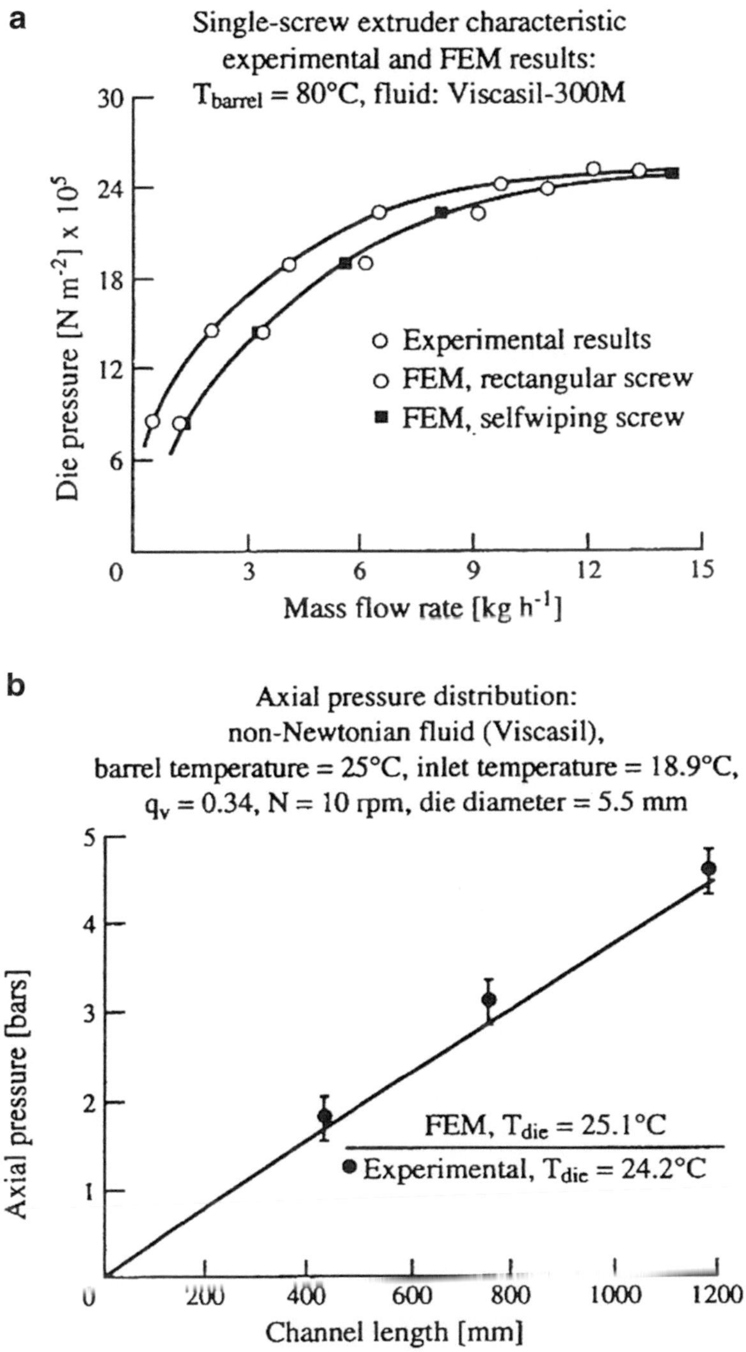

Fig. 6.13 Comparisons between 3D (FEM) predictions and experimental results for a single screw extruder, using Viscasil-300M as the fluid. (**a**) Variation of die pressure with mass flow rate for $T_i = T_b = 80$ °C and cylindrical die of 2.4 mm diameter; (**b**) axial variation of pressure for $T_i = 18.9$ °C, $T_b = 25.0$ °C, $N = 10$ rpm, 5.5 mm diameter die, and $q_v = 0.34$. (Sastrohartono et al. 1995)

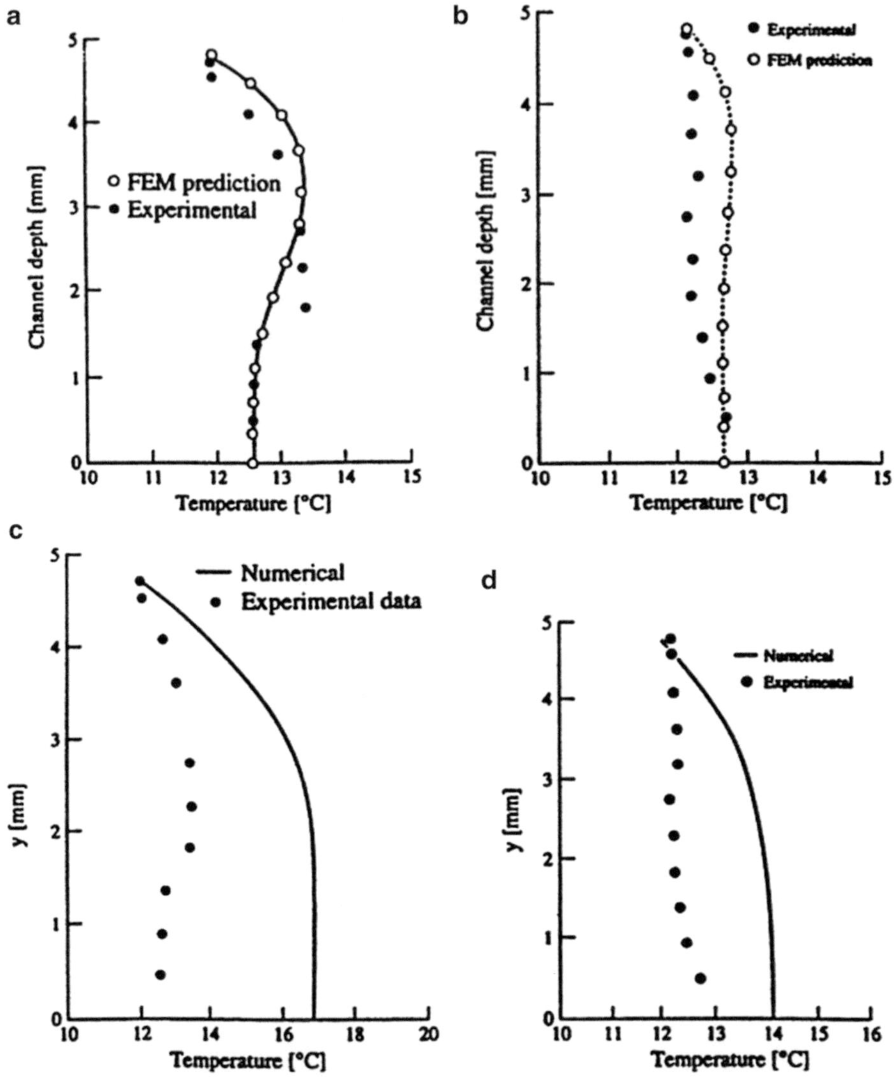

Fig. 6.14 Comparisons between numerical and experimental results on temperature profiles for Viscasil-300M, with (**a**) and (**c**) from the 3D (FEM) model and (**b**) and (**d**) from the 2D (FDM) model. For (**a**) and (**b**): $T_i = 20.3$ °C, $T_b = 12.2$ °C, $N = 20$. For (**c**) and (**d**): $T_i = 18.8$ °C, $T_b = 22.3$ °C, $N = 35$. (Sastrohartono et al. 1995)

are shown in Fig. 6.14 for Viscasil-300M, which is, as mentioned earlier, a non-Newtonian fluid. The effect of flow recirculation in the screw channel is seen in terms of the temperature near the screw root being greater in the 3D case, than that predicted by the two-dimensional model. This indicates that a three-dimensional model is needed to capture the fluid recirculation in the channel. There is good agreement between the numerical and experimental results on the temperature profiles for 3D simulations, as compared to the 2D simulations.

6.6 Twin Screw Extrusion

Various types of twin screw extruders have been developed in recent years to improve the mixing, control, and stability characteristics of the extrusion process, as well enhance the production rate. They are finding growing interest and usage in polymer thermal processing involving operations like pumping, polymer blending, and distribution of pigments and reinforcing materials in molten polymers. They are of particular interest in reactive polymers like food in which good mixing is of paramount importance. Two screws are located adjacent to each other in a barrel casing whose cross section has a figure of eight patterns; see Figs. 1.3d and 6.2. Twin screw extruders are of many types, depending on the screw configuration, rotation, and separation between screws. These include intermeshing, non-intermeshing, co-rotating, counter-rotating, and so on. If the screws rotate in the same direction, they are called co-rotating, and when they rotate in opposite directions, they are known as counter-rotating twin screw extruders. The extruders are classified as intermeshing or non-intermeshing extruders depending upon the separation between the two screw axes. If the distance between the axes is less than the diameter at the tip of the screw flight, then one screw intermeshes with the other. This is then an intermeshing twin screw extruder. Otherwise, it is known as non-intermeshing. If the distance between the screw axes is equal to twice the radius at the screw root and the flights of one screw wipe the root of the other screw, then the extruder is known as a fully intermeshing, self-wiping, twin screw extruder. This is the one commonly used for food materials because the wiping action in co-rotating extruders cleans the screw and ensures that material does not stay in the extruder and get over-processed or spoilt.

As seen in Fig. 1.3d, the flow domain of a twin screw extruder is a fairly complicated one, and the rotation of the screws makes the simulation very involved and challenging (Wang and White 1989). The numerical simulation may be simplified by considering two regions, the translation, or *T* region, and the intermeshing, or *M* region, as shown in Fig. 1.3d. This figure schematically shows a section normal to the screw axis of tangential counter- and co-rotating twin screw extruders and the two regions in the two cases. The flow in the translation region may be analyzed using the approach outlined earlier for a single screw extruder, approximating the region by a channel flow. The intermeshing, or mixing, region is represented by the geometrically complex central portion of the extruder, between the two screws. The two translation regions are separated by a hypothetical boundary used for the numerical calculations. The simulations in the two regions, *T* and *M*, are coupled at this boundary to yield the simulation for the entire region. Further details on this model for twin screw extruders and on the numerical scheme are given in the literature (Kalyon et al. 1988; Wang and White 1989; Sastrohartono et al. 1994; Chiruvella et al. 1996b).

The flow in the intermeshing region is three-dimensional. The governing equations in terms of the velocity components u, v, and w in the three coordinate directions x, y, and z are obtained, with approximations similar to those made earlier for the two-dimensional flow in a single screw extruder, as

$$\frac{\partial p}{\partial x} = \frac{\partial}{\partial y}\left(\mu \frac{\partial u}{\partial y}\right) + \frac{\partial}{\partial z}\left(\mu \frac{\partial u}{\partial z}\right) \tag{6.12}$$

$$\frac{\partial p}{\partial y} = \frac{\partial}{\partial y}\left(\mu \frac{\partial v}{\partial y}\right) + \frac{\partial}{\partial z}\left(\mu \frac{\partial v}{\partial z}\right) \tag{6.13}$$

$$\frac{\partial p}{\partial z} = \frac{\partial}{\partial y}\left(\mu \frac{\partial w}{\partial y}\right) + \frac{\partial}{\partial z}\left(\mu \frac{\partial w}{\partial z}\right) \tag{6.14}$$

$$\rho C_p\left[\frac{\partial}{\partial y}(vT) + \frac{\partial}{\partial z}(wT)\right] = \frac{\partial}{\partial y}\left(k \frac{\partial T}{\partial y}\right) + \frac{\partial}{\partial z}\left(k \frac{\partial T}{\partial z}\right) + \mu(\dot{\gamma})^2 \tag{6.15}$$

$$\dot{\gamma} = \left[\left(\frac{\partial u}{\partial y}\right)^2 + \left(\frac{\partial u}{\partial z}\right)^2 + 2\left(\frac{\partial v}{\partial y}\right)^2 + 2\left(\frac{\partial w}{\partial z}\right)^2 + \left(\frac{\partial v}{\partial z} + \frac{\partial w}{\partial y}\right)^2\right]^{1/2} \tag{6.16}$$

The finite-difference or finite volume method may be used to solve these equations for different geometries. Solutions are obtained for the translation and intermeshing regions and linked at the interface or overlapping region. The screw channels are assumed to be completely filled, and leakage across the flights is neglected. The flow moves from one channel to the other, with mixing arising due to the change in the geometry.

The finite element method is particularly well suited to the complex domain of the mixing region. Figure 6.15 shows the finite element mesh used and some typical results on the transport in the mixing or nip region of the extruder. It is seen that large gradients in pressure, velocity, and shear rate arise in the nip region, resulting in substantial fluid mixing, unlike the small recirculation seen earlier in single screw extruders. Figure 6.16 shows a comparison between experimental and numerical simulation results for a counter-rotating twin screw extruder. A rotating two-cylinder system was specially fabricated for these experiments (Sastrohartono et al. 1995). The flow division ratio is the fraction of the mass flow that crosses over from one channel to the other. Clearly, good agreement between simulation and experimental results is seen. Additional results are given in the literature for different operating conditions and materials. Viscous dissipation was found to increase the temperature in the fluid above the barrel temperature for typical operating conditions. The pressure rise, residence time, and energy transfer were also calculated.

Velocity measurements are quite involved because of the complex geometry and rotating screws. Some measurements of the fluid velocity field for heavy corn syrup, which is transparent, have been carried out (Karwe and Sernas 1996; Bakalis and Karwe 2002). A Plexiglas window was used for visual access to the mixing region. A two-component laser Doppler anemometer (LDA) in the back-scatter mode was

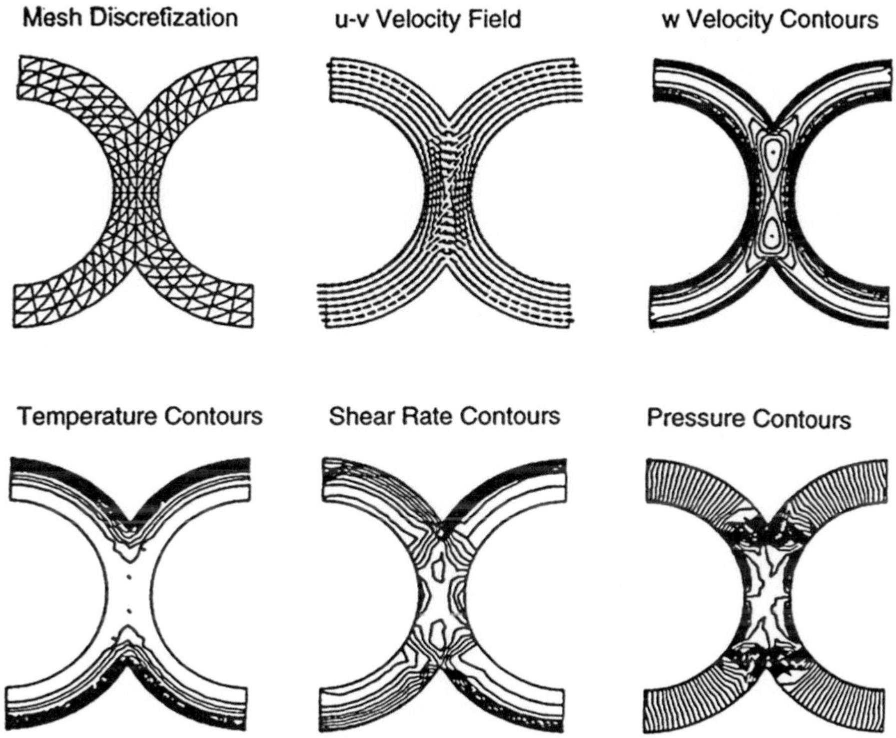

Fig. 6.15 Mesh discretization for the mixing region in a co-rotating tangential twin screw extruder, along with typical computed results for low-density polyethylene (LDPE) at $n = 0.48$, $T_b = 320\ ^\circ$C, $T_i = 220\ ^\circ$C, $N = 60$ rpm, $q_v = 0.3$

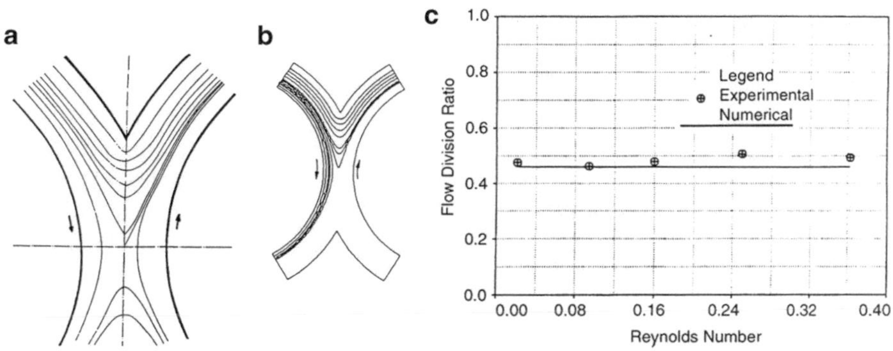

Fig. 6.16 Streamlines in the nip region for CMC solution at 16 rpm. (**a**) Experimental results; (**b**) numerical predictions; (**c**) comparison of flow division ratio obtained from experimental and numerical results for CMC solution. (Sastrohartono et al. 1995)

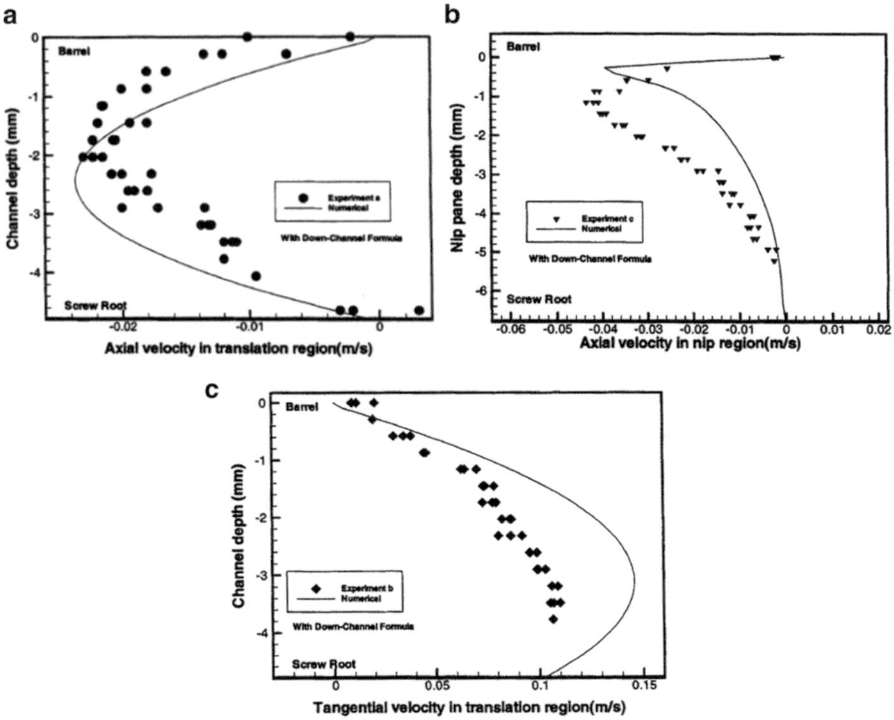

Fig. 6.17 Comparison between calculated and measured velocity profiles in a twin screw extruder for isothermal heavy corn syrup at 26.5 °C, with mass flow rate of 6 kg/h and screw speed of 30 rpm. (Karwe and Sernas 1996)

used to measure the local velocities in the extruder. As expected, the 3D flow was found to be quite complicated. The measured velocity distributions were compared with the numerical results, and the results are shown in Fig. 6.17. Thus, a fairly good agreement is observed in the translation and nip regions, lending support to the model. The results indicated that leakage across the flights was significant and should be accounted for in an improved model. Considerable work has been done on the modeling and experimental study of twin screw extruders for a wide variety of polymeric materials; see, for instance, Kohlgruber (2007) and White and Kim (2010).

6.7 Combined Heat and Mass Transfer and Chemical Reactions

Combined thermal and mass transport mechanisms are important in many extrusion processes, particularly those dealing with food, reactive polymers, and several other materials with multiple species. Many polymers of interest are chemically reactive.

For instance, food is a polymeric material, and chemical reactions are of particular importance in the extrusion cooking process. As mentioned earlier, extrusion is an important manufacturing technique in thermal processing of food materials. It is used in the processing of many snack foods, cereals, pasta, confectionary, pet foods, and bread substitutes. Various starches, wheat, rice flour, and other materials, along with a chosen amount of water, are fed into the hopper and cooked through the input of shear and heat to obtain different extruded products (Harper 1981; Mercier et al. 1989; Kokini et al. 1992). Chemical reactions can substantially alter the structure and characteristics of the material. Even though the following discussion considers the extrusion of food materials, the basic approach and trends are similar for other chemically reactive materials.

6.7.1 Moisture Transport

The transport of moisture in the extruder is very important in the extrusion of food materials both during the extrusion process and after the material emerges from the extrusion die. The viscosity is a strong function of moisture concentration, as given by an equation such as Eq. (6.2). Therefore, the transport of moisture in the extruder affects the flow and the heat transfer process. As the material comes out of the extruder, unbound or free water evaporates and flashes off due to the much lower ambient pressure, as compared to that in the die. This can cause a large amount of extrudate expansion and, thus, considerable reduction in the density of the extrudate. This in turn affects the nature and quality of the final product. Therefore, it is important to study the transport of moisture in the extruder and also the effect of chemical reactions.

For mass transfer in the extruder, for example, when moisture transport in the extrusion process for food materials is considered, the governing equation may be written, for small concentration levels, as

$$w\frac{\partial c}{\partial z} = D_{\mathrm{m}}\frac{\partial^2 c}{\partial y^2} + S\,c^{\mathrm{m}}; \quad S = 0 \text{ for } T < T_{\mathrm{gel}} \tag{6.17}$$

where c is the concentration of the species under consideration, D_{m} is the mass diffusivity, S is the reaction rate, and m is the order of the reaction, being zero if the reaction rate is constant and equal to 1.0 if the reaction rate varies linearly with concentration. Here, Sc^{m} represents a source term. However, for moisture it behaves as a sink, with S taking on negative values, to indicate moisture removal from the flow due to the chemical reaction. The concentration may be nondimensionalized with the inlet concentration of the species c_{i}. The ambient concentration may be taken as zero, and the sink term comes into effect only when the temperature T is greater than a threshold temperature T_{gel} needed for gelatinization to occur. Thus, this equation may be solved by the same numerical procedure adopted for heat transfer, with D_{m}, m, and S as additional variables.

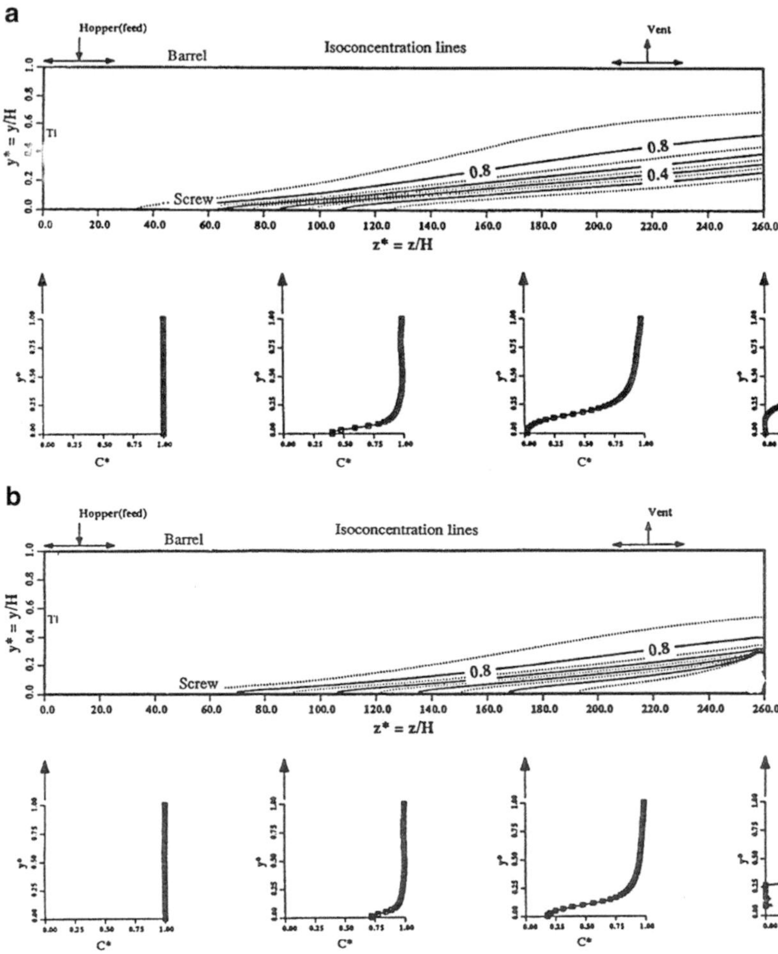

Fig. 6.18 Lines of constant moisture concentration c^* and concentration profiles at four downstream locations in a single screw extruder for a non-Newtonian fluid with $n = 0.3$, $\phi = 16.54°$, $q_v = 0.25$, $Pe = 7050$, $\theta_{gel} = 0.5$, for viscosity varying with both temperature and moisture. (**a**) $S = -500$; (**b**) $S = -1000$

Figure 6.18 shows the moisture concentration contours and profiles for $m = 0$ for two different values of S. As seen here, the effect of the sink term, S, due to the reaction, is manifested in the form of a decrease in the moisture concentration due to bonding of water for gel formation as the material reaches the dimensionless gelatinization temperature θ_{gel}. This loss of moisture occurs at a rate specified by the sink term S and is obviously more rapid for larger S. For the chosen conditions, the screw gets hotter than the barrel, and moisture removal occurs there first. The dimensionless pressure at the die was found to be lower for higher sink strengths. The temperature rises at higher sink strengths and causes the viscosity to decrease.

This results in the material flowing more easily and causing smaller pressure rise. However, the results are generally very sensitive to the material properties and operating conditions, making it difficult to predict the trends. The convergence of the numerical scheme is also not assured for widely varying materials and conditions. The effects of changing the throughput q_v on the pressure and the bulk temperature are similar to those discussed earlier for thermal transport in single screw extruders. The results are strongly influenced by the throughput. For a q_v value of 0.3, there is essentially no pressure variation in the extruder channel, indicating a situation with no obstruction because of a die. At lower values of q_v, the pressure rises due to the obstruction, simulating the circumstance in typical extrusion processes. At q_v values larger than 0.3, the pressure drops in the axial direction, indicating that the material must be pushed by a higher pressure at the inlet to maintain such a high flow rate. The bulk temperature rises faster for lower q_v since the material stays in the extruder channel for longer time, i.e., residence time is larger, allowing the material to heat up.

6.7.2 Chemical Reactions

It is also important to model the chemical conversion process in reactive materials, such as food, to determine the nature and characteristics of the extruded material. A simple approach to analyze these processes is discussed here. The governing equation for chemical conversion may be given as (Wang et al. 1989)

$$\frac{d}{dt}(1 - X) = -K(1 - X)^{\mathrm{m}} \tag{6.18}$$

where t is time and X is the degree of conversion, defined as

$$X = \frac{M_o - M_t}{M_o - M_f} \tag{6.19}$$

Here, M_o is the initial amount of unconverted material, such as starch, corn meal, etc., M_f is the final amount of unconverted starch, and M_t is the amount of unconverted starch at time t. The order of the reaction is m, and K is the reaction rate.

The order of the reaction m in Eq. (6.18) has been shown to be zero and the rate of the reaction K given as (Wang et al. 1989)

$$K = K_T + K_S \tag{6.20}$$

where

$$K_T = K_{To} \exp\left(-E_T/RT\right) \tag{6.21a}$$

$$K_S = K_{\mathrm{So}} \exp(-E_S/\tau\eta) \tag{6.21b}$$

Thus, conversion is due to shear stress τ, denoted by subscript S, and thermal effects, denoted by subscript T. Here, E is the activation energy, subscript o denotes

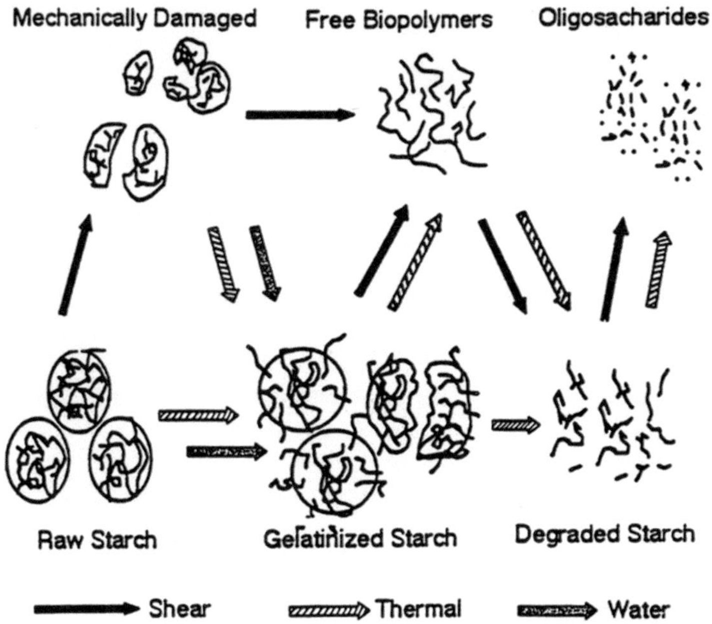

Fig. 6.19 A sketch indicating shear- and temperature-driven conversion of starch

reference conditions, R is the universal gas constant, and η is a constant which is obtained experimentally for the material. Figure 6.19 shows a schematic indicating the two components, shear and temperature, for the conversion of starch. Detailed experiments were carried out by Wang et al. (1989) to obtain the relevant constants for the chemical kinetics of starch and other food materials.

A one-dimensional approximation may be applied to model the degree of conversion defined in Eq. (6.19), as (Chiruvella et al. 1996c)

$$w\frac{dX}{dZ} = K \tag{6.22}$$

Here, w is the velocity in the down-channel z-direction. Thus, numerical results on conversion are obtained by integrating this equation. Typical results on the conversion of Amioca, which is a pure form of starch, at constant screw speed and throughput, are presented in Fig. 6.20. The degree of conversion depends upon the down-channel velocity and the reaction rate constant K which varies with the local temperature and shear rate. For smaller velocities, the degree of conversion up to a given axial location is higher. This is clear in the region near the screw root in all of the conversion contours. As the temperature increases along the down-channel direction, the degree of conversion increases, though the down-channel velocity does not change much. At a barrel temperature of 150 °C, the local temperature is higher than that at 125 °C. Since the reaction rate constant is higher when the local

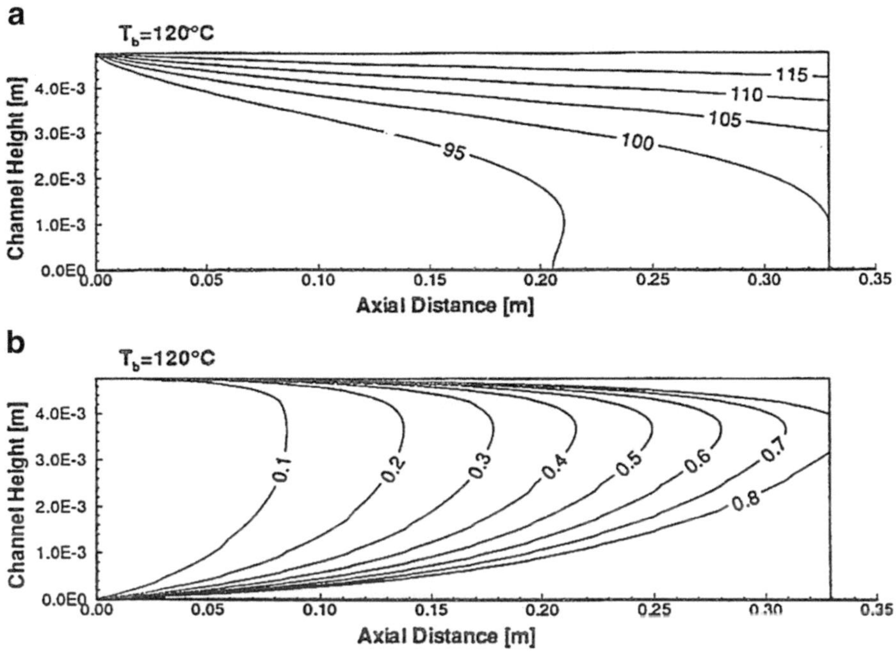

Fig. 6.20 Isotherms and constant conversion contours at (**a**) an inlet temperature of 100 °C and a barrel temperature of 125 °C for pure starch (Amioca) and (**b**) inlet temperature of 100 °C and barrel temperature of 150 °C

temperature is higher, the material gets converted in a shorter distance than that in the case where the temperature is lower. Several other results were obtained under different temperature levels and throughputs. This figure shows the typical trends observed. Results were also obtained for a tapered single screw extruder and for other conditions. With increasing flow rate, the degree of conversion decreases since the residence time is smaller allowing less time for the chemical reactions to occur, as seen in Fig. 6.21. A comparison with experimental results also shows good agreement.

Using the microscale conversion mechanisms, given in Eq. (6.20), the flow, heat transfer, and conversion in a twin screw extruder were investigated by Zhu and Jaluria (2001a, b). Residence time was also determined for different operating conditions. Also considered was the feasibility of the extrusion process. The feasible domain for a twin screw extruder with Amioca, or pure starch, as the extruded material is shown in Fig. 6.22. The feasibility of the process is determined largely by the flow and the pressure and temperature rise in the extruder. An upper limit is obtained for the mass flow rate, beyond which it is necessary to impose a favorable pressure gradient to push the material down the channel. Therefore, a negative pressure gradient along the axial direction will occur in the channel, and that is not physically acceptable for an extruder. A lower limit on mass flow rate is also obtained because of recirculation of the flow, which makes the process unstable

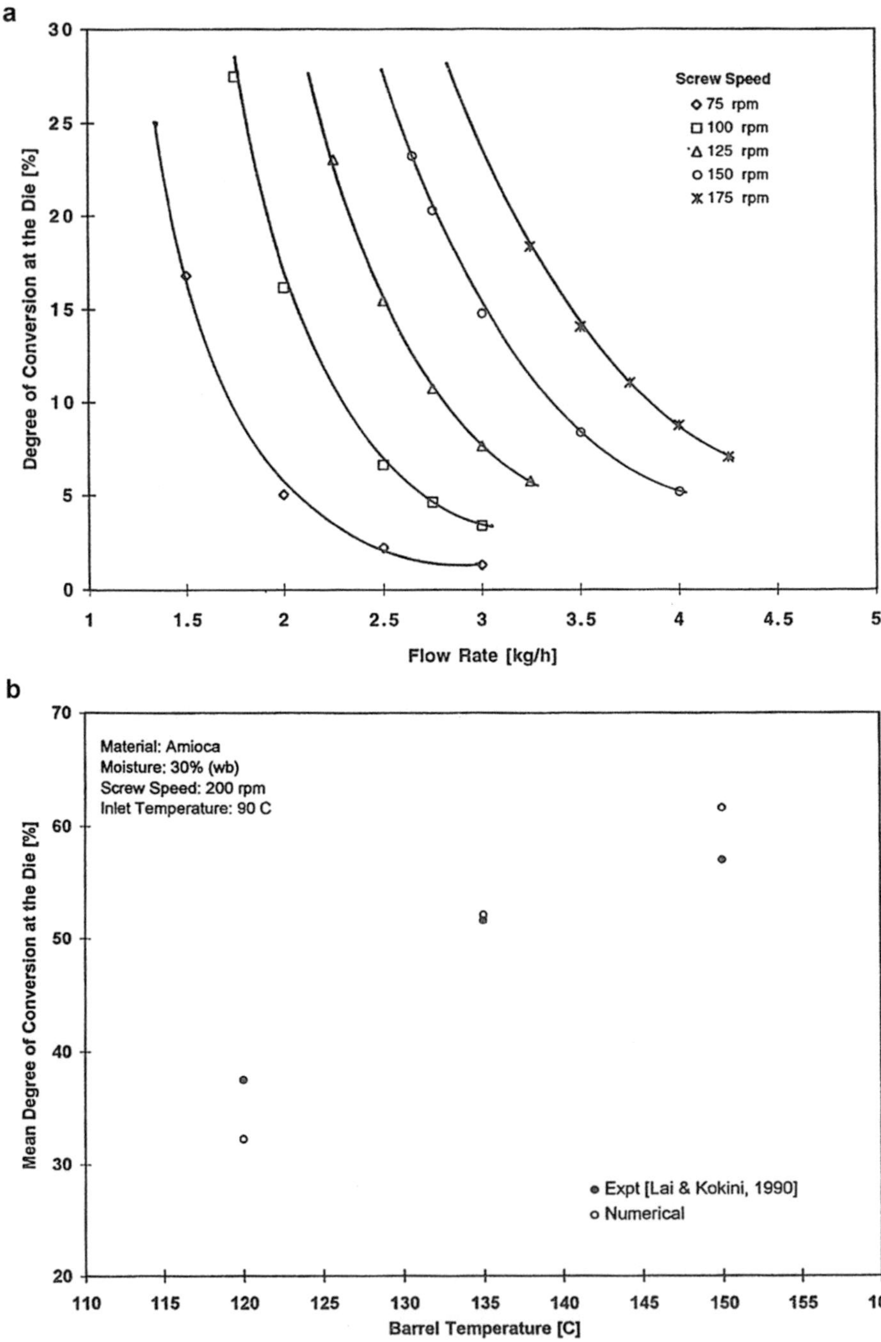

Fig. 6.21 (**a**) Predicted variation of mean degree of conversion at the die with the flow rate at different screw speeds for Amioca with 30% moisture, $T_b = T_i = 90\ ^\circ C$; (**b**) comparison between numerical predictions and experimental results for Amioca at 30% moisture at various barrel temperatures, with $\dot{m} = 3.0$ kg/h, $N = 200$ rpm, $T_i = 90\ ^\circ C$. (Chiruvella et al. 1996c)

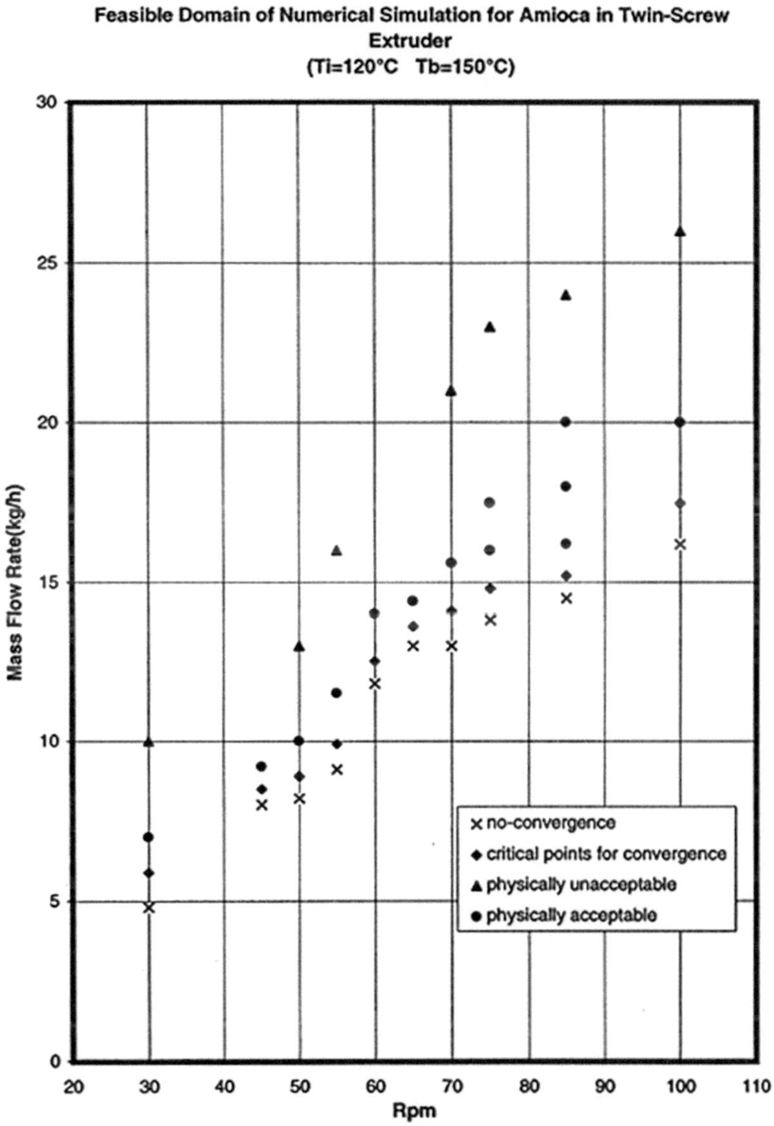

Fig. 6.22 Numerically determined feasible domain for twin screw extrusion of Amioca, with $T_i = 120\ °C$ and $T_b = 150\ °C$

and leads to high pressures and material degradation. Thus, a fairly narrow range of mass flow rates yields a feasible extrusion process.

It must be noted that, in these cases, the extruder screw channel is not completely filled with a rheological fluid throughout its length. Only a fraction of the channel length, as given in the figure, is completely filled, with the remaining portion either partially full or containing powder which cannot be treated as a fluid. The changes in the chemical structure due to chemical conversion can also be included as a factor in these equations if experimental data are available. One such equation multiplies the viscosity equation, such as the one given earlier, Eq. (6.2), by $(\mathrm{DG})^a$, where DG is the fraction of starch that has been converted, or gelatinized, and a is a constant obtained from a curve fit of the experimental data. There are several other such practical aspects that may be included in the analysis.

6.8 Die Flows

The heat transfer and flow in an extrusion die are very important in determining the characteristics of the extrudate and the operating conditions of the extruder since the flow rate through the die is determined by the pressure difference across the die. The pressure generated in the extruder must match the pressure needed to force the fluid through the die at the given flow rate. Therefore, the flow rate, or throughput, which is an important parameter in extruder analysis is determined by the die. At a given screw speed, a smaller diameter die causes a greater obstruction to the flow. It, thus, reduces the flow rate and increases the pressure rise so that the resulting pressure can force the fluid through the narrower die. Different geometries of dies are used in industry to obtain different extrudate characteristics, such as shape, density, homogeneity, and expansion after extrusion. In addition, complicated dies are used for coextrusion to combine materials, for making composites, and for extruding different materials. A considerable amount of literature exists on different types of dies, their heating and cooling, design, and analysis (Michaeli 2003). A few important aspects are outlined here.

6.8.1 *Coupling of Extruder with Die*

The analysis of the flow and heat transfer in the extruder channel yields the pressure at the entrance to the die, whereas an analysis of the die yields the pressure needed for the given flow rate. A correction procedure may be used to match the two, as done by Chiruvella et al. (1995) using a Newton-Raphson correction scheme. Different dimensions and geometries were used. As given earlier in Chap. 3, the relationship between the pressure drop Δp across a cylindrical region and the flow rate for a non-Newtonian fluid is given by Kwon et al. (1986) as

$$\Delta p = \frac{2L}{R}\widehat{C}\,(T)\,\left[\frac{3n+1}{4n}\,\frac{4\dot{m}}{\rho\pi R^3}\right]^n \tag{6.23}$$

where R is the radius and L the length of the cylindrical pipe, n is the power-law index of the fluid, and $\widehat{C}\,(T)$ is a temperature-dependent coefficient in the viscosity expression, which is given as

$$\mu = \hat{C}(T)(\dot{\gamma})^{1-n} \tag{6.24}$$

Using these expressions, Chiruvella et al. (1995) investigated the characteristics of an extruder channel with a die. Figure 6.23 shows some typical results obtained for two fluids and several dies of different dimensions and geometries. Die 1 is the most restrictive one and die 6 the least. As expected, the pressure at the die increases with flow rate for a given die, with the largest value obtained for the most restrictive die. At a given screw speed N, a decrease in flow rate implies going to a smaller diameter die, and the pressure increases. All these trends are physically expected. Therefore, for a given die and given screw speed, the operating point of the extruder may be determined from the intersection of the two characteristics in these graphs. This information is useful in the design of a system for the extrusion process.

6.8.2 Transport in Complex Dies

For more complicated dies, a numerical analysis may be carried out to obtain the pressure drop across the die for a given throughput. This may then be coupled with the analysis of the extruder channel, as discussed above. The flow and temperature fields may also be determined for improving the design to obtain better control on the process and uniformity of the extrudate. A three-dimensional finite element model was developed by Gupta et al. (1993) to simulate isothermal flow in an extrusion die with the rotating conical screw of the extruder on the left-hand side and the cylindrical outflow on the right, as shown in Fig. 6.24a. This is a fairly complicated flow geometry due to the rotating screw and the changing channel geometry as the fluid flows toward the outlet. The finite element method is particularly useful in the simulation of such a flow. The number of revolutions/min (rpm) affects the flow rate which in turn affects the pressure drop, being highest at the highest speed. The viscosity is lower at higher temperature, resulting in smaller pressure drop across the die, as shown in Figs. 6.24b and c. A comparison with experimental results showed good agreement over the range of flow rates considered; see Fig. 6.24d. Detailed results are given by Gupta et al. (1993).

The simulation of a practical extrusion die system is shown in Fig. 6.25. In this die, a long cylindrical region is used as flow path of the fluid to help in eliminating unsteadiness in the flow that arises due to the rotating screw and for improving the uniformity of the material. A conjugate heat transfer problem results here because of conduction in the thick wall of the die and convection in the flow (Lin and Jaluria 1996). Numerical modeling is used for obtaining the velocity and temperature fields, as well as the pressure variation, under a range of operating conditions. Most of the temperature variation arises near the boundaries due to the low thermal conductivity of the material. The flow is very well behaved with no separation or recirculation arising from the step changes in the cross section. This is expected from the large fluid viscosity and resulting low Reynolds number in the flow. Most of the pressure drop occurs in the final region of the die where the diameter is the smallest. Very

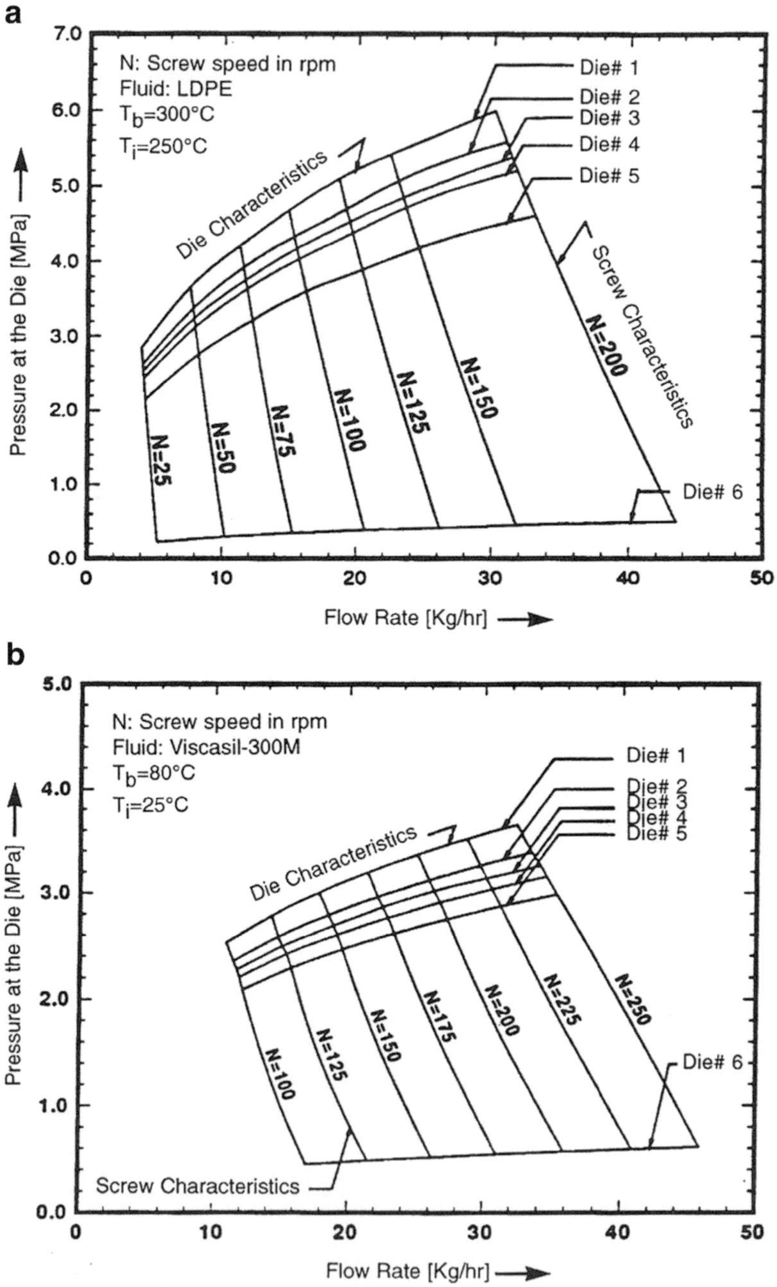

Fig. 6.23 Variation of pressure at the die with mass flow rate at different screw speeds using different dies during extrusion of (**a**) LDPE and (**b**) Viscasil-300M, at the operating conditions shown in the figure

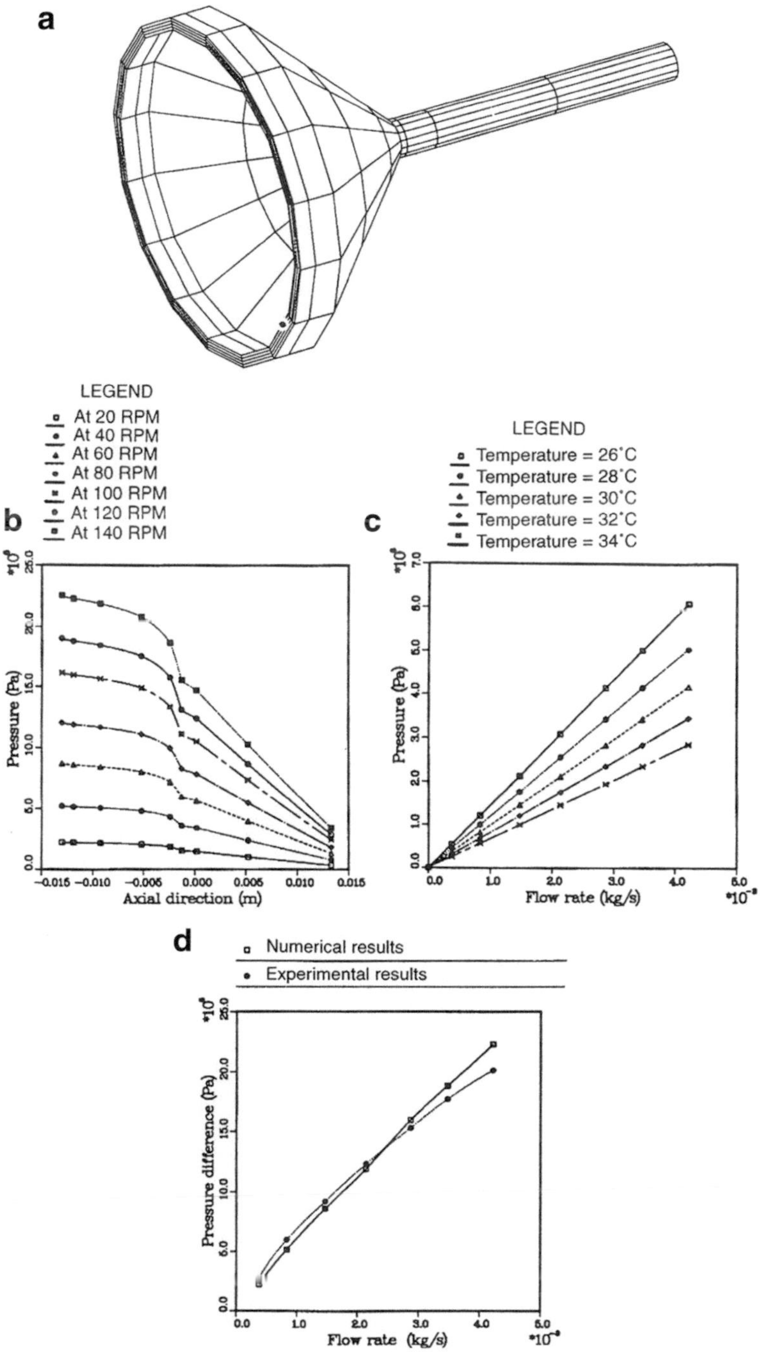

Fig. 6.24 (**a**) Finite element discretization of a circular die with the conical end of the screw rotating in the entrance region of the die; (**b**) pressure variation in the axial direction; (**c**) variation of pressure at the die with the flow rate; (**d**) comparison between numerical and experimental results. The fluid for these results is heavy corn syrup

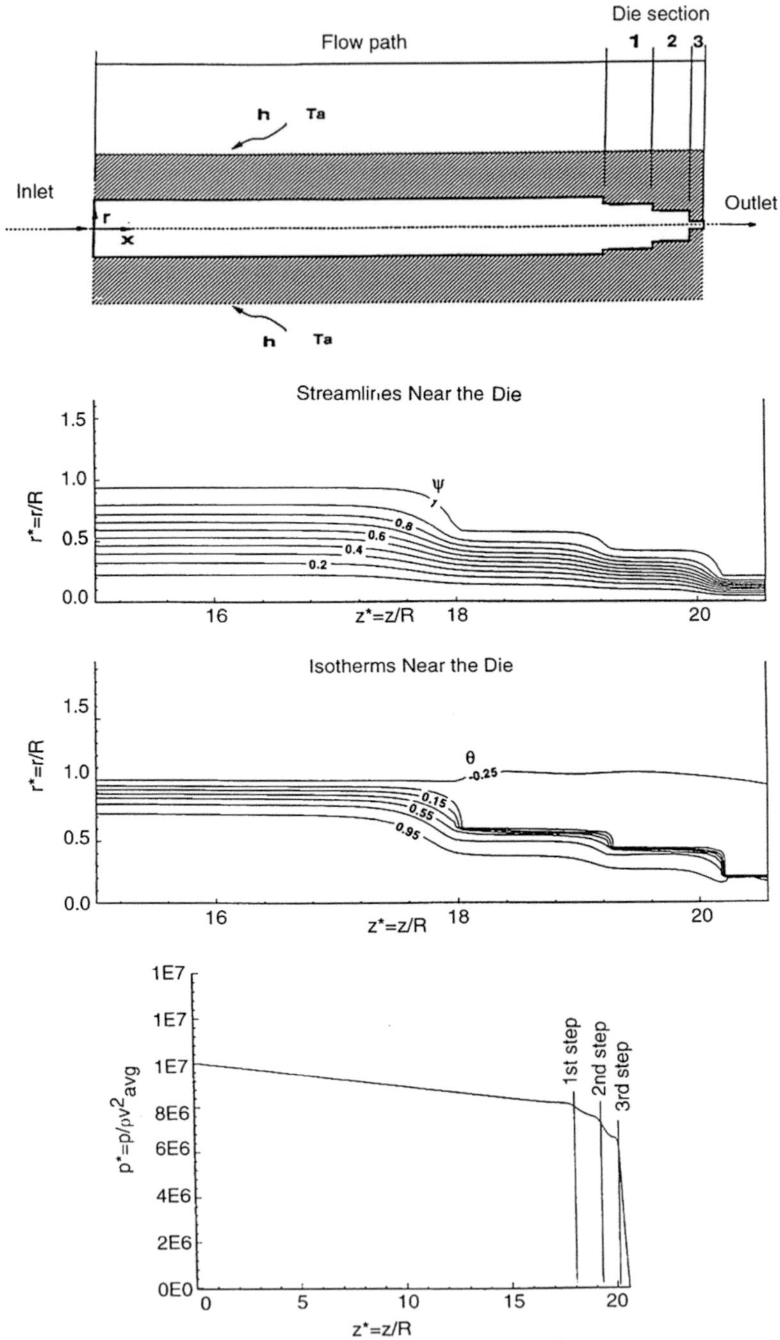

Fig. 6.25 Numerical modeling of a particular extrusion die system to obtain the flow and temperature fields, as well as the pressure variation in the axial direction, using corn meal at 21% moisture as the fluid

Table 6.1 Comparison between predicted and experimental results on the pressure drop for the flow of corn meal in a food extrusion die. Case 1: $m_c = 21\%$, $T_i = 382\ K$, $\dot{m} = 51.8$ kg/h. Case 2: $m_c = 17.3\%$, $T_i = 394\ K$, $\dot{m} = 59.4$ kg/h, where m_c gives the moisture percentage

Case	Simulation Δp (MPa)	Experiment Δp (MPa)	Error(%)
1	6.275	6.611	5.35
2	6.204	6.263	1.02

little pressure variation occurs in the relatively large diameter cylindrical flow path region. Thus, the flow and heat transfer in the die may be simulated to determine the pressure drop which is then matched with the pressure rise in the extruder channel. The flow and temperature fields are used to indicate if any stagnation or recirculation arises in the flow and to improve the design to obtain better quality of the extrudate. A comparison with experimental results on the total pressure drop across the die is shown in the following table, indicating close agreement (Table 6.1).

Clearly, dies are a critical component of the extrusion system. They determine the dimensions, geometry, and production rate of the final product. Since the die also strongly influences the pressure rise and residence time in the extruder, it has a substantial effect on the quality and characteristics of the product. Therefore, it is important to design the die used for a given application with great care. The die swell due to viscoelastic effects of the material, if important, must also be taken into consideration. The die may be cooled or heated as well to control the temperature of the flow and thus influence the pressure, shear, and conversion in the die channel.

6.9 Additional Aspects in Extrusion

The polymer extrusion process is an important and widely used technique for the processing of a wide range of polymers. However, as seen here, it is a complicated process due to material, geometry, non-Newtonian behavior, viscous dissipation, and other aspects that characterize polymer processing. Though most important and basic issues have been outlined here, there are several other aspects that need to be considered in practical extrusion processes. The material is generally fed into the extruder as solid pieces or as powder. This material is conveyed by the rotating screw, compacted, and then melted or chemically converted due to the energy input. Figure 6.26 shows a schematic of the overall process. The solid conveying region can often be modeled as plug flow with friction and slip at the boundaries, as shown. Friction factors for different materials have been measured and are available in the literature. The force balance yields the pressure variation in this region (Tadmor and Gogos 1979; Jaluria et al. 1999). For modeling the compaction process, information is needed on the variation of density or porosity of the material with pressure. Work has been done on the flow and compaction of such powders and the associated thermal effects. The melting and solidification of the material, effects of different screw configurations, leakage across screw flights, flow stability, conjugate transport

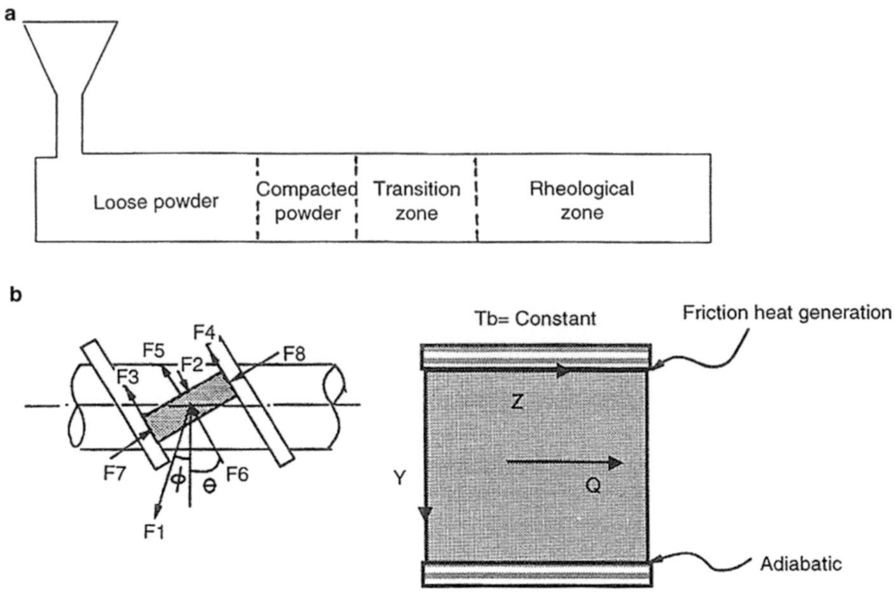

Fig. 6.26 (**a**) Schematic of the various regions in extrusion; (**b**) modeling of powder flow in a single screw extruder

due to conduction in the barrel, and transient effects are other aspects that are important and need detailed consideration. Several relevant results, particularly from experimental studies, have been reported in the literature, and the references given here may be consulted for additional information.

6.10 Injection Molding

The injection molding process for plastics and other polymeric materials is an important manufacturing process. It is extensively used for the fabrication of a variety of polymer-based products, ranging from plastic cups and toys to bathtubs, car bumpers, and molded parts made of composite materials (Bryce 1996; Rubin 2013; Yang et al. 2016). It is similar to the casting process discussed in Chap. 4, and, therefore, heat transfer, fluid flow, and phase change are the important basic mechanisms involved. The polymer is melted and injected under pressure into a mold cavity (Lord and Williams 1975). After the mold is filled, the material is allowed to cool and thus solidify. The pressure may be applied by means of a plunger or a rotating screw, as shown in Fig. 6.27, or by other methods. As the polymer starts to solidify, additional amounts of melt may be injected to fill the gaps left due to shrinkage during solidification. The mold is held together by a clamping unit, which opens and closes the mold and also ejects the final solidified product. Solidification

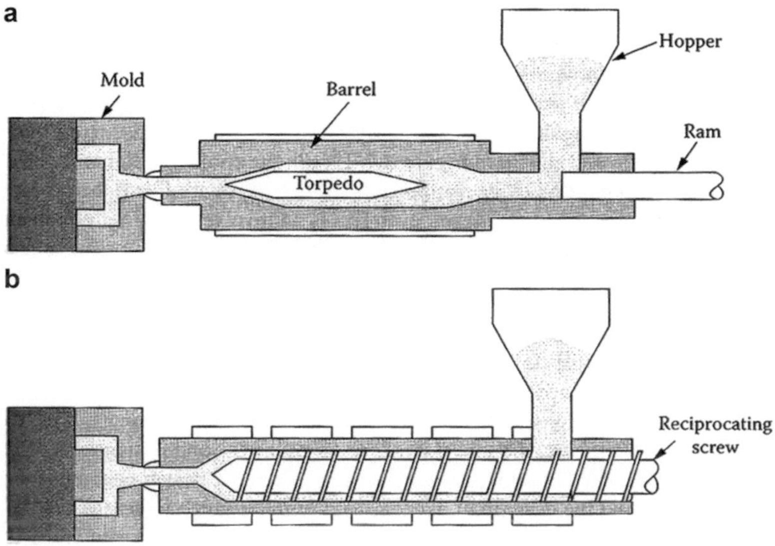

Fig. 6.27 (**a**) Ram-fed injection molding machine; (**b**) screw-fed injection molding machine

in channels leading to a mold cavity must be avoided since it affects the polymer flow and the pressure due to the resulting blockage.

Many of the issues that arise in injection molding are similar to those discussed earlier with regard to polymer extrusion. For instance, combined heat and mass transfer processes arise and significantly affect the thermal transport and the product quality. The fluids are generally non-Newtonian and highly viscous, making viscous dissipation an important consideration. The materials generally have low thermal conductivity and get damaged at relatively low temperatures, compared to materials like metals and alloys. Complex geometries, conjugate transport, and high pressures are commonly encountered. Phase change is obviously an important aspect in injection molding. Channel flow with solidification was considered in Chap. 4; see Fig. 4.25. A solidified layer was obtained near the boundaries, with flow in the central core. It is important to determine the blockage due to this solidified layer and if complete blockage might occur in some cases. In the circumstance shown in Fig. 4.25, complete blockage did not occur because of increasing viscous dissipation effects as the flow region decreased. The interface between the solid and melt regions, the velocity and temperature distributions, and pressures may thus be determined to form the basis for the design of injection molding systems.

Because of the industrial importance of the injection molding process, substantial amount of information exists on it, including polymer flow in channels, mold filling, and solidification. The effect of the flow on the interface and on the microscale structure has also been investigated. Of particular interest are flow in channels with phase change and mold filling. Williams and Lord (1975) and Lord and Williams (1975) carried out a detailed study of these processes and presented the basic

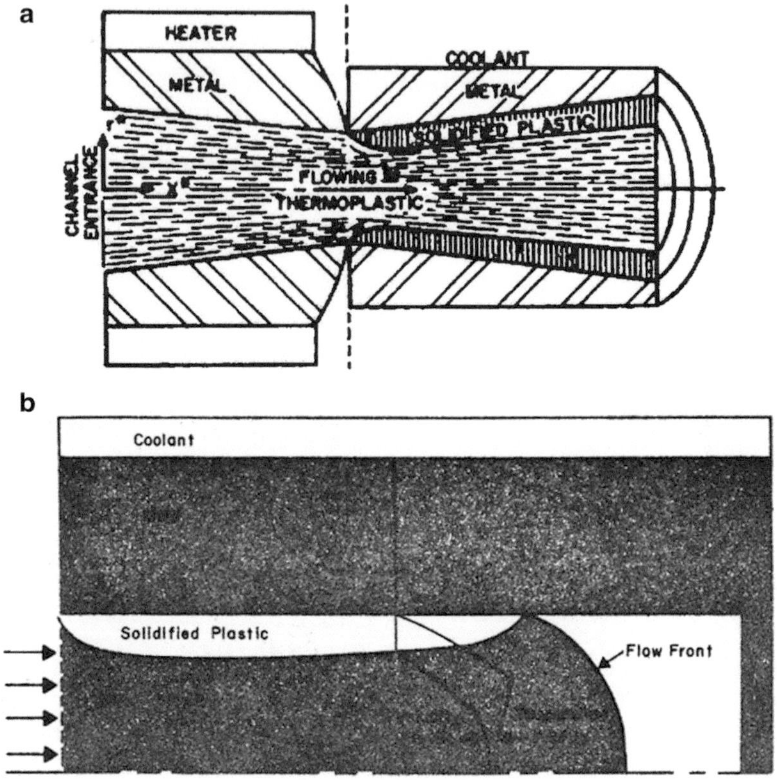

Fig. 6.28 (**a**) Typical flow channel geometry showing hot and cold flow channel walls and solidified plastic layer. (**b**) Plastic flow into a mold cavity

transport phenomena as well as important results on the injection molding process; see Fig. 6.28. Since solidification in the mold is an important consideration in the rate of production and in the quality of the product, considerable attention has been directed at the modeling and design of the cooling system. Many studies have focused on the characteristics of solidification under different cooling designs and operating conditions, as well as on temperature and stress distributions in the mold; see, for example, the papers by Park and Kwon (1998) and Chiang et al. (1991). Because of the practical importance of this manufacturing process, extensive commercial software is available to design, predict, control, and optimize the system (Himasekhar et al. 1992). The process can also be extended to other materials, such as composite materials, metals, and ceramics (Lee et al. 2011; Gupta and Wang 1993). The injection molding process is widely used in manufacturing, and several industries are based on products obtained by this process. Many manufacturing systems have been developed in recent years to address the needs for new composite materials and other polymeric products.

6.11 Thermoforming

Thermoforming is an important manufacturing process for thermoplastics. Basically, it involves heating the material to a point beyond its glass transition temperature T_g, at which temperature the material transitions from a solid to a viscous state, forming or shaping the material and then cooling it to make the new shape or configuration permanent. The temperature should not exceed a maximum value T_{max}, which could be the melting point or a temperature at which the material is damaged. Thus, the temperature T_0 of the plastic is maintained between a minimum T_{min}, which is essentially the glass transition temperature, and T_{max}. This temperature level is maintained for a certain time in order that the desired thermal process may be completed. There may be constraints on rate of heating of cooling as well in order to avoid damage to the material. The temperature variation for such a process is shown qualitatively in Fig. 6.29.

Generally, thermoforming refers to a manufacturing process where a sheet of plastic is heated to a forming temperature, beyond T_g, formed to a specific shape in an appropriately designed mold and cooled to a finished shape to obtain a usable product. Figure 6.30 shows some common thermoforming processes for converting a plastic sheet into desired shapes (Schey 1999). Thus, the mold is designed and fabricated with the desired shape in mind, and the heating process, as well as the forming operation, can be varied for different thermoplastic materials, thickness of the sheet used, and desired product characteristics. The products range from disposable cups, containers, and trays, obtained from thin sheets, to vehicle dash panels, refrigerator liners, and utility beds, obtained from thick sheets. The main considerations relate to the temperature and time needed, as well as the forming process using the mold. Since plastics generally have low thermal conductivity, it is necessary to design a heating system that leads to uniformity across the sheet thickness. This is particularly important for thick sheets.

Even though thermoforming is often focused on plastic sheets, the process of heating a given plastic part and forming or shaping it into a desired configuration or dimension can be applied to a variety of applications. For instance, consider the

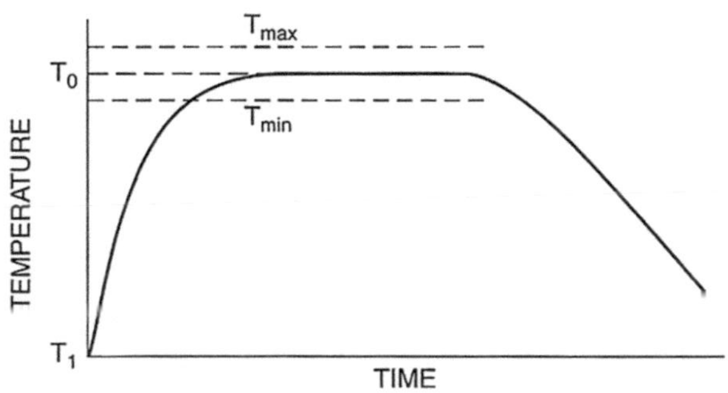

Fig. 6.29 Typical temperature cycle for a thermoforming process, with T_1 as the initial temperature, T_0 as the required temperature for the process, T_{min} being the minimum or the glass transition temperature, and T_{max} the allowable maximum temperature beyond which the material is damaged

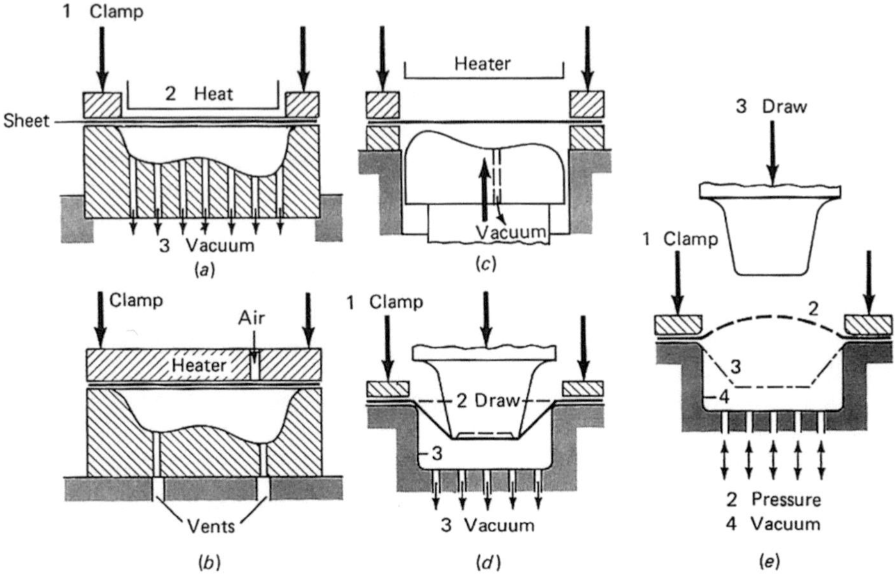

Fig. 6.30 Thermoforming processes to convert a plastic sheet into various shapes by (**a**) vacuum, (**b**) pressure, (**c**) drape-vacuum, (**d**) plug-assist, and (**e**) pressure-bubble plug-assist methods. Numbers refer to the sequence of the process. (Schey 1999)

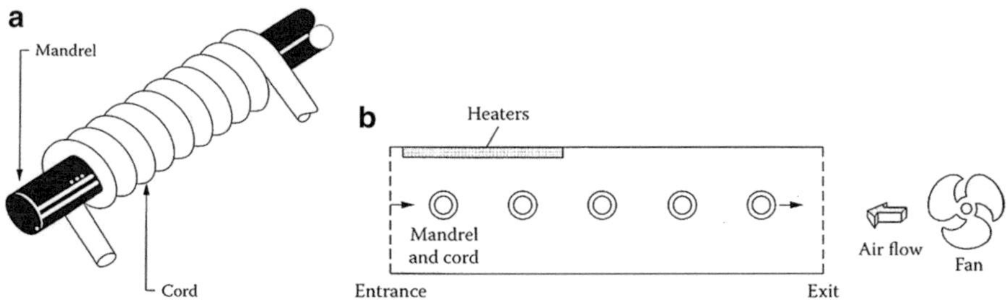

Fig. 6.31 (**a**) Plastic cord wound on a metal mandrel for forming it into a coil; (**b**) sketch of a system that may be used for continuous manufacture of coiled cords

manufacture of coiled plastic (PVC) cords, such as those used for telephone connection between the receiver and the housing. Straight cords are wound over a stainless steel mandrel and heated beyond the glass transition temperature of 121.1 °C, without exceeding the maximum temperature of 160 °C, followed by cooling to about 48 °C to make the shape permanent. It is important to achieve the temperature needed, maintain this level for the time needed for the process, and cool the material to a temperature low enough to permanently hold the shape. As mentioned above, it is necessary to keep the material below the maximum temperature to avoid damage. But it is also important to obtain temperature uniformity over the diameter or thickness of the cord to obtain satisfactory forming over the entire cord. This becomes challenging due to the low thermal conductivity of the material. Figure 6.31 shows a sketch of the basic

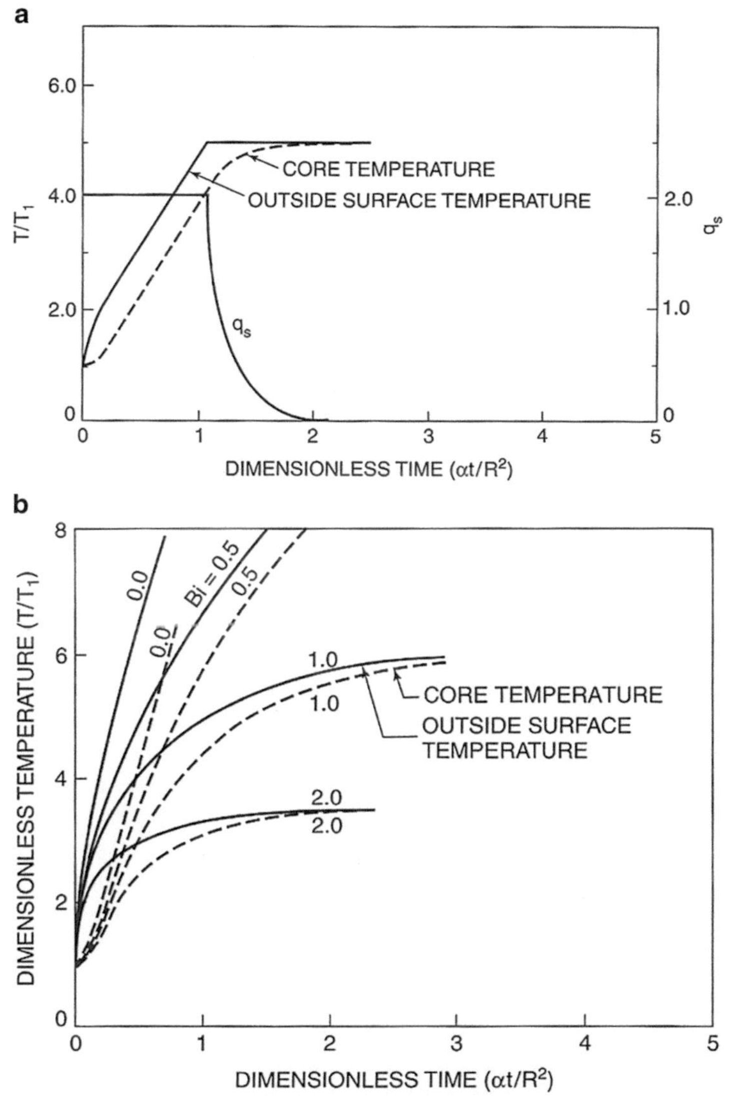

Fig. 6.32 (**a**) Variation of temperature and dimensionless heat flux q_s required to heat the outer surface to a given temperature at constant flux and then maintain it at that value; (**b**) temperature variation of the plastic cord with time due to combined convection, given by the Biot number *Bi*, and radiation heating with dimensionless heat flux $q_s = 5$

problem and a possible system design for continuous manufacture of coiled cords. Jaluria (2008) has discussed this problem in detail, giving the mathematical model, numerical results, and design considerations.

Figure 6.32a shows the time variation in dimensionless heat flux q_s needed to heat the outer surface of the plastic cord at constant heat flux to a given temperature and then maintain it at that level, if heat loss from the outer surface is neglected. Due to the thermal capacity of the mandrel and since heat flux is imposed at the outer surface, the core temperature, which is the inner surface temperature, lags behind the surface temperature. Large differences between the two can arise due to the low thermal conductivity of the material, so that the outer surface may reach the maximum temperature, while the inner temperature has hardly changed. The heat flux variation shown in this figure is difficult to obtain in actual practice. However, various parameters may be adjusted to vary the heating of the cord as well as the temperature across its diameter. Figure 6.32b shows the numerical results for a constant heat flux input with convective loss for different convective coefficients. A high heat flux helps in heating the cord rapidly, so that the production rate is high. But this also leads to a large temperature difference between the outer and inner surfaces. So, the time variation of the temperature and the difference between the two surfaces can be varied by adjusting the heating (and cooling) parameters. This situation is typical of many manufacturing processes that employ shaping and forming of plastics to obtain different products.

6.12 Summary

This chapter discusses the processing of polymers, which include a wide range of materials such as plastics, food, rubber, resins, silicones, and biopolymers. Because of the extensive use of these materials, many important processing techniques are available and are widely used to manufacture products for many different applications. Substantial literature exists on the fundamental aspects involved in polymer processing and on the design and operation of the relevant systems. It is thus not possible to cover all the techniques, the current practice, and existing systems in detail. The focus is on the three major processes of extrusion, injection molding, and thermoforming. Extrusion is considered in significant detail in order to discuss the characteristics of the materials involved, major concerns, requirements and constraints, process modeling, and typical simulation and experimental results that may be used for the control, design, and optimization of the processes. Extrusion also serves as the feeding mechanism in several injection molding systems. The discussions on phase change, transport phenomena in polymers, and channel flow can also be applied to injection molding processes. Thermoforming is a very different process, since it depends on forming and shaping after adequately heating the plastic beyond the glass transition temperature. A few examples are taken to illustrate this process as well as other similar ones.

There are several other processing methods that may be mentioned here. Coating is an important process and will be discussed in later chapters, particularly for optical fiber coating. Polymer casting is another important process, which is quite similar to the casting processes discussed in Chap. 4, though the material characteristics lead to important differences. There are several other similar processes, such as compression, transfer, rotational, and blow molding, with their specific procedures and requirements. Melt spinning is a process used for manufacture of fibers and is similar to wire drawing, which was considered in the preceding chapter. Polymer foam processing is another interesting method to manufacture different foams, such as Styrofoam, that are used in automotive, acoustic, medical, and other applications. The manufacture of composite materials also generally involves the processing of polymers in the matrix. Overall, polymer processing is an important area in materials and manufacturing. It continues to be expanded to include new materials and meet new challenges posed by emerging applications. Extensive literature is available on it, and many of the references given in the chapter may be consulted for further information.

References

S. Bakalis, M.V. Karwe, Velocity distributions and volume flow rates in the nip and translational regions of a co-rotating, self-wiping, twin-screw extruder. J. Food Eng. **51**, 273–282 (2002)

R.B. Bird, R.C. Armstrong, O. Hossager, *Dynamics of Polymeric Liquids, Vol. I: Fluid Dynamics* (Wiley, New York, 1987)

D.M. Bryce, *Plastic Injection Molding: Manufacturing Process Fundamentals* (Society of Manufacturing Engineers, Dearborn, 1996)

H.H. Chiang, C.A. Hieber, K.K. Wang, A unified simulation of the filling and post-filling stages in injection molding: part 1. formulation; part 2. experimental verification. Polym. Eng. Sci. **31**, 116–139 (1991)

R.V. Chiruvella, Y. Jaluria, A. Abib, Numerical simulation of fluid flow and heat transfer in a single screw extruder with different dies. Polym. Eng. Sci. **35**, 261–273 (1995)

R.V. Chiruvella, Y. Jaluria, M. Esseghir, V. Sernas, Extrusion of non-Newtonian fluids in a single-screw extruder with pressure back flow. Polym. Eng. Sci. **36**, 358–367 (1996a)

R.V. Chiruvella, Y. Jaluria, M.V. Karwe, V. Sernas, Transport in a twin-screw extruder for the processing of polymers. Polym. Eng. Sci. **36**, 1531–1540 (1996b)

R.V. Chiruvella, Y. Jaluria, M.V. Karwe, Numerical simulation of the extrusion process for food materials in a single-screw extruder. J. Food Eng. **30**, 449–467 (1996c)

M.J. Crochet, A.R. Davies, K. Walters, *Numerical Simulation of Non-Newtonian Flow* (Elsevier Science Publication, Amsterdam, 1984)

M. Esseghir, V. Sernas, A cam-driven probe for measurement of the temperature distribution in an extruder channel. SPE ANTEC Tech. Papers **37**, 54–57 (1991)

M. Esseghir, V. Sernas, Experiments on a single screw extruder with a deep and highly curved screw channel. in *Food Extrusion Science and Technology,* ed. by J.L. Kokini, C.T. Ho, M.V. Karwe (Marcel Dekker, New York, 1992) p. 21–40

R.T. Fenner, *Principles of Polymer Processing* (Chemical Publishing, New York, 1979)

S. Gopalakrishna, Y. Jaluria, M.V. Karwe, Computational study of heat and mass transfer in a single screw extruder for non-Newtonian materials. Int. J. Heat Mass Transf. **35**, 221–237 (1992)

M. Gupta, K.K. Wang, Fiber orientation properties of short-fiber-reinforced-molded composites: simulated and experimental results. Polym. Compos. **14**, 367–382 (1993)

M. Gupta, Y. Jaluria, V. Sernas, M. Esseghir, T.H. Kwon, Numerical and experimental investigation of three-dimensional flow in extrusion dies. Polym. Eng. Sci. **33**, 393–399 (1993)

C.D. Han, *Rheology in Polymer Processing* (Academic Press, Orlando, 1976)

J.M. Harper, *Extrusion of Foods: Volume I* (CRD Press, Boca Raton, 1981)

K. Himasekhar, J. Lottey, K.K. Wang, CAE of mold casting in injection molding using three-dimensional numerical simulation. J. Engg. Industry **114**, 213–221 (1992)

Y. Jaluria, Heat and mass transfer in the extrusion of non-Newtonian materials. Adv. Heat Tran. **28**, 145–230 (1996)

Y. Jaluria, *Design and Optimization of Thermal Systems*, 2nd edn. (CRC Press, Boca Raton, 2008)

Y. Jaluria, K.E. Torrance, *Computational Heat Transfer* (Taylor & Francis, New York, 2003)

Y. Jaluria, Y. Liu, R.V. Chiruvella, Modeling and simulation of the solids conveying and unfilled regions in polymer extrusion. J. Reinf. Plast. Compos. **18**, 15–26 (1999)

D.M. Kalyon, A.D. Gotsis, U. Yilmazer, C. Gogos, H. Sangani, B. Aral, C. Tsenoglou, Development of experimental techniques and simulation methods to analyze mixing in co-rotating twin screw extrusion. Adv. Polym. Technol. **8**, 337–353 (1988)

M.V. Karwe, Y. Jaluria, Numerical simulation of fluid flow and heat transfer in a single screw extruder for non-Newtonian fluids. Num. Heat Transf. **17**, 167–190 (1990)

M.V. Karwe, V. Sernas, Application of laser Doppler anemometry to measure velocity distribution inside the screw channel of a twin screw extruder. J. Food Process Eng. **19**, 135–152 (1996)

M.V. Karwe, R.V. Chiruvella, Y. Jaluria, Coordinate system independence of shear rate during isothermal single screw extrusion of a Newtonian fluid. J. Food Process Eng. **18**, 55–69 (1995)

K. Kohlgruber, *Co-Rotating Twin Screw Extruders* (Hanser, New York, 2007)

J. L. Kokini, C.-T. Ho, M. V. Karwe (eds.), *Food Extrusion Science and Technology* (Marcel Dekker, New York, 1992)

T.H. Kwon, S.F. Shen, K.K. Wang, Pressure drop of polymeric melts in conical converging flow: Experiments and predictions. Polym. Eng. Sci. **26**, 214–224 (1986)

T.H. Kwon, J.W. Joo, S.J. Kim, Kinematics and deformation characteristics as a mixing measure in the screw extrusion process. Polym. Eng. Sci. **34**, 174–189 (1994)

L.S. Lai, J.L. Kokini, The effect of extrusion operating conditions on the online barrel viscosity of 98% amylopectin (Amioca) and 70% amylose (Hylon-7) corn starches during extrusion. J. Rheol. **34**, 1245–1266 (1990)

S.W. Lee, S. Ahn, C.J. Whang, S.J. Park, S.V. Atre, J. Kim, R.M. German, Effects of process parameters in plastic, metal and ceramic injection molding processes. Korea-Australia Rheo. J. **23**, 127–138 (2011)

P. Lin, Y. Jaluria, Numerical approach to model heat transfer in polymer melts flowing in constricted channels. Num. Heat Transf. **30**, 103–123 (1996)

H.A. Lord, G. Williams, Mold-filling studies for the injection molding of thermoplastic materials. Part II: the transient flow of plastic materials in the cavities of injection-molding dies. Polym. Eng. Sci. **15**, 569–582 (1975)

C. Mercier, P. Linko, J. M. Harper (eds.), *Extrusion Cooking* (Am. Assoc. Cereal Chem, St. Paul, 1989)

W. Michaeli, *Extrusion Dies for Plastics and Rubber*, 3rd edn. (Hanser Publication, Munich, 2003)

S.J. Park, T.H. Kwon, Optimal cooling system design for the injection molding process. Polym. Eng. Sci. **38**, 1450–1462 (1998)

J. R. A. Pearson, S. M. Richardson (eds.), *Computational Analysis of Polymer Processing* (Applied Sci. Pub, London, 1983)

C. Rauwendaal, *Polymer Extrusion*, 5th edn. (Hanser Pub, New York, 2014)

I.I. Rubin, *Injection Molding: Theory and Practice* (Wiley-Interscience, New York, 2013)

T. Sastrohartono, T.H. Kwon, Finite element analysis of mixing phenomena in tangential twin screw extruders for non-Newtonian fluids. Int. J. Num. Meth. Engg. **30**, 1369–1383 (1990)

T. Sastrohartono, Y. Jaluria, M.V. Karwe, Numerical coupling of multiple region simulations to study transport in a twin screw extruder. Num. Heat Transf. **25**, 541–557 (1994)

T. Sastrohartono, Y. Jaluria, M. Esseghir, V. Sernas, A numerical and experimental study of three-dimensional transport in the channel of an extruder for polymeric materials. Int. J. Heat Mass Transf. **38**, 1957–1973 (1995)

J.A. Schey, *Introduction to Manufacturing Processes*, 3rd edn. (McGraw-Hill, New York, 1999)

Z. Tadmor, C. Gogos, *Principles of Polymer Processing* (Wiley, New York, 1979)

Y. Wang, J.L. White, Non-Newtonian flow modeling in the screw region of an intermeshing co-rotating twin screw extruder. J. Non-Newtonian Fluid Mech. **32**, 19–38 (1989)

S.S. Wang, C.C. Chiang, A.I. Yeh, B. Zhao, I.H. Kim, Kinetics of phase transition of waxy corn starch at extrusion temperatures and moisture contents. J. Food Sci. **54**, 1298–1301 (1989)

J.L. White, E.K. Kim, *Twin Screw Extrusion: Technology and Principles*, 2nd edn. (Hanser, New York, 2010)

G. Williams, H.A. Lord, Mold-filling studies for the injection molding of thermoplastic materials. Part 1: the flow of plastic materials in hot- and cold-walled channels. Polym. Eng. Sci. **15**, 553568 (1975)

B. Xu, Y. Liu, H. Yu, L.-S. Turng, Chaotic mixing in a single screw-extruder with a moving internal baffle. Polym. Eng. Sci. **54**, 198–207 (2014)

Y. Yang, X. Chen, N. Lu, F. Gao, *Injection Molding: Process Control, Monitoring and Optimization* (Hanser Pub, New York, 2016)

W. Zhu, Y. Jaluria, Residence time and conversion in the extrusion of chemically reactive materials. Polym. Eng. Sci. **41**, 1280–1291 (2001a)

W. Zhu, Y. Jaluria, Transport processes and feasible operating domain in a twin-screw polymer extruder. Polym. Eng. Sci. **41**, 107–117 (2001b)

Chapter 7
Thin Film Deposition: Micro-/Nanoscale Fabrication

The fabrication of microscale devices and nanoscale materials and structures, which refer to length scales of micrometer and nanometer, respectively, has been of considerable interest in recent years as the development of smaller and smaller integrated electronic systems has grown. Such small devices have also been of interest in other areas such as biosystems, diagnostics, microfluidics and controls. This drive toward reducing size has been largely possible due to new fabrication techniques and better control on the quality and reliability of the microscale devices obtained. Among the earliest techniques was crystal growing that was discussed in Chap. 4. Electronic wafers or chips were obtained from the pure crystal and subjected to a variety of processes, such as doping, etching, heat treatment, and epitaxial growth, which involves the deposition of a crystalline layer on a substrate. The microfabrication processes lead to the desired electronic circuitry to be generated in individual electronic chips, which are then used as components in electronic systems for various applications. Substantial work has been done in recent years on wafers, deposition, microfabrication, integrated circuits, and nanosystems.

One of the important processes used in the manufacture of electronic devices is chemical vapor deposition (CVD). It is used for depositing materials in various forms, such as monocrystalline, polycrystalline, and amorphous. Even though CVD is the main focus of this chapter, the basic aspects of deposition, materials involved, product quality, modeling, and optimization are of interest in many other micro-/nanoscale fabrication processes. Details on the process and characteristic results are given for several materials, systems, and techniques.

7.1 Introduction

The deposition of thin films onto a solid substrate or susceptor is an important materials processing technique. It is of interest in many applications such as those involved with microelectronic circuits, optical and magnetic devices, high-

Y. Jaluria, *Advanced Materials Processing and Manufacturing*, Mechanical Engineering Series, https://doi.org/10.1007/978-3-319-76983-7_7

Table 7.1 Different materials fabricated and applications for CVD reactors

Different CVD applications	Common materials
Semiconductors	Silicon (Si), gallium arsenide (GaAs)
Dielectrics	Silicon dioxide (SiO_2), silicon nitride (Si_3N_4)
Metallic conductors	Tungsten silicide (WSi_2), molybdenum silicide ($MoSi_2$), tungsten (W), aluminum (Al), molybdenum (Mo), polysilicon (Si)
Protective coatings	Titanium nitride (TiN), tungsten (W), molybdenum (Mo), gold (Au), platinum (Pt)
Ceramics	Aluminum oxide (Al_2O_3), titanium carbide (TiC), silicon carbide (SiC), boron carbide (B_4C), titanium diboride (TiB_2)
Anticorrosive coatings for turbine blades	Boron nitride (BN), molybdenum disilicide ($MoSi_2$), silicon carbide (SiC), boron carbide (B_4C)
Powers for sintering and hot pressing	Silicon nitride (Si_3N_4), silicon carbide (SiC)
Fibers for composite materials	Boron (B), boron carbide (B_4C), silicon carbide (SiC)
High-purity monolithic materials for infrared optics	Zinc selenide (ZnSe), zinc sulfide (ZnS), cadmium sulfide (CdS), cadmium telluride (CdTe)
Synthetic diamonds	Carbon (C)

performance cutting, polishing and grinding tools, corrosion-resistant coatings, and solar cells. This fabrication method is relatively new, but it has grown substantially in interest and activity in recent years because of its application to many important areas, high quality of material generated, good control of the process, and overall efficiency of the process. There are many physical and chemical deposition processes in use today. Mechanical, electromechanical, or thermodynamic methods are used to obtain thin films in physical deposition, through processes such as sputtering, molecular beam epitaxy, and pulsed laser deposition. Chemical processes are used in chemical vapor deposition (CVD) to obtain the desired thin films. A few of the applications of CVD and materials involved are given in Table 7.1.

In CVD, thin films are deposited from a gas phase on to a solid surface. Chemical reactions take place in the gases as well at the substrate on which deposition occurs. The products of the reactions form a solid crystalline or amorphous layer on the substrate. The activation energies needed for the various chemical reactions are provided by an external heat source, such as heat input to the substrate. After the deposition, the by-products of the reactions are removed by carrier gases (Mahajan 1996). The sequence of events involved in a horizontal CVD process is shown schematically in Figs. 1.3c and 7.1. The inflow of gases, chemical reactions in the gases and at the surface, and the removal of the remaining mixture are shown. Film thicknesses generally range from a few nanometers to tens of microns. The quality of the deposited film is characterized in terms of its purity, composition, thickness uniformity, adhesion, surface morphology, and crystalline structure. The level of quality needed depends on the intended application, with electronic and optical materials imposing the most stringent demands. In order to improve the quality of the film deposited and to increase the production rate, it is necessary to understand

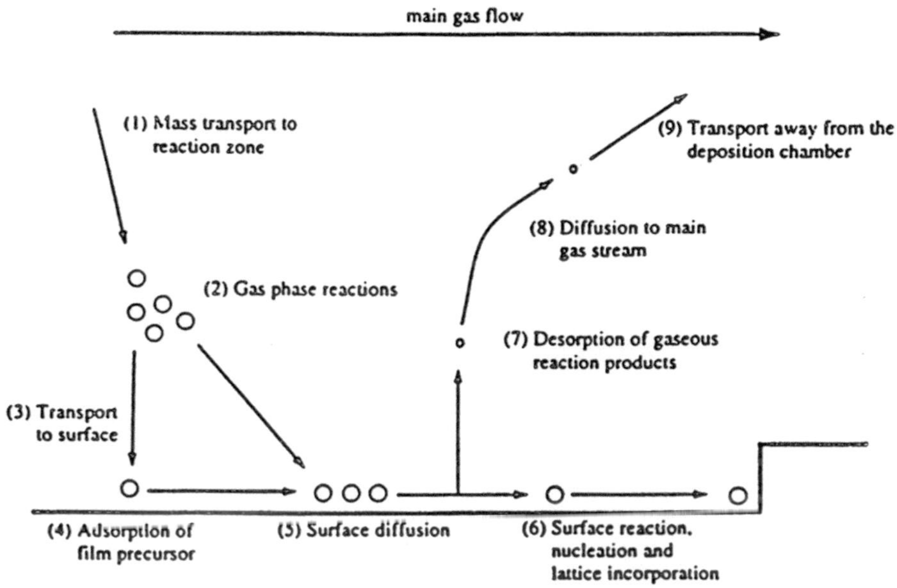

Fig. 7.1 A schematic of the various steps in the chemical vapor deposition process

the basic mechanisms that govern the reactions leading to the deposited material as well as access of the appropriate chemical species to the substrate. These aspects obviously depend on the flow, the associated heat and mass transfer, and chemical reactions (Chiu et al. 2002; Lin et al. 2009).

Many different types of CVD reactors have been developed and employed for different applications. The reactor configurations can be classified into two general categories: the horizontal and the vertical reactors. In the horizontal reactor, the heated susceptor or surface on which deposition occurs is generally placed at an angle to the incoming flow. The vertical flow reactor has the susceptor positioned perpendicular to the downward flow. Both types of reactors are commonly used. A few practical CVD reactor systems are shown in Fig. 7.2. In most cases, a batch process is used, with the susceptor stationary or rotating, and after the process is completed, the susceptor is removed and a new charge undertaken. Continuous processing has also been considered with moving substrate in the manufacture of photovoltaic cells and other devices. Different flow configurations have been considered to reduce the loss of reactive gases and to achieve uniform deposition over the susceptor. The CVD reactor may be at atmospheric or low pressure, including ultrahigh vacuum. Much of the initial effort was directed at silicon deposition because of its traditional importance to the semiconductor industry. However, the recent work has been directed at the deposition of a wide range of materials such as gallium arsenide, titanium nitride, silicon carbide, diamond, graphene, gallium nitride, and metals like titanium, tungsten, aluminum, and copper. Metal-organic precursors are used in several cases such as gallium nitride and silicon carbide. These

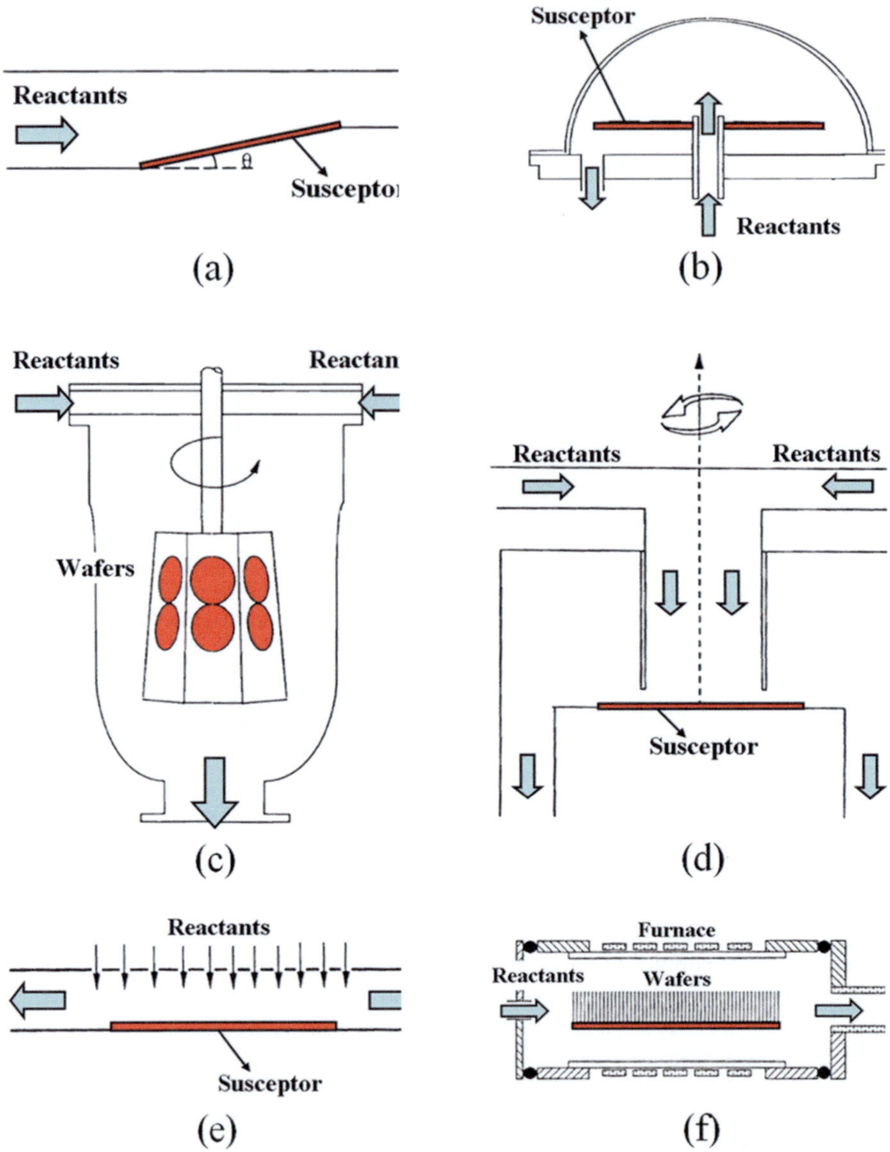

Fig. 7.2 Different configurations of CVD reactors: (**a**) horizontal reactor; (**b**) pancake reactor; (**c**) barrel reactor; (**d**) vertical impinging reactor with rotating susceptor; (**e**) vertical impinging reactor; and (**f**) tubular reactor (Mahajan 1996)

CVD systems are generally classified as metal-organic chemical vapor deposition or MOCVD systems. Various other systems, including plasma assisted or enhanced, have been developed for different materials and applications.

7.2 Basic Aspects

The chemical vapor deposition process involves the flow of gases in the reactor, along with the associated heat and mass transfer and chemical reactions. The susceptor may be stationary or moving, as in continuous deposition systems. It may also be rotating, as is the case with many vertical CVD reactors to achieve greater film uniformity. Therefore, the following basic considerations arise in typical CVD reactors:

Fluid flow
Heat transfer
Mass transfer
Chemical reactions
Buoyancy effects
Conjugate transport
Complex 2D and 3D domains
Dufour and Soret effects
Variable properties

Most of these aspects have been considered earlier. The Dufour effect is the energy flux due to a concentration gradient, and the Soret effect is the inverse, being mass diffusion caused by a temperature gradient. Both these effects are often important in CVD processes (Mahajan and Wei 1991).

Substantial information exists in the literature on the heat and mass transfer, flow, and chemical reactions for many of the materials mentioned earlier. These studies have focused on the flow and temperature and concentration levels for complex geometries encountered in common CVD processing systems. Rotating susceptors, three-dimensional flow, and experimental measurements have also been considered in various investigations, such as Jensen et al. (1983, 1991), Fotiadis et al. (1990), Karki et al. (1993), Chiu and Jaluria (2000) and Yoo and Jaluria (2002). Experimental studies have also been carried out on the transport processes in channels that simulate horizontal CVD systems such as the one shown in Fig. 7.1 (Jensen et al. 1991; Chiu et al. 2001).

The governing equations for the transport processes in a CVD reactor can be obtained from the basic equations given earlier in Chap. 2. Besides the conservation of mass, momentum, and energy, species conservation equations are also needed. It is often assumed that the gas mixture is ideal gas, the flow is subsonic and laminar, and the turbulence and viscous dissipation effects in the gas mixture are negligible. Then the following equations are obtained:

Conservation of mass:

$$\frac{\partial \rho}{\partial t} + \nabla.\left(\rho \bar{V}\right) = 0 \tag{7.1}$$

Momentum force balance:

$$\frac{\partial \rho \bar{V}}{\partial t} + \nabla.(\rho \bar{V}\,\bar{V}) = -\nabla p + \nabla.\tau + \rho\,\bar{g} \tag{7.2}$$

Conservation of energy:

$$\frac{\partial \rho h}{\partial t} + \nabla.(\rho h \bar{V}) = \nabla.(k\nabla T) + \nabla.(\tau.\bar{V}) + S_h \tag{7.3}$$

Conservation of species:

$$\frac{\partial \rho \phi_i}{\partial t} + \nabla.(\rho \phi_i \bar{V}) = \nabla.(\rho D_i \nabla \phi_i) + M_i \dot{\omega}_i + \nabla.j_i \tag{7.4}$$

with

$$\sum_{i=1}^{N} \phi_i = 1 \tag{7.5}$$

Here, ϕ_i is the mass fraction of each of the N species, D_i the mass diffusivity of species i, M_i its molecular weight, ω_i the molar production rate per unit volume due to the gas phase reactions, and j_i the species diffusive flux. The boundary conditions are based on the usual no-slip conditions at the walls, inlet/outlet conditions, and imposed thermal conditions at the boundaries. Initial conditions are important for the transient problem. A sudden change in heat input at start-up or other scenarios can be considered to study the time-dependent deposition.

The governing equations are coupled through buoyancy, variable properties, Dufour and Soret effects, and energy release or absorption due to chemical reactions. Consequently, the problem is generally quite complicated. Though the flow and heat transfer problem, without reactions, is relatively straight forward, the chemistry involved with the gas and surface reactions can complicate the problem substantially, depending on the deposited material. In several cases, it is possible to simplify the problem by assuming axisymmetric or two-dimensional, planar, steady flow. However, transient effects are often important, particularly at start-up and shutdown, and to study unsteadiness in the process. A three-dimensional problem has to be solved for various practical situations, such as with a rotating susceptor. The flow and heat transfer results do not yield quantitative information on the deposition, but they do provide significant information on the flow characteristics and on the heat transfer coefficient at the surface. These results can be employed to determine the dependence of the expected characteristics of the deposition, particularly the uniformity of the film, on the flow of reacting gases. Other aspects such as fluid loss from the system, which is an important consideration due to high costs of precursor gases, also depend on the flow. The results can be used in the design of CVD system with respect to the heating arrangement, inflow and outlet channels, rotating mechanism, and so on.

As expected, chemical reactions and, therefore, the chemical kinetics and properties play a critical role in the deposition of material from the gas phase. The chemical kinetics for several materials is available in the literature and is often complicated by the large number of intermediate reactions and species involved. In addition, there are reactions at the surface as well as within the gas. These are considered in detail later in connection with metal-organic CVD (MOCVD) systems.

For the fairly well-known case of the deposition of silicon from silane (SiH_4), with hydrogen as the carrier gas, the overall reaction is given by

$$\mathrm{SiH_4(gas)^4} \rightarrow \mathrm{SiH_4(absorbed)} \rightarrow \mathrm{Si(Deposited)} + 2\mathrm{H_2} \tag{7.6}$$

The chemical kinetics were proposed in a simple form by Claasen et al. (1982) as

$$K = \frac{K_o p_{\mathrm{SiH_4}}}{1 + K_1 p_{\mathrm{H_2}} + K_2 p_{\mathrm{SiH_4}}} \tag{7.7}$$

where p's are the partial pressures of the two species in the reactor, K is the surface reaction rate in mole of Si/m^2 s, $K_o = A \exp(-E/RT)$, E being the activation energy, and A, K_1, and K_2 are constants which are obtained experimentally and given by Claasen et al. (1982). Similarly, the chemical equation and reaction rate for TiN are given by Buiting et al. (1991) for titanium tetrachloride ($TiCl_4$) and ammonia (NH_3) as the reacting gases and argon (Ar) as the carrier gas as

$$6\mathrm{TiCl_4} + 8\mathrm{NH_3} \Leftrightarrow 24\mathrm{HCl} + \mathrm{N_2} + 6\mathrm{TiN} \tag{7.8}$$

$$K = K_o p_{\mathrm{T}}^{0.8} X_{\mathrm{TiCl_4}}^{-0.5} X_{\mathrm{NH_3}}^{1.3} e^{-E/RT} \tag{7.9}$$

where X is the mole fraction, p_{T} is the total pressure, and the constants are given in the reference.

Even though, these simple chemical equations, which are generally obtained experimentally from the overall deposition in CVD systems, give the general behavior of the deposition process, they do not give the desired level of accuracy, particularly with respect to local behavior and movement of different species. Generally, several intermediate reactions are involved that must be included in order to accurately predict the deposition rate. This is particularly true for depositing materials like gallium nitride (GaN) and silicon carbide (SiC), as discussed later. Also, accurate values of the material properties, such as density and diffusivity, are needed for valid and accurate results from the modeling. Commercial software, such as Chemkin, are often used for solving such complex chemical kinetics problems.

7.3 Characteristic Results

A commonly used CVD system is the horizontal reactor, as shown in Fig. 7.2. The susceptor, on which deposition occurs, may be placed horizontally or at an angle with the flow. Figure 7.3 shows the predicted and experimental results in a horizontal

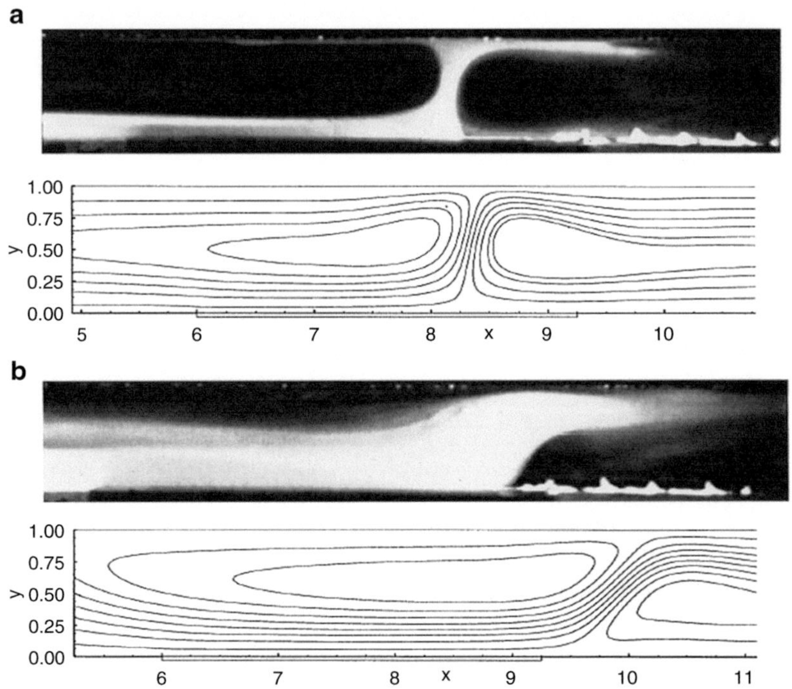

Fig. 7.3 Comparison between experimental observations and numerical predictions of streamlines at (**a**) $Re = 9.48$ and (**b**) $Re = 29.7$ for a ceramic susceptor in a horizontal CVD reactor

CVD reactor for flow of air in a straight channel over a susceptor heated by a uniform heat flux source, for two flow rates. As the gas flows over the heated susceptor, heating causes significant buoyancy effects to arise. In the case of the lower *Re*, a plume develops above the susceptor. This flow pattern generates two transverse rolls, with their axis of rotation perpendicular to the flow direction. These rolls can significantly influence the deposition rate and film uniformity. As *Re* is increased, the plume shifts downstream due to larger bulk gas flow. The transverse rolls become smaller, and flow entrainment from the outside is reduced. The appearances of oscillatory flow, turbulent flow, and transverse and longitudinal rolls, which would also affect the deposition, were found to be dependent on the channel cross-sectional aspect ratio, flow rate, and heat input. The plume location is quantified by calculating the flow separation distance from the susceptor's leading edge and compared to experimentally measured values for a range of mixed convection parameter. There is good agreement between numerical and experimental results.

Significantly different flow regimes are observed when the same parametric space is examined for a converging channel with a 8° tilted heating section. Flow in the converging channel remains steady and laminar over a larger range of conditions since the reduction in channel height increases the local gas velocity and diminishes the relative effect of the buoyancy force. When the flow rate and heating rate are

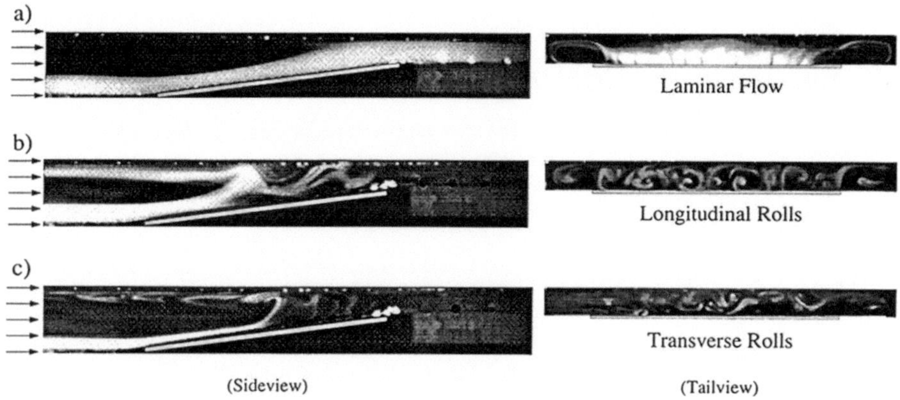

Fig. 7.4 Side view and tail view of the flow pattern in a converging channel with 8° tilt. Tail views are located at the end of the heated section with a light sheet oriented perpendicular to the main flow (**a**) $Re = 9.48$, $Gr = 4.3 \times 10^5$, (**b**) $Re = 29.7$, $Gr = 4.3 \times 10^6$ and (**c**) $Re = 9.48$, $Gr = 4.3 \times 10^6$

increased, the transition to a new flow regime, composed of longitudinal rolls with an axis of rotation parallel to the main flow, occurs. The side view of this regime shows oscillatory flow. The tail view reveals the presence of rolls along the heated plate. These rolls are unsteady in nature, and their presence promotes mixing within the channel. Buoyancy effects increase as *Re* is decreased or as *Gr* is increased, and a thermal plume appears above the heated plate, as shown in Fig. 7.4. The corresponding side view shows the breakup of longitudinal roll structures. When this condition occurs, the flow shifts to a regime dominated by transverse rolls. The plume generated above the heated plate creates an upstream transverse roll which is observed to be two-dimensional near the center of the channel. Near the exit, the flow structure breaks up into turbulent flow. The transition from longitudinal rolls to transverse rolls is defined by fitting a curve across the transition zone, yielding a critical mixed convection parameter of $Gr/Re^2 = 6000$. In the design of CVD reactors, entrainment at the exit and transverse rolls are undesirable. Flow entrainment introduces by-products and other forms of contaminants onto the deposition surface, thereby lowering product quality. Rolls may act like stagnation zones, preventing the desired species from migrating to the deposition surface. These issues must be considered in detail when horizontal CVD reactors are designed.

Figure 7.5 presents a comparison between the numerical predictions and experiments for the preceding horizontal reactor configuration. The temperature distributions at various heights above the heat source are compared. A fairly good agreement is observed. The temperature field is an important consideration since the chemical reactions are strongly affected by the local temperature. Similarly, heat transfer rate at the surface is important in indicating the uniformity in the transport processes acting at the depositing surface. Figure 7.6 shows the typical results from simulations of a horizontal reactor for depositing TiN. The results obtained include temperature and concentration differences as functions of location on the surface. The observed trends are as expected. The deposition rate, which is obviously of particular interest, was also calculated.

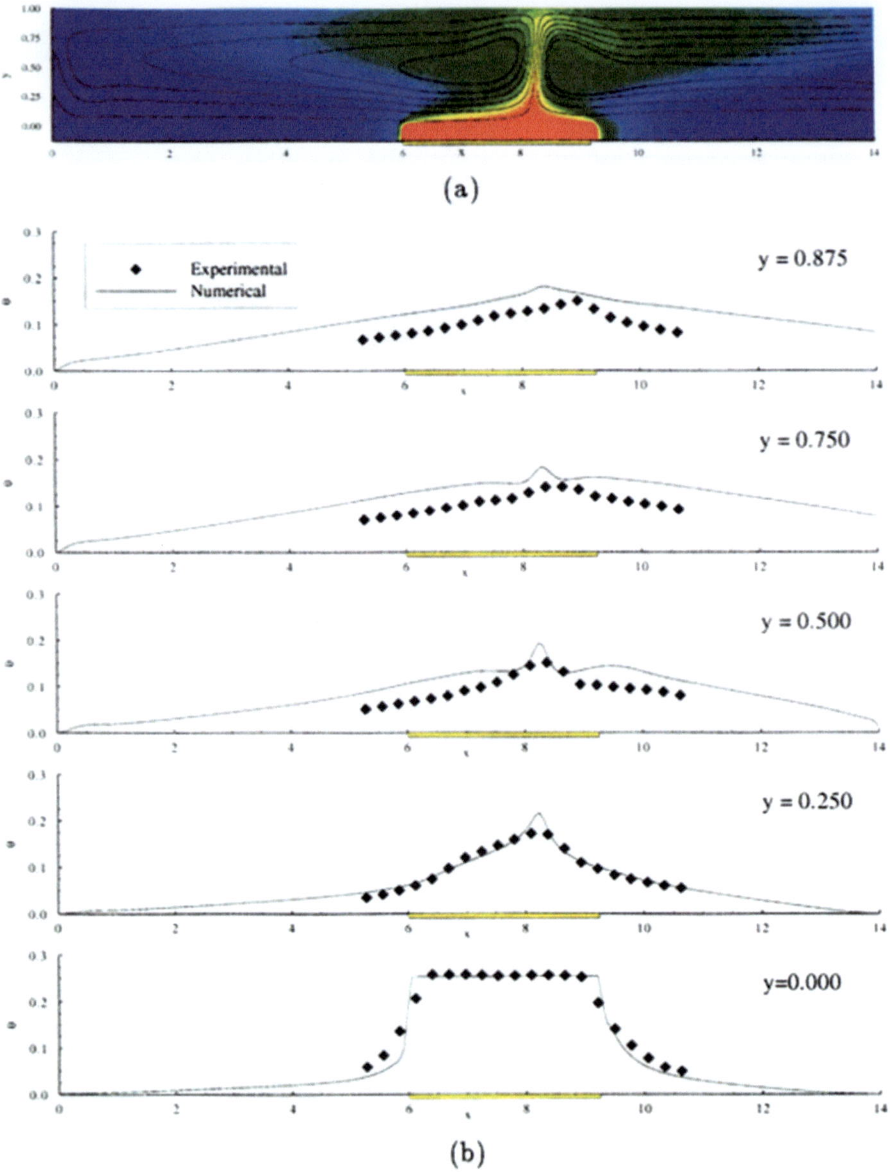

Fig. 7.5 (**a**) Calculated streamlines and temperature contours and (**b**) experimental temperature data (symbols). Comparison with numerically predicted results for an aluminum heated region in a horizontal channel

Figure 7.7a shows a comparison between the computed silicon deposition rate in a horizontal reactor obtained by different models, using the chemical kinetics expression given by Eq. (7.7), with the measurements of Eversteyn et al. (1970). It

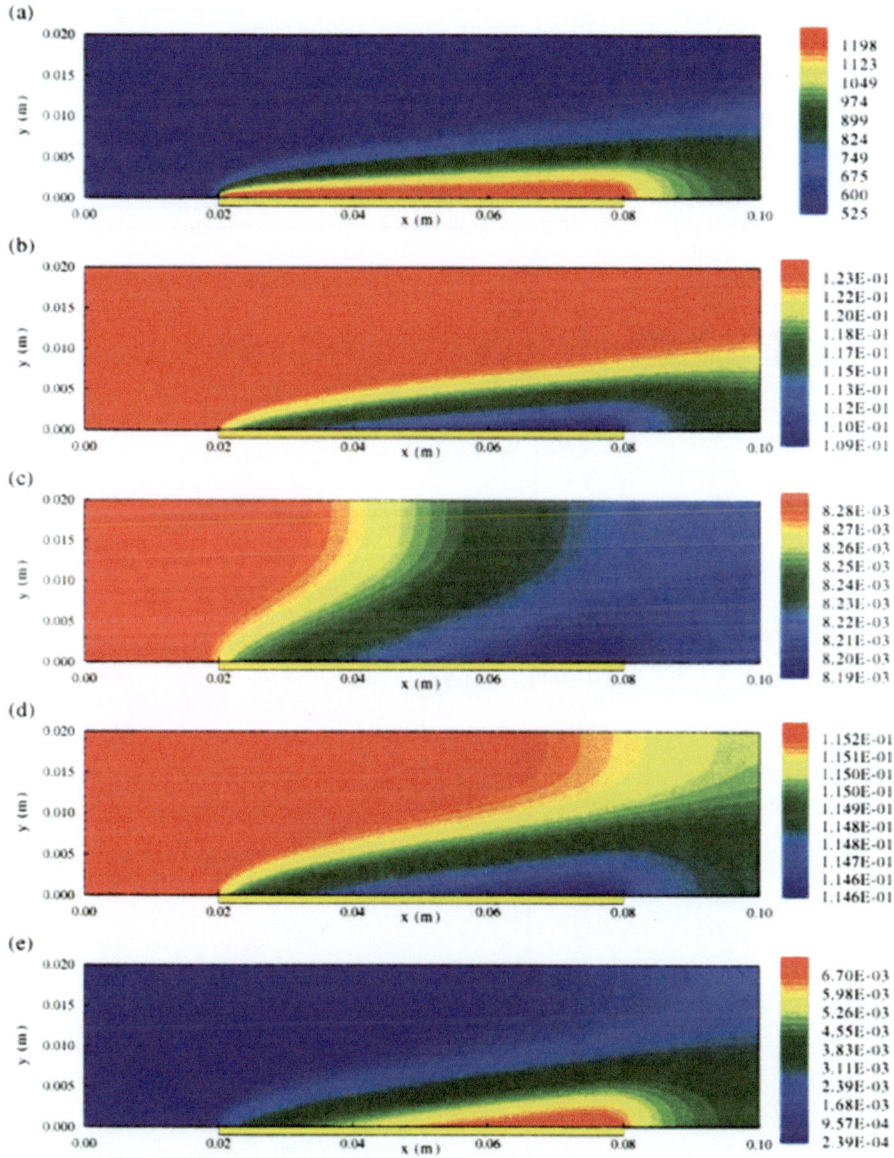

Fig. 7.6 Calculated results for a horizontal CVD reactor at atmospheric pressure, with a horizontal heated susceptor, showing the temperature distribution and mass fraction contours for the deposition of TiN (**a**) Temperature (K), (**b**) $TiCl_4$ Mass fraction, (**c**) H_2 Mass fraction, (**d**) N_2 Mass fraction, (**e**) HCl Mass fraction

is seen that the basic trends are similar, and a fairly good agreement is observed in most cases. A larger discrepancy is seen near the susceptor's leading edge, mainly because of the uncertainty in locating the edge in the experiment. The significant

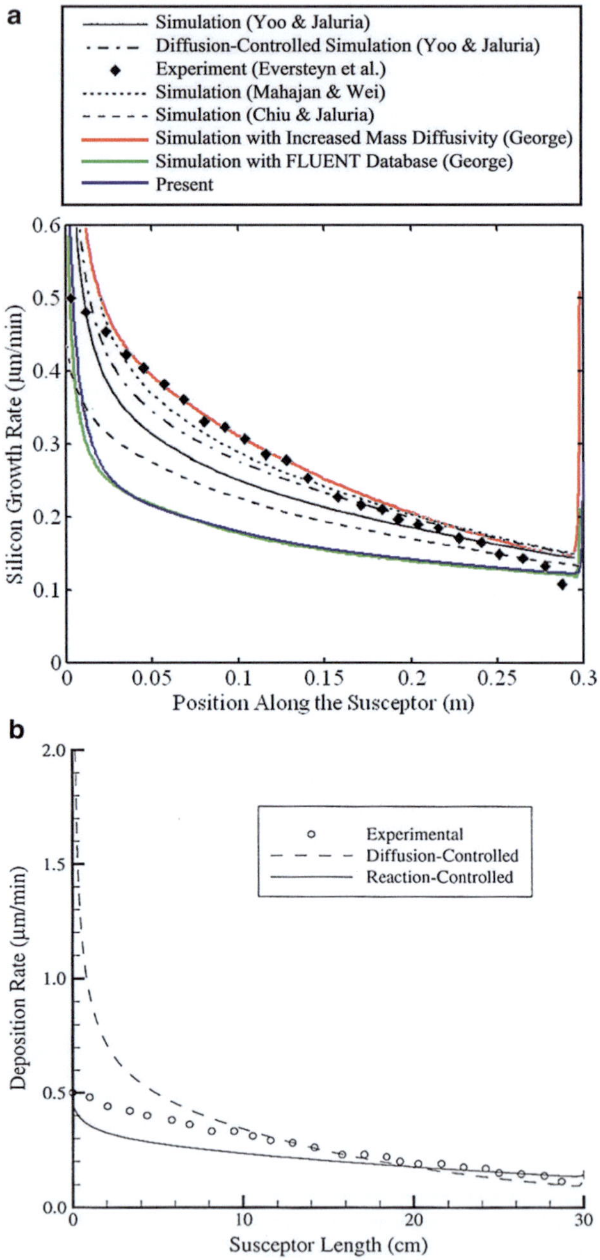

Fig. 7.7 (**a**) Comparison between numerical results and measurements for a horizontal CVD reactor for silicon deposition with different models. In this figure, present refers to Lin et al. (2010) and George to George et al. (2006); (**b**) comparison between the diffusion-controlled approximation and the reaction-controlled chemical modeling. The experimental results were obtained by Eversteyn et al. (1970)

effect of the mass diffusivity used indicates the importance of accurate property values. A chemical kinetics limited deposition may be assumed if the transport processes allow easy access of the species to the surface, and a diffusion limited case may be assumed if the deposition is largely restricted by the transport process.

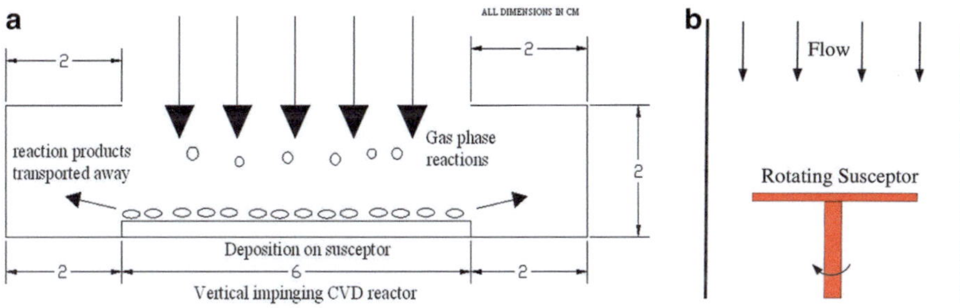

Fig. 7.8 (**a**) An impingement-type CVD reactor; (**b**) Vertical CVD reactor with a rotating susceptor

In the present example, the reaction-controlled deposition model yields much better agreement with experimental results, see Fig. 7.7b. The reaction-controlled deposition model accounts for the deposition rate dependence on temperature and species concentration at the deposition surface. However, the full problem is fairly complicated, and the transport processes as well as the chemical reactions must be considered at the surface and in the gas phase to satisfactorily model the overall deposition process. Even though chemical kinetics data are available for a few common materials like silicon, the lack of accurate data for deposition of metals and other materials is a major problem in accurate prediction of the deposition rate (Wang et al. 2006).

Two commonly used vertical CVD reactor systems are shown in Fig. 7.8. One is an impinging type vertical reactor, with dimensions given for a small system, and the other one is with a rotating susceptor. The former may be studied as a 2D or 3D problem, depending on the geometry and dimensions. The second one may be considered as an axisymmetric problem if the susceptor is stationary or as a 3D one with a rotating susceptor. Typical results on the former system in terms of temperature, flow, and deposition are shown in Fig. 7.9. The green region in Fig. 7.9c represents acceptable uniformity and the red region unacceptable nonuniformity in film thickness. The nonuniformity of the deposited film is a major concern, and efforts are made to achieve better uniformity by varying the operating conditions like susceptor temperature and inflow velocity distribution and concentration. Similar considerations arise in the deposition of other materials like TiN, SiC, GaN, and GaAs, needed for different new and emerging applications like laser diodes, light-emitting diodes, and high-power transmitters. This also points to an optimization problem where the process has to balance between film quality and deposition rate, as discussed later.

7.4 Metal-Organic Chemical Vapor Deposition (MOCVD)

Many new materials have been developed in recent years to meet the challenges and needs of new and emerging applications, such as power amplifiers and optoelectronic devices (Steigerwald et al. 1997; Chattopadhyay et al. 2009; Mokkapati and

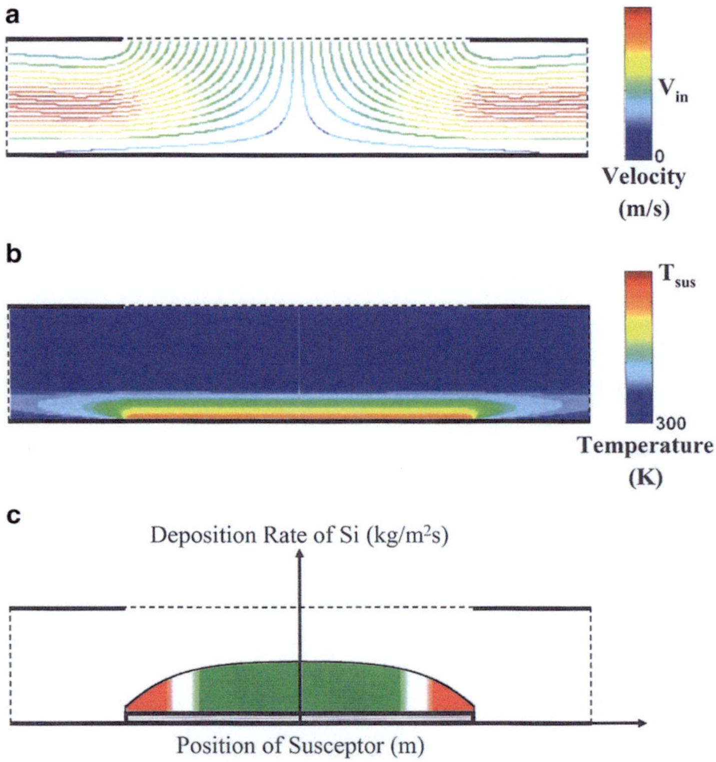

Fig. 7.9 Typical results on flow, temperature distribution, and deposition in a CVD reactor

Jagadish 2009; Amorim et al. 2012; Justice et al. 2013). Chattopadhyay et al. (2009) stated that "Group III-nitride semiconductors have been pivotal in technology enrichment, especially in the optoelectronics field." This field includes technology that involves light-emitting diodes (LEDs), photodiodes in solar cells and collectors, photosensors, and other technologies involving the use of a light source. Chemical vapor deposition using metal-organic precursors has been the manufacturing process used to develop and produce many of these advanced materials. These manufacturing systems form a class of CVD reactors, generally known as metal-organic CVD or MOCVD reactors. The deposited materials include GaN, GaAs, InGaN, AlGaN, and ZnO, among many others. As one of the CVD methods, MOCVD is used to produce single or polycrystalline thin films, which involves a highly complex process for growing crystalline layers to create semiconductor multilayer structures. Due to its versatility and capability for large-scale production, MOCVD has become a major process in the manufacture of thin films of many of these materials.

7.4.1 Basic Process

As with a typical CVD process, four types of gases flow into the reaction chamber. The first type is called the carrier gas. It is typically composed of nitrogen, hydrogen, or argon. The purpose of the carrier gases is the transportation of the other gases and vapors into the chamber due to the fact they do not react with the precursor gases. The second type is the precursor gas. Precursor gases are the raw material gases that will undergo chemical reactions to produce the new material. The third type is the reactive gases, and they react with the precursor gases. The fourth type is a dopant gas, and it is used to enhance the growth by having the dopant gas react with the unwanted radical gases from the by-products of the chemical reactions. For example, to produce SiO_2 or silicon dioxide, films used for electrical isolation in semiconductors, the gases required are SiH_4 or silane for precursor gas, O_2 or oxygen for reactive gas, H_2 or hydrogen for the carrier gas, and carbon fluoride gas for dopant.

As briefly discussed earlier, the energy source, which may include microwave, induction coil, or resistive heating, heats up the chamber, and the precursor gas and the reactive gas molecules begin to split apart from the excessive thermal energy. This is called the adsorption phase. After molecules are broken apart, radical molecules react with other molecules and recombine to form the new materials. This is where the chemical reactions occur, indicating the reaction phase. At the same time, as the new molecules form, excessive molecules and energy are released from the stable substances. This is called desorption phase. The new material deposits on to the substrate, and it begins to grow in thickness as more and more chemical reactions occur. The substrate is the crystalline material on which the deposition occurs, and sapphire is used in many cases. Other substrates are also used, depending on the application. The by-products from the reaction exit the chamber as more precursor and reactive gases enter the chamber. The following sections focus on the deposition of gallium nitride to bring out the major concerns and considerations in these processes. For other materials, the relevant chemical kinetics, properties, and operating conditions are obviously employed.

7.4.2 Gallium Nitride

Gallium nitride (GaN) is one of the most important semiconductor materials, with wide bandgap (3.4 eV), high breakdown field properties, and chemical and thermal stability. Since the first p-n junction GaN light-emitting diodes (LEDs) were demonstrated by means of the low-energy electron beam irradiation (LEEBI) treatment (Amano et al. 1989), GaN has been shown to have tremendous potential in electronic and optoelectronic devices. The mixture of GaN with In (InGaN) or Al (AlGaN), with a bandgap dependent on the ratio of In or Al to GaN, makes it suitable for manufacturing green, blue, ultraviolet (UV), and eventually white light-emitting diodes (LEDs), which have important advantages over traditional incandescent or fluorescent light. As a starting point for the application of GaN, the film growth of GaN provides a base for improving the technology. In the literature, GaN thin films have been successfully deposited onto several substrates made of sapphire and silicon carbide, by using the metal-organic chemical vapor deposition (MOCVD) and molecular beam epitaxy (MBE) techniques. While the latter involves a very simple process and the growth process can be monitored as the film is built up one atomic layer at a time, it has limitations for commercial application. First, the need for ultrahigh vacuum (UHV) conditions makes it expensive and essentially unaffordable. Second, frequent shutdown and opening of the UHV apparatus need considerable preparation time, before returning to the film growth process, which leads to a waste of valuable production time. Compared with MBE, MOCVD has several major attractions such as versatility and capability for large-scale production. MOCVD is considered as one of the most versatile and economical techniques, because it is suitable for producing virtually all V/III semiconductor compounds and alloys and devices requiring large areas, such as LEDs and solar cells. At present, the preferred industrial method for GaN film growth is MOCVD using trimethylgallium (TMGa) and ammonia (NH_3) as precursors and hydrogen (H_2) or nitrogen (N_2) as the carrier gas (Acharya 2014).

7.4.3 Chemistry for GaN Deposition

It was seen earlier that chemical kinetics and material properties play critical roles in the deposition of solid material from the gas phase in CVD. The chemical kinetics is generally quite complicated due to the large number of intermediate reactions and species involved. In addition, there are reactions at the surface as well as within the gas. In some cases, simplified overall kinetics have been developed, as given earlier. But for greater accuracy, intermediate reactions and relevant chemical species need to be taken into account. A closer look in the chemical pathway process of GaN would help understand the basis of GaN production. In a general pathway, GaN is produced by the following

$$2\mathrm{Ga} + 2\mathrm{NH_3} = 2\mathrm{GaN(s)} + 3\mathrm{H_2} \quad (7.10)$$

However, the typical pathways that arise with the precursors used in actual practice are much more complicated. Typically, the corresponding comprehensive chemistry model contains many intermediate species and gas phase and surface reactions. Gas phase reactions for the GaN growth process are composed of a series of complicated chemical reactions. It was reported that a stable adduct $TMGa{:}NH_3$ was formed between TMGa and NH_3 during the GaN MOCVD process. Then, it was transformed into a ring compound, which was thought to affect the film growth and its properties (Thon and Kuech 1996). However, by using density functional and reaction rate theories, it was reported that this ring compound did not affect the main pathways for the GaN growth and could be taken as negligible (Sengupta 2003). The main pathways involved in the GaN growth process are the decomposition of TMGa to dimethylgallium (DMGa), monomethylgallium (MMGa), and Ga via successive elimination of methyl groups and their reactions with NH_3:

$$
\begin{aligned}
&Ga(CH_3)_3 \rightarrow Ga(CH_3)_2 + CH_3 \rightarrow GaCH_3 + 2CH_3 \rightarrow Ga + 3CH_3 \\
&Ga(CH_3)_{x+1} + NH_3 \rightarrow Ga(CH_3)_{x+1}.NH_3 \\
&\qquad \rightarrow (CH_3)_x.GaNH_2 + CH_4 (x = 0 - 2)
\end{aligned}
\tag{7.11}
$$

According to these main pathways, a gas phase chemistry model has been proposed, and the details can be obtained from Meng and Jaluria (2013, 2015). Although surface reactions focus on the physical and chemical processes on the surface, including gas adsorption and desorption, diffusion, formation and etching of solid film, surface site occupation, etc., it is a challenge to consider all of these processes in one surface reaction model. Therefore, none of the GaN surface reaction mechanisms reported in the literature includes all of these factors together. A detailed kinetic surface reaction model, given by Sengupta et al. (2005), considers gaseous species absorption, surface reactions, surface species desorption, and formation of the solid film. The mathematical calculation scheme of the deposition rate can be obtained from the work of Mazumder and Lowry (2001).

A comprehensive chemistry model for deposition of GaN from TMGa and NH_3 in excess H_2 was used in the calculations of Meng and Jaluria (2015). This mechanism, which includes 17 gas phase reactions and 52 surface reactions, has been validated to be fairly universal and of good adaptability by comparison with

Table 7.2 List of detailed gas phase homogeneous reactions involved in the GaN growth process

No.	Gas phase reactions			A	α	E_a/R_g
G1	TMGa	<->	DMGa + CH_3	1.6×10^{17}	0	30057
G2	DMGa	<->	MMGa + CH_3	2.5×10^{15}	0	17883
G3	MMGa	<->	Ga + CH_3	4.3×10^{13}	0	24519
G4	TMGa + NH_3	->	TMGa·NH_3	2.28×10^{31}	−8.31	1568
G5	TMGa + NH_3	->	$(CH_3)_2GaNH_2$ + CH_4	17	2.0	10050
G6	DMGa + NH_3	->	DMGa·NH_3	4.08×10^{28}	−7.03	1628
G7	DMGa + NH_3	->	CH_3GaNH_2 + CH_4	530	1.56	10440
G8	MMGa + NH_3	->	MMGa·NH_3	7.95×10^{21}	−5.21	1054
G9	MMGa + NH_3	->	$GaNH_2$ + GH_4	8.1×10^{2}	1.3	8919
G10	NH_3 + CH_3	<->	NH_2 + CH_4	3.31	2.51	4962
G11	CH_3 + H_2	<->	CH_4 + H	1.2×10^{9}	0	6300
G12	TMGa + H	<->	DMGa + CH_4	5.0×10^{10}	0	5051
G13	DMGa + H	<->	MMGa + CH_4	5.0×10^{10}	0	5051
G14	TMGa·NH_3	<->	MMGa + $2CH_3$ + NH_3	1.33×10^{44}	−8.24	39150
G15	CH_3 + H + M	<->	CH_4 + M	2.4×10^{19}	−1	0
G16	$2CH_3$	<->	C_2H_6	2.0×10^{10}	0	0
G17	2H + M	<->	H_2 + M	1.0×10^{13}	0	0

Rate constants for unimolecular and bimolecular reactions are expressed as s^{-1} and m^3/(kmol.s), respectively. Backward rates are calculated from equilibrium by minimizing the Gibbs free energy experimental data for both laboratory-scale and commercial reactors. Table 7.2 shows the model for the gas reactions. Here, the rate constant K is given by the equation

$$K = AT^{\alpha}\exp\left(-\frac{E_a}{R_g}T\right) \tag{7.12}$$

where A and α are constants, E_a the activation energy, and R_g the universal gas constant. There are additional reactions at the surface (Wu et al. 2003; Kadinski et al. 2004 ; Sengupta et al. 2005). Because of the complexity of this chemistry model, other simpler models have also been developed by considering the activation energies involved and comparing the results obtained with more complex models. Several such simpler models are available and can be used for the deposition of GaN and other materials (Hu et al. 2016, 2017).

The chosen chemistry model can be applied at the micro- or nanoscale level to obtain the reaction rates and the deposition, while the boundary conditions, in terms of inlet velocity, susceptor temperature, inlet concentration, rotational speed, etc., are imposed at the system level. In the numerical scheme, the grid near the surface is much finer than that in the main flow, resulting in control volumes that would allow the capture of deposition which is in microns. The chemistry is applied to each control volume and the results linked with the transport processes at the larger scales to obtain the final deposition results. Some characteristic results for silicon were discussed earlier.

In order to solve the equations for the flow and heat and mass transfer, with chemical reactions, material properties of gaseous species need to be implemented into the model. Typically, either constant values or certain functional dependent on temperature are prescribed. The mixture density can be calculated using the ideal gas law, viscosity evaluated using Sutherland's law, specific heat computed using NIST-JANAF thermochemical tables (Chase 1998), and both thermal conductivity and binary diffusion coefficients evaluated by using transport theory. Most of the material properties can also be found from the material database of several commercial software such as CFD ACE+, Fluent, and Chemkin.

It may be mentioned that the preceding detailed discussion of the chemistry of GaN deposition is presented to indicate the complexity of the process, the importance of accurate property and chemical kinetics data, and the large number of chemical equations and species that arise in typical cases. Many of these equations are stiff and require very fine grids and step sizes to obtain the solution, making the computational solution quite challenging.

7.5 Validation

It is very important to verify and validate the numerical model because of the complexity of the governing equations. Verification is achieved by grid refinement and ensuring that the numerical results do not vary significantly with arbitrary

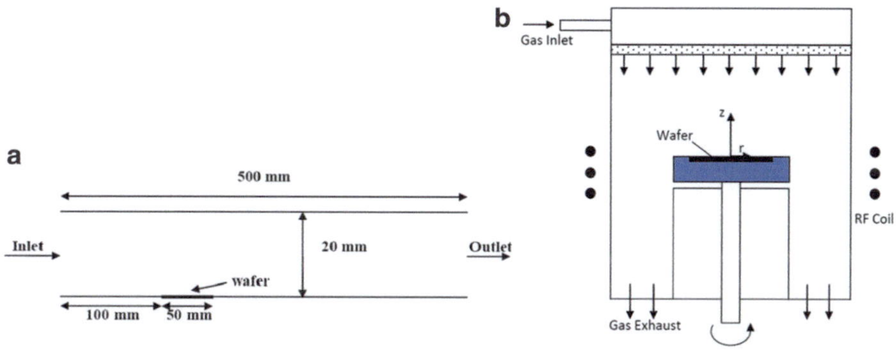

Fig. 7.10 MOCVD reactors considered for validation of the modeling of the GaN deposition process. (**a**) Horizontal MOCVD reactor; (**b**) Rotating disk reactor

parameters chosen to obtain the solution. For validation, it is necessary to ensure that the results obtained are physically reasonable and that they compare well with available benchmark numerical results and experiments. Two experimental studies are considered, and the results presented in the literature are compared with numerical predictions.

The numerical model with the preceding chemistry is applied to the two MOCVD systems shown in Fig. 7.10. The experimental data for the reactor in Fig. 7.10a were obtained by Chen et al. (1996). This is a simple horizontal reactor with the precursors entering from the left. A 50 mm sapphire substrate is placed on the wafer, 100 mm away from the inlet of the reactor. Its height is fixed at 20 mm, but the total length is not available, so some reasonable assumptions had to be made. The range of wafer temperature variation is set from 700 to 1400 K, based on which the proposed numerical approach was applied and the simulation carried out. A comparison between the simulation results and experimental data is shown in Fig. 7.11a. It is seen that the two results agree quite well, lending support to the model.

The experimental study of the other reactor shown in Fig. 7.10b and investigated by Theodorpoulos et al. (2000) is also considered. It consists of a quartz tube with an inner diameter of 11.43 cm. A screen is installed at the entrance to ensure uniform flow at the inlet. The temperature of the outer walls of the tube is fixed at 300 K due to water cooling. The distance between the inlet screen and the exhaust outlet is 30 cm. The substrate with a diameter of 8.9 cm is installed on the rotating disk in the center of the tube, and the distance from the inlet screen and the substrate surface is 10 cm. One wafer is placed on the susceptor, and its diameter is 5.08 cm. The carrier gas H_2 is supplied into the reactor at a flow rate of 7.1 standard liter/min (slm), and the flow rates of TMGa and NH_3 are kept at 2.98 slm and 6.0 slm, respectively. The operating pressure is kept at 140 torr, and the rotating speed of the wafer is 1200 rpm. The proposed model was used to perform the numerical calculations over a range of the substrate temperature, 1123–1273 K. The simulation results are compared with the experimental data, as shown in Fig. 7.11b. Again, a good agreement between the modeling results and experimental data is observed. The temperature range of

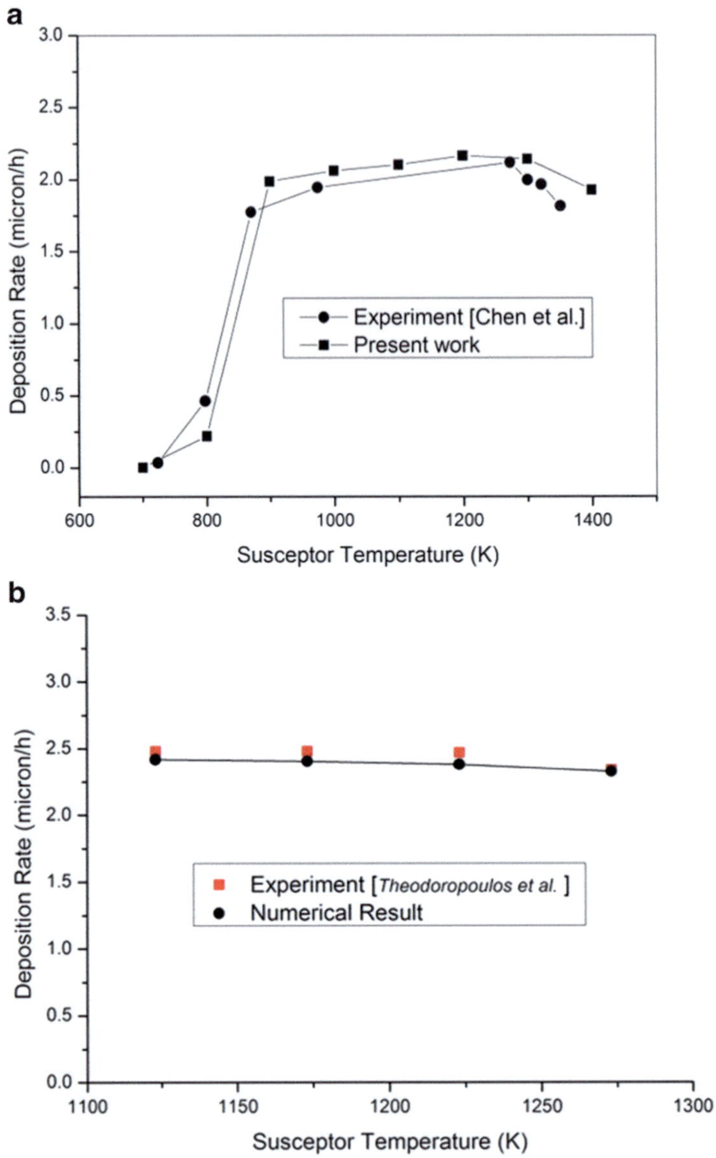

Fig. 7.11 Comparisons between numerical and experimental results for the two reactors shown in Fig. 7.10a, b respectively

1123–1273 K is not broad enough to cover all the growth conditions. However, it is significant for the GaN MOCVD process. Therefore, the proposed model is believed to be valid and accurate.

7.6 Flow and Temperature Distributions

As mentioned earlier, the flow and heat transfer mechanisms in a reactor can be investigated to determine the flow behavior and heat transfer coefficients at the surface. The characteristics of the flow, particularly any instability, recirculation, or chaotic behavior, can be used to qualitatively predict the uniformity and quality of the deposition. Similarly, heat transfer at the surface could be used to explore the mass transfer and deposition characteristics. Once the flow has been simulated, the chemistry model could be initiated to determine the deposition rates. As discussed earlier, many different CVD reactors are used for a wide range of materials. Two different reactors with impingement flow are considered here. The first is a two-dimensional vertical reactor shown in Fig. 7.12, and the other is a three-dimensional, rotating susceptor, reactor to be discussed later. For the 2-D reactor, a 120 mm sapphire substrate is placed in the middle of the bottom wall of the reactor, and other walls are assumed to be adiabatic. Inflow gases which contain TMGa, NH_3, and H_2 enter the chamber from the top at a uniform temperature and exit from both left and right. The flow rates, concentration ratio (V/III), which is defined as the ratio of volumetric flow rate of NH_3 to that of TMGa, the susceptor temperature, and the reactor pressure are fixed at typical values used in practical reactors. Due to high temperature of the susceptor, it is necessary to consider thermal radiation in the energy conservation equation. A surface-to-surface radiation method may be applied in this case.

Since the reactor is perfectly symmetric, one half of the entire domain may be employed for numerical simulation. Figure 7.13 shows the computed temperature and velocity distributions for a typical circumstance, with susceptor temperature 1073 K, pressure 100 torr, inlet velocity 4.18 cm/s, and concentration ratio 8064. The flow field shows the effect of buoyancy, which provides a vertical upward force, near the center. This would lead to the gases not reaching the central region and thus lack of adequate deposition there. The flow exits the reactor after flowing along the wafer and the bottom wall, as well as near the top. A recirculation is seen to arise near the mid-height, with ambient fluid brought in due to entrainment into the outflow. The temperature distributions show the decreasing temperature above the

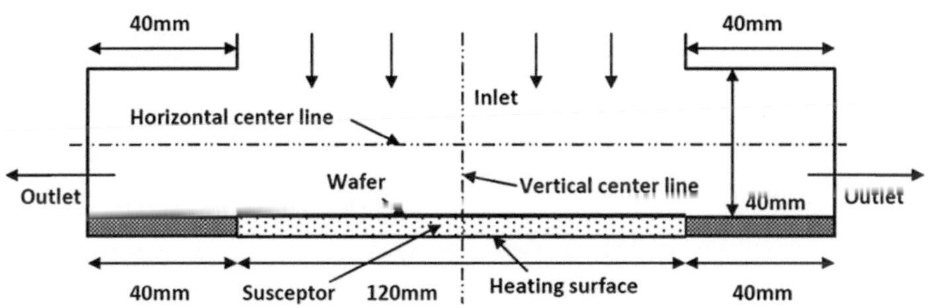

Fig. 7.12 An impinging two-dimensional MOCVD reactor for GaN deposition

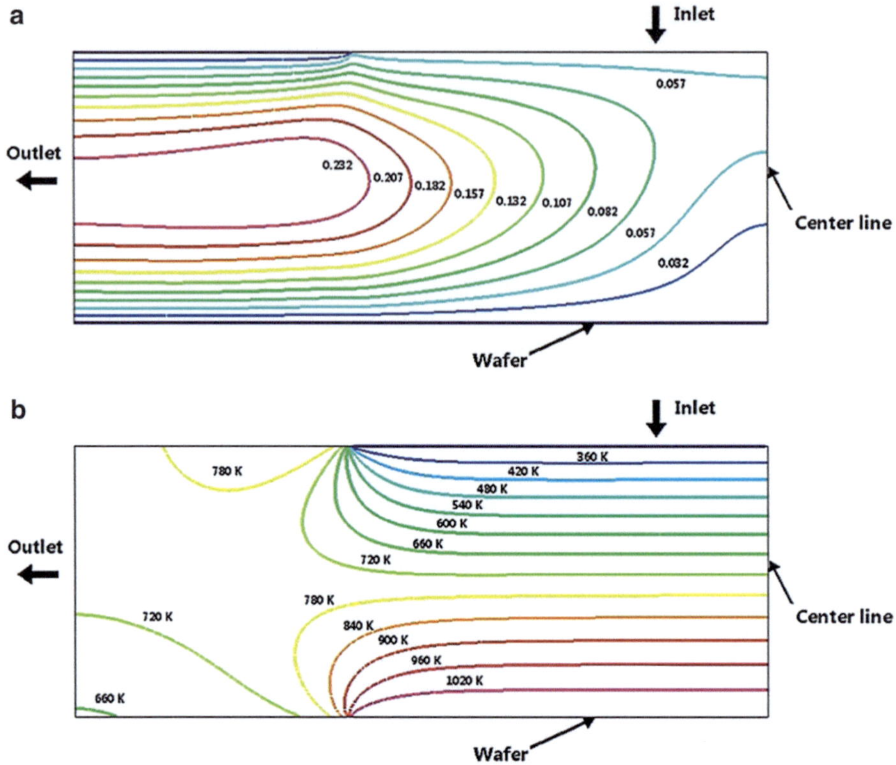

Fig. 7.13 Results obtained with the parameters fixed at the initial values given in the text. (**a**) Temperature field; (**b**) velocity distribution in m/s

heated wafer and the spreading out of the thermal field due to conduction in the bottom wall. Fairly uniform conditions are seen at the wafer, indicating that conditions are good for uniform deposition.

An impingement MOCVD reactor with a rotating base is commonly used in GaN deposition, as well as for various other materials. Figure 7.14 shows a typical wafer distribution on the rotating base for a large reactor, which is typically of diameter around 2 m. However, many smaller reactors have been developed, with much fewer wafers, including a small reactor with just one wafer. The typical diameter of each wafer is around 3–4 cm. The inflow may be spread uniformly over the inlet at the top, or a particular velocity distribution may be imposed. Many inlet designs have been developed and employed to obtain better deposition characteristics. An upper flow guide (UFG) may also be incorporated to channelize the flow over the rotating base. Such a flow guide could provide better control on the flow, as well as reduce the loss of reactive gases that are often very expensive.

Figure 7.15a shows a schematic of an impingement MOCVD reactor with a rotating base and an upper flow guide. This configuration has been considered in detail by Meng et al. (2015), George et al. (2015) and Wong and Jaluria (2016).

Fig. 7.14 A typical wafer loading in an industrial MOCVD reactor for GaN

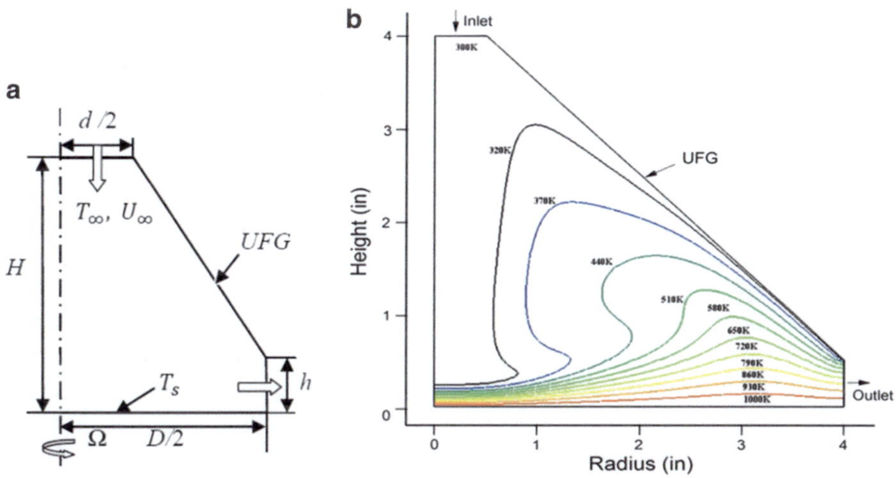

Fig. 7.15 (**a**) MOCVD reactor chamber with a rotating susceptor and upper flow guide (UFG); (**b**) calculated temperature contours on one cross section for typical conditions

Different flow guide configurations, including the conical one shown in the figure, were considered, along with different size reactors. Both experimental and numerical studies were carried out. Figure 7.15b shows the computed thermal field for typical operating conditions of 100 torr pressure, 1000 K susceptor temperature, 6 m/s inlet velocity, 600 rpm rotational speed, and concentration ratio of 2000. A three-dimensional numerical simulation was performed, and the flow and temperature

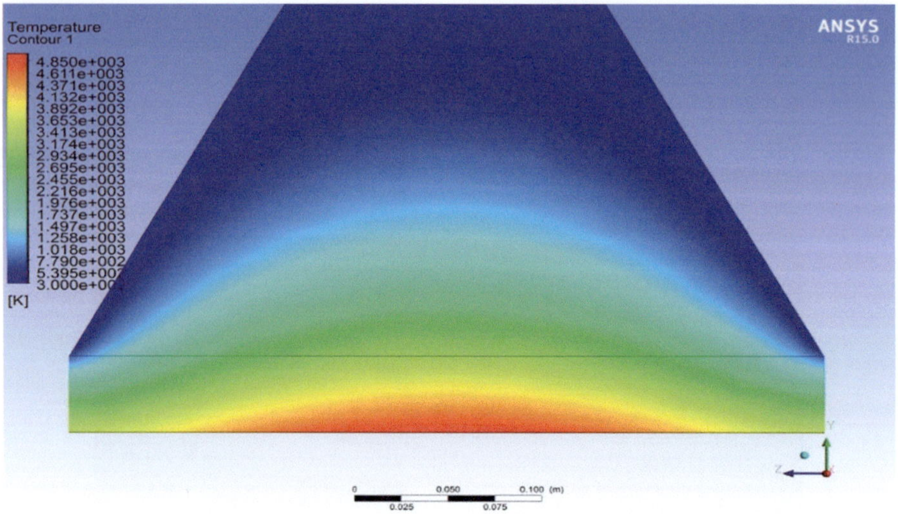

Fig. 7.16 Temperature distribution above the susceptor with no flow

contours indicated that gas recirculation arises near the reactor outlet. This could affect the uniformity of the thin film deposited in this region.

Figure 7.16 shows the thermal field when there is no inlet flow and no rotation. This case serves as the experimental control case, since there are many results available in the literature for transport from a horizontal heated plate. These results are used to further validate the computational model. This figure shows that the heated susceptor heats up the reaction chamber and the temperature distributions are symmetric about the central axis, as expected. The temperature profiles shown also indicate that buoyancy effects could be important at the generally low inlet flow rates and high temperatures employed in practice. Higher temperature in the center and tapering off at the edge are seen. Such temperature variation can affect the uniformity and local growth rates of the thin film.

Figure 7.17a, c shows the temperature distributions and flow streamlines above the heated susceptor, for an inlet velocity of 0.25 m/s, no rotation, and nitrogen as the fluid. As soon as the cool air enters the chamber, the overall temperature decreases. The center is seen to be cooler than the edge because the flow is impinging the center area first before the flow turns and exits the chamber. Figure 7.17c shows recirculation caused by the outflow and the wide cone-shaped flow guide. The recirculation is clearly seen on the two sides where the average velocity changes. Recirculation in a CVD reactor causes thermal variations and influences uniformity and deposition. The effect of rotation is seen in Fig. 7.17b, d. The flow and temperature inside the chamber change due to the rotation dramatically, even at a relatively small rpm of 60. The temperature field above the susceptor becomes much more uniform. The overall temperature decreases, as more thermal energy is drawn out of the reaction chamber due to the rotation. Overall flow velocities increase due to centrifugal acceleration, and flow streamlines are flatter near the susceptor compared with the no-rotation case. The flow near the edge of the susceptor is being drawn out by the centrifugal effect, hence the high-velocity area. The recirculation increases due to increase in overall fluid velocities, and a second eddy occurs near the top of the flow. This effect causes higher particle residence time inside the chamber. Longer residence time indicates that the molecules are not undergoing reaction and deposition. It causes thermal hotspots where the temperature is too high and precursor gas either turn into unwanted radicals or it burns up completely. Many such cases were investigated to study the effect of various parameters, such as inlet flow rate, susceptor temperature, and rotation on the resulting flow and temperature distributions. The effect of these parameters on deposition is discussed in the next section.

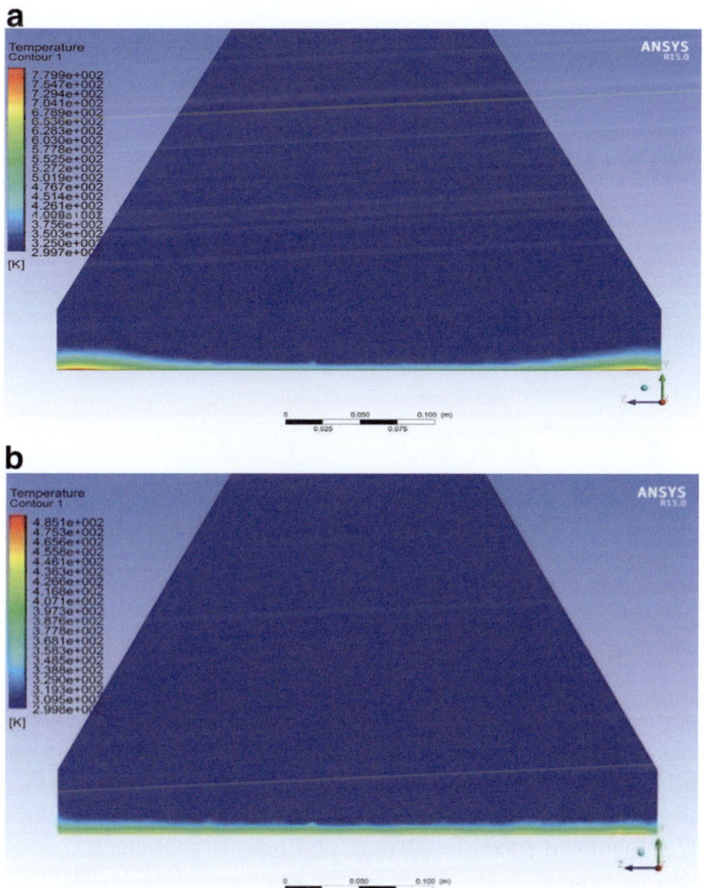

Fig. 7.17 Temperature and flow in the reactor for air at inlet velocity of 0.25 m/s. (**a**, **c**) No rotation; (**b**, **d**) Rotation at 60 rpm

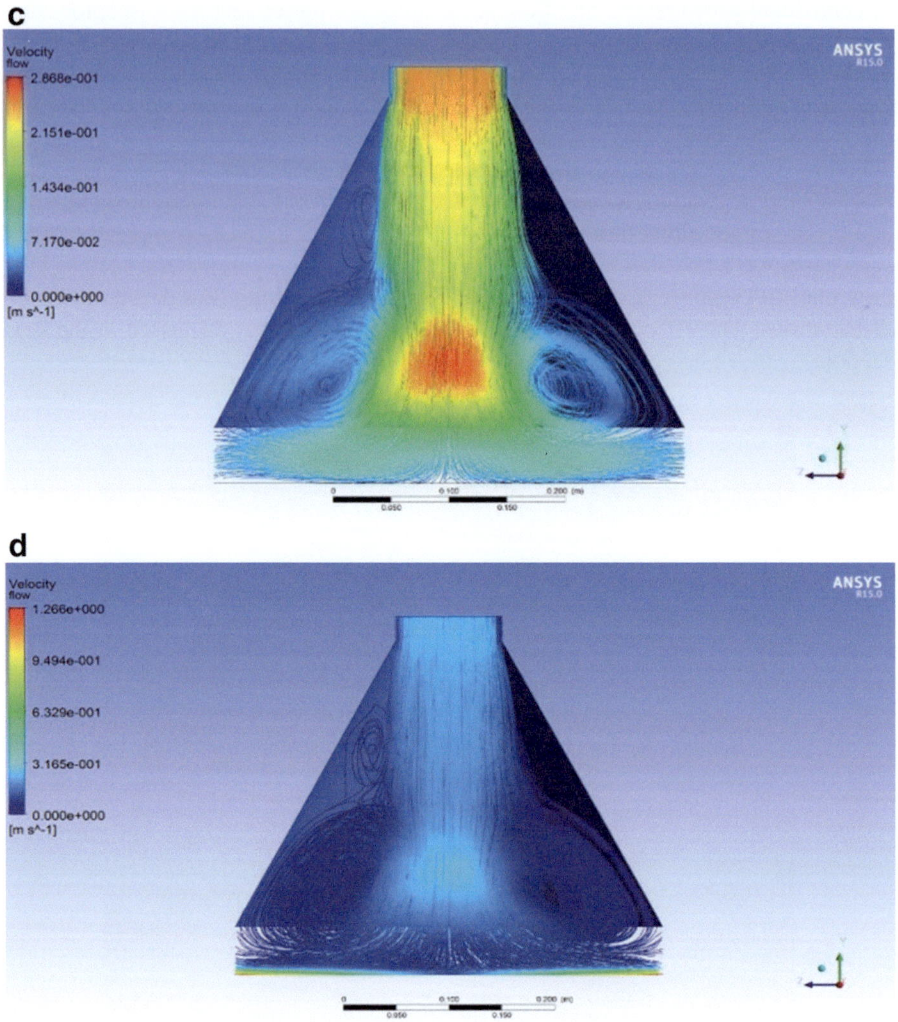

Fig. 7.17 (continued)

An important practical issue is maintaining an isothermal surface for deposition. Generally, heat input is provided by electrical heaters located below the susceptor. A complicated conjugate heat transfer problem arises that determines the temperature distribution at the surface. A typical scenario was modeled, in which inlet velocity is 1 m/s, wafer rotational speed is 60 rpm, and reactor pressure is 1 atm. The gas supplied into the reactor is air. Based on the analysis of heat transfer, considering conduction, convection, and radiation, the surface temperature for given heat flux at the bottom was calculated. Figure 7.18a shows the comparison between the numerical results and experimental data for this circumstance and Fig. 7.18b for another

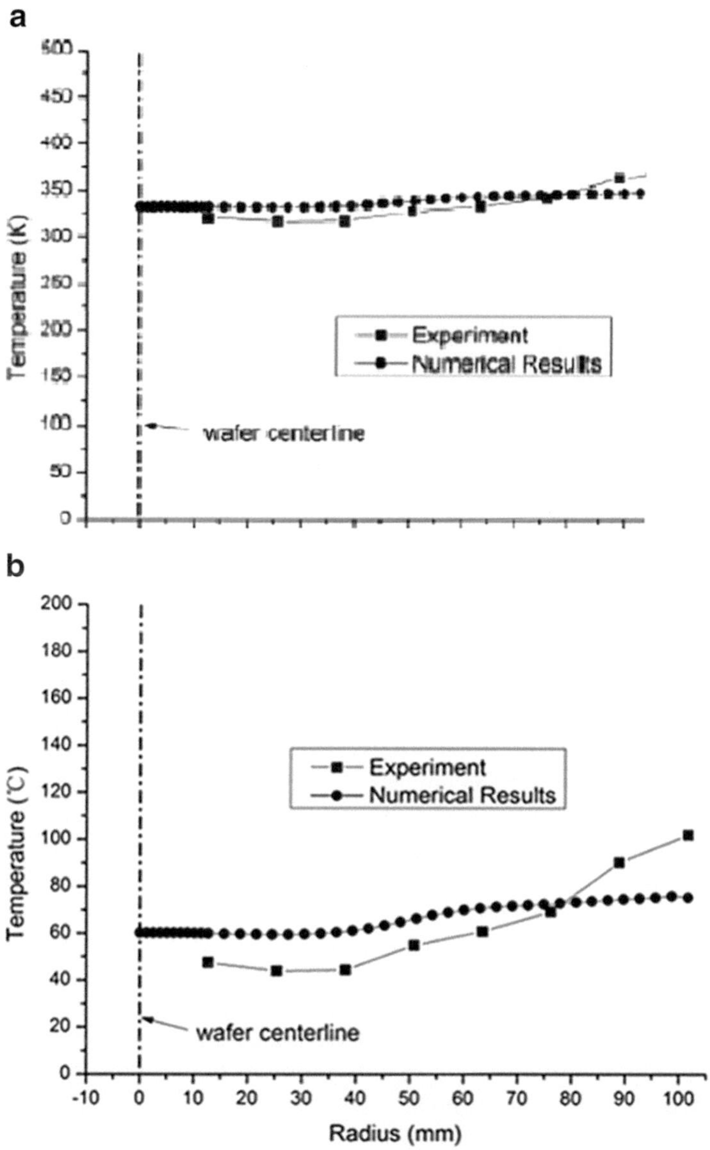

Fig. 7.18 Comparison between simulation results and experimental data for two operating conditions. Fluid temperature is measured at the height of 0.8 mm above the wafer surface

case with different operating conditions. A good agreement between the experimental and numerical results is observed. As seen here, the trends are similar, but better temperature uniformity is observed from the numerical results due to the assumed heat flux input. In practice, it is difficult to maintain the desired temperature or uniform heat flux condition at the susceptor. Therefore, measured temperature distribution at the susceptor is often used in modeling and design, rather than solving the conjugate problem for given heat inputs.

7.7 GaN Deposition

Though the flow and temperature distributions give qualitative information about the deposition process, it is necessary to link the simulation with the chemistry model in order to obtain deposition rates, from which the average and local film growth may be determined. Steady-state conditions are generally assumed, and the thickness of the film after a specified time is obtained with the growth rate taken as constant with time. Though only a few characteristics results are presented here, the references given earlier may be consulted for further details and additional results. Similarly, the deposition of other materials has been investigated in various studies that may be employed for specific information regarding these processes.

Consider first the reactor shown in Fig. 7.12. Figure 7.19a shows the GaN growth rate as a function of the reactor pressure. In these simulations, all the parameters, such as susceptor temperature, inlet velocity, and concentration ratio, are fixed at typical values except for the reactor pressure, and its range of variation is 25–760 torr. According to this figure, a basic conclusion that can be drawn is that the deposition rate increases with the reactor pressure, with all other variables being fixed. The main reason for this result is that higher reactor pressure leads to more active nitrogen atoms on the wafer surface. Although the velocity is fixed, due to the ideal gas behavior, a higher pressure leads to a higher density, followed by a higher mass flow rate, which provides more precursor molecules in the reactor. The effect of the reactor inlet velocity on average GaN growth rate is shown in Fig. 7.19b. The GaN growth rate is found to increase with the inlet velocity, again because larger mass of reacting gases enters the chamber. Similarly, the effects of other variables such as susceptor temperature and concentration ratio were investigated by Meng and Jaluria (2013). The effect on film nonuniformity was also studied. Overall, the effects of concentration ratio, reactor pressure, and mass inlet flow were found to be most significant. Several conditions that led to higher deposition rate, such as higher pressure, generally gave rise to greater nonuniformity in film thickness, thus leading to an optimization problem, as discussed in Chap. 10.

Figure 7.20 shows the calculated results for the deposition of GaN on the rotating susceptor, with a flow channel to guide the downward flow of precursor and carrier gases, shown in Fig. 7.15a. The reference conditions are taken as reactor pressure of 100 torr, wafer temperature of 1000 K, inlet velocity of 6 m/s, rotational speed of 600 rpm, and concentration ratio (V/III) of 2000. For varying a given condition, the others are held constant at these reference values. Buoyancy effects give rise to a vertically rising flow that opposes the downward flow, resulting in the reacting gases not reaching the surface and lower deposition near the center. The flow arising in the reactor was discussed earlier. Figure 7.20 shows the effect of susceptor rotation on deposition. A higher rotational speed tends to make the deposition more uniform and

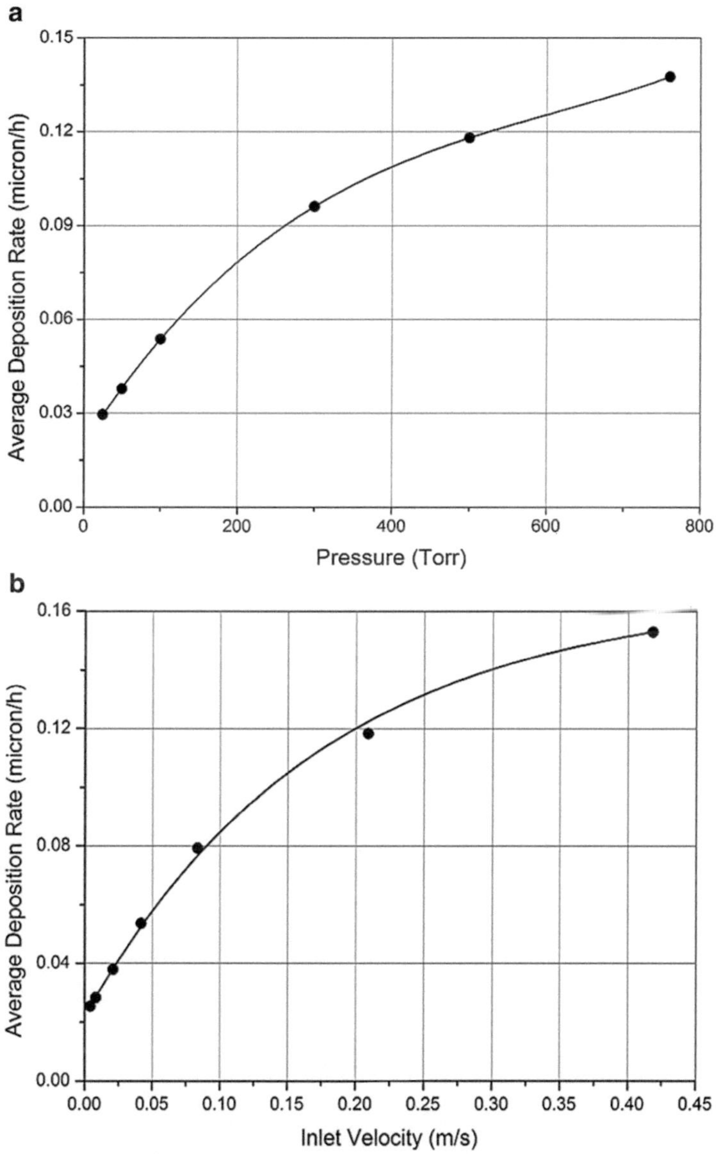

Fig. 7.19 The effect of the (**a**) reactor pressure and (**b**) inlet velocity, on the average growth rate of GaN in the reactor shown in Fig. 7.12. The susceptor temperature is maintained at 1073 K, and ratio of volumetric flow rate of NH_3 to that of TMGa is fixed at 8064

also increases the rate of deposition. The effects of other parameters such as inlet velocity, susceptor temperature, and the inlet composition of gases have been investigated in detail. Figure 7.21 shows some of the characteristic results. Again, the concentration ratio, inlet mass flow rate, and reactor pressure are the dominant

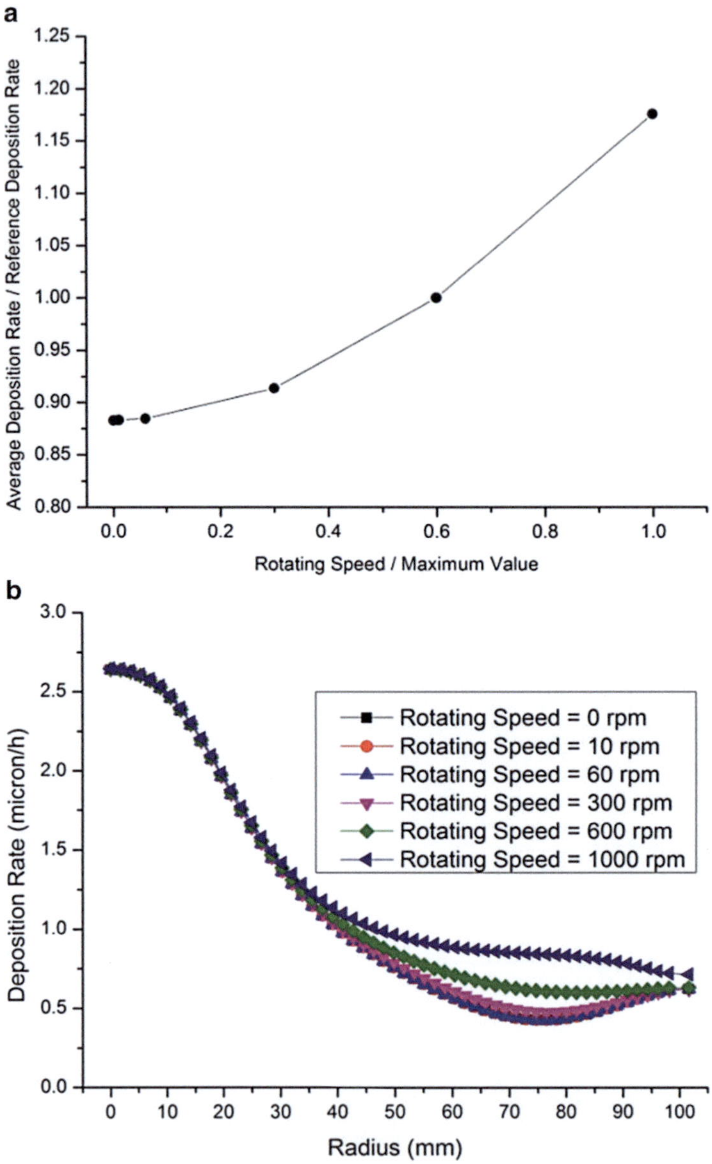

Fig. 7.20 Dependence of (**a**) the average rate of deposition and (**b**) the radial distribution of the deposition rate, on the rotational speed for a vertical CVD reactor with a rotating susceptor

variables. The susceptor temperature range is largely limited by the material to be deposited and thus by the chemistry. References, such as Kadinski et al. (2004), Meng et al. (2015), Wong and Jaluria (2016), and others given earlier may be consulted for further information on the deposition of GaN, which is considered

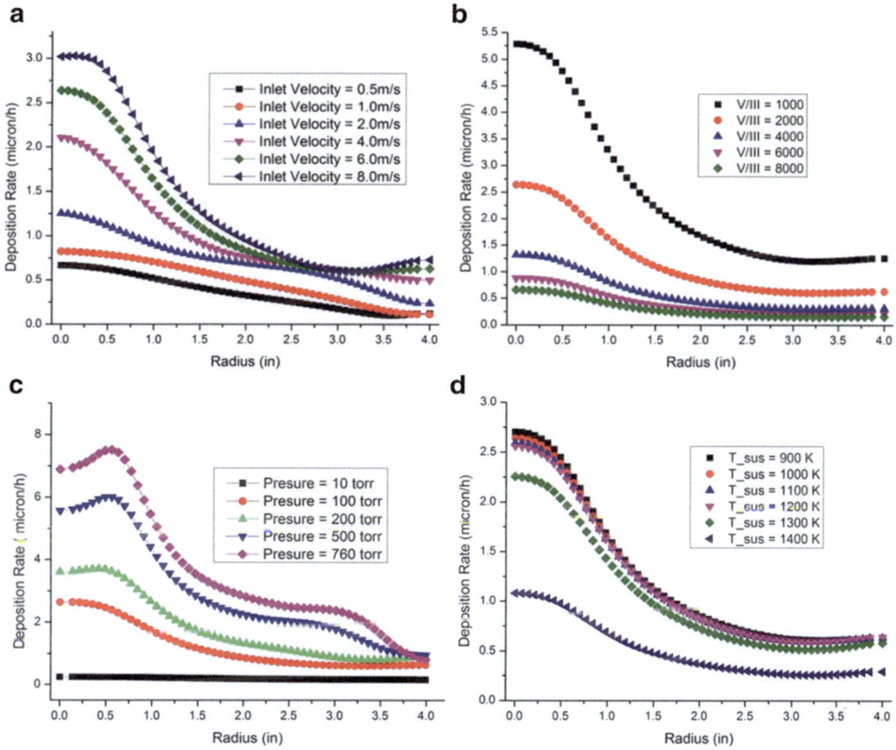

Fig. 7.21 Radial distribution of the deposition rate for variations in (**a**) inlet velocity; (**b**) inlet concentration; (**c**) pressure; and (**d**) susceptor temperature

here. Other materials like gallium arsenide, silicon carbide, and graphene have also been investigated, and the earlier references, as well as others in the literature, may be employed for relevant details.

7.8 Transient Effects

In the preceding sections, the main aspects involved in the fabrication of thin films by chemical vapor deposition have been discussed. Even though these processes are time dependent due to the growth of the film and the initiation and termination of the application, most of the studies referenced earlier have focused on steady-state conditions for the simulation in both two- and three-dimensional cases. Experiments have obviously involved transient effects, but the results presented and the focus have been on the steady-state situation. These studies thus lacked the capability of dealing with film initiation and grain nucleation, as well as instability and fluctuations in operating conditions. There is reason to believe that the optimal conditions for film initiation and grain nucleation are different from those for optimum steady growth. For one single process operating in both regimes, it is necessary to concentrate the efforts on the investigation of the transition from one to the other, which can only be done by performing transient numerical simulation of the entire process. Nevertheless, not much work has been reported on time-accurate transient simulation of the entire process, including steady-state deposition and the system start-up and shutdown, in order to find effective ways to reduce the input power and reduce the loss of precursors, especially trimethylgallium (TMGa), as well as to control the production time.

Meng and Jaluria (2015) carried out a numerical study on the time-dependent process for the reactor considered earlier and shown in Fig. 7.12, using the chemistry model for the GaN MOCVD process given earlier. The transient terms were added to the governing equations, as given in earlier chapters, and the model discussed earlier for steady state was extended to transient processes. Validation was based on comparing the numerical results with the available experimental data, which are given only for steady deposition. Of particular interest were changing from one steady-state condition to another and the initiation process.

Several different scenarios exist for one complete MOCVD process, depending on the time-dependent variations employed in practice for a given system. Typically, the entire procedure consists of three stages: system start-up, steady-state deposition process, and system shutdown. During the system start-up, the reactor chamber is initially filled with hydrogen at the ambient temperature, and heat is turned on. Natural convection occurs above the wafer surface. In order to save the valuable precursors, only after the flow and temperature fields reach steady state, the first stage ends. As the inlet gas valves are turned on, the second stage starts. Then precursors as well as the carrier gas H_2 are supplied into the reactor chamber, and gas phase and surface reactions occur, gradually reaching steady-state conditions. Finally, in order to terminate the steady-state deposition process, the inlet gas valves are turned off. However, heat is still kept on initially to ensure that the remaining reacting gases are completely consumed or removed from the reactor. Once the steady state is obtained again, heat is terminated so that the gas temperature in the reactor chamber finally reaches the ambient temperature.

This typical case is considered, and the calculated time-dependent temperatures are shown in Fig. 7.22 at a few chosen locations during the process. The time intervals 0 to t_1, t_1 to t_2, and t_2 to the end correspond to the three stages of the entire procedure. The chosen locations are placed along the vertical center line of the reactor. Taking the average wafer temperature as an example, it is seen from the figure that the average wafer temperature is initially at 300 K, and then it rises to 1153 K after the heat is turned on. It takes time Δt for the system to reach steady state. At time t_1, the inlet gas valves are turned on. Time Δt_1 is taken for the system to reach the steady state again, and the average wafer temperature becomes 1228 K. At time t_2, the inlet gas valves are turned off, and the temperature rises to 1258 K at steady state before heat is turned off at time t_3. After that, it drops down to the initial temperature of 300 K. The values of Δt and Δt_1 are obtained as 6590 s and 13,356 s, respectively, in this example. In order to ensure the steady state to be reached after

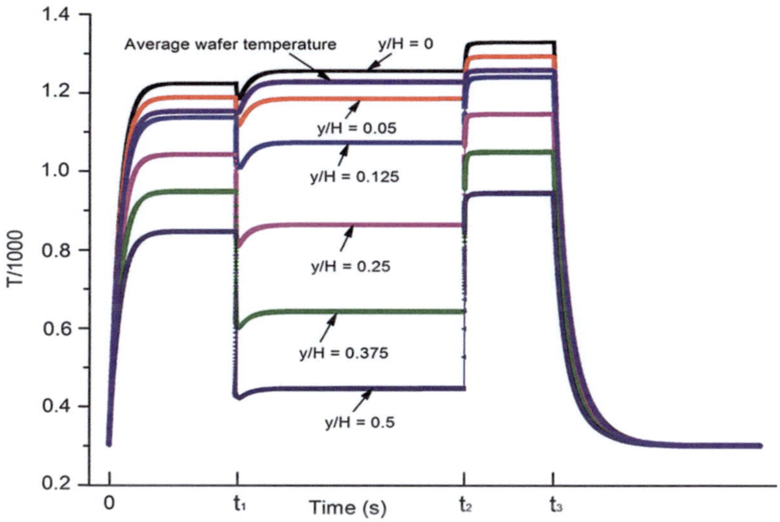

Fig. 7.22 Temperature versus time at selected locations during the entire MOCVD process for GaN deposition

each action of system operation, such as turning on or off the heat, etc., the values of t_1, t_2, and t_3 can be chosen according to the values of Δt and Δt_1.

It is seen that, when the inlet gases start entering the reactor chamber, the temperatures at all the chosen locations first decrease sharply and then gradually approach steady state. This is due to gases at 300 K entering the reactor chamber and arriving in the vicinity of the chosen locations, which leads to abrupt temperature decline. As the flow field approaches steady state, the temperatures rise slowly and reach the constant steady-state values. Similarly, at time t_2, when the inlet gas valves are closed, the temperatures at all the chosen locations rapidly rise to higher temperatures. It is a different gas composition, from the initial H_2, that affects the steady-state temperature.

The calculated mass fraction profiles of the valuable precursor TMGa at some typical times along the horizontal and vertical center lines are shown in Fig. 7.23. The times t_4, t_5, t_6, and t_7 lie, respectively, in the steady-state portions of the three regions and in the region after heat is turned off. The curve at t_5 in Fig. 7.23a shows a uniform mass fraction distribution of TMGa near the vertical center line and lower values near the reactor outlet, due to the consumption of TMGa in the reactions. As shown by the curve at t_5 in Fig. 7.23b, the mass fraction of TMGa stays constant near the reactor inlet since gas phase reactions do not occur there due to the low temperature (around 300 K). As the temperature rises along the vertical center line from top to bottom, gas phase reactions occur gradually, and the mass fraction of TMGa decreases steadily. However, one peak value is observed on this curve, due to natural convection and recirculation. When the inlet gas valves are turned off, all the

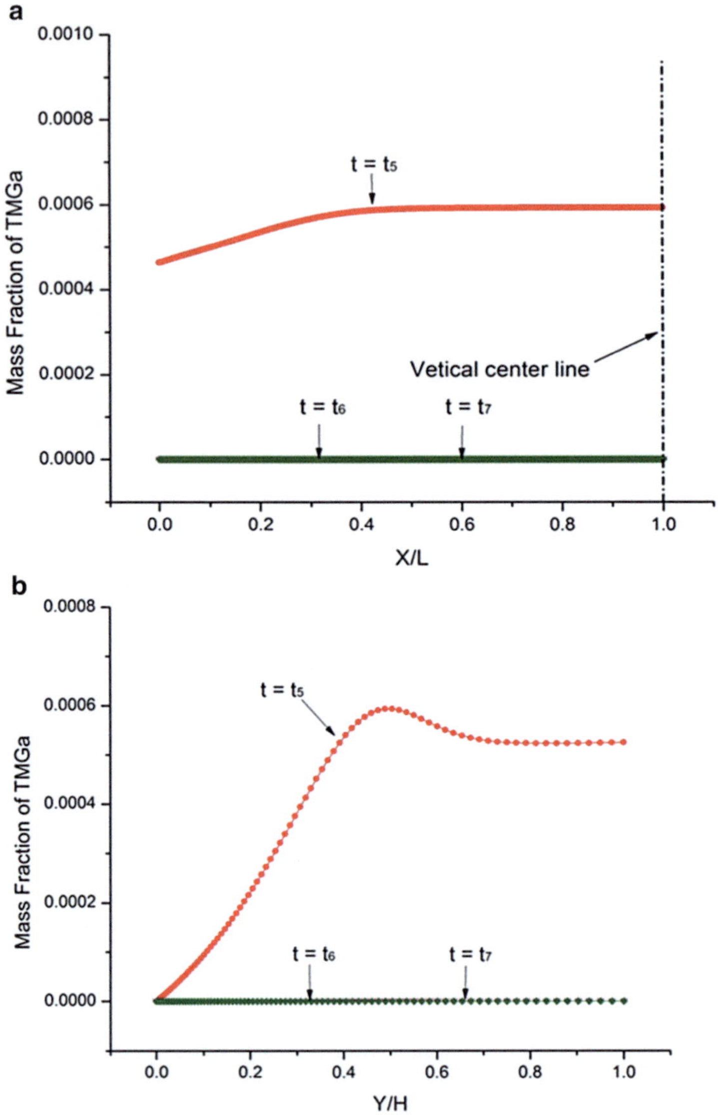

Fig. 7.23 Mass fraction profiles of TMGa at given times along (**a**) the horizontal center line; (**b**) the vertical center line

remaining reactants are eventually completely consumed due to chemical reactions. This feature is also observed from the curves at t_6 and t_7 in these figures.

Some of the corresponding temperature results are shown in Fig. 7.24a.No matter what time is considered, the highest temperature occurs in the center of the wafer due to a constant heat flux applied to the heating surface. The curve at t_5 is much flatter than those at t_4 and t_6, since gas flow pushes the reactants of higher temperature to

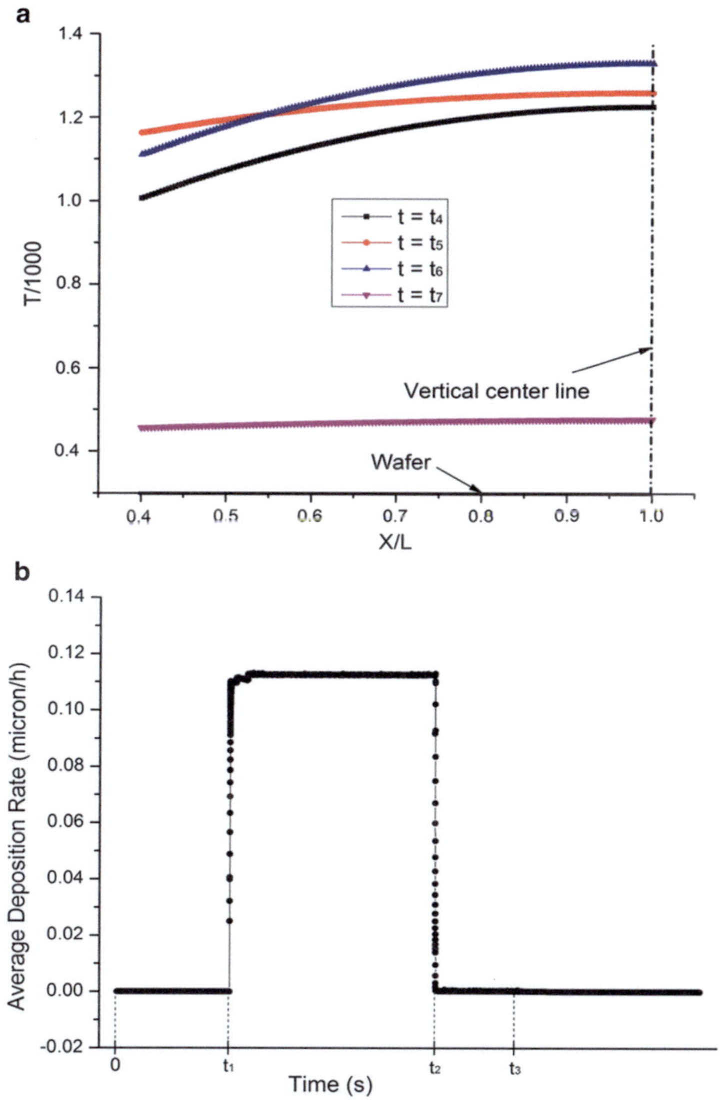

Fig. 7.24 (**a**) Temperature profiles at given times on the wafer; (**b**) Average deposition rate versus time during the entire MOCVD process

the wafer edge, leading to a more uniform temperature on the wafer. The curve at t_7 is also relatively flat because the heating has been turned off, causing the temperature within the entire domain to drop sharply. All these concentration, temperature, and flow results allow one to study the transient behavior within the reactor. For this particular time-dependent scenario used for the GaN deposition process, the transient deposition rate is shown in Fig. 7.24b. Before time t_1, the average deposition rate is zero since no reacting gases are introduced into the reactor chamber. After the inlet gas valves are turned on, both gas phase and surface reactions occur inside the reactor, and consequently the average deposition rate increases rapidly before approaching a constant steady-state condition. A sharp drop occurs at time t_2, when the inlet gas valves are turned off. After that, very little deposition is observed due to the reactions between the remaining reacting gases. The deposition rate thus decreases and eventually drops back to zero. After time t_2, only a short time is required for this level to become negligible. This short interval, i.e., the response time for estimation of the appropriate time interval t_2–t_3 is necessary in regard to the aim of reducing the input power without wasting the valuable reactant gases.

Clearly, transient effects are important with respect to the overall efficiency and cost. Different time-dependent sequence of operations are used for different applications, with the one described here being one of the common ones used in practice. A detailed discussion is given to indicate the basic trends and response times. The transient solutions can also be used to determine instabilities in the transport processes that may lead to deterioration of the quality of the deposited film. Fluctuations in the operating conditions may also be considered and the results used in system design.

7.9 Characteristics of Deposited Thin Films

A critical aspect in the fabrication of thin films by chemical vapor deposition, as is the case for all micro/nanoscale fabrication, is the quality and characteristics of the final product. For thin films, this refers to the uniformity of deposition and thickness of the film, as well as the deposition characteristics as seen by imaging of the surface. The numerical model yields the deposition rate as a function of location and the average deposition rate. These aspects can be linked with the system design and operating conditions to optimize the process to achieve high and uniform deposition rates. Many parameters that lead to high deposition rates also result in greater nonuniformity. Some of these considerations have been mentioned earlier and are discussed later in terms of optimization of the process.

Several experimental techniques are available to study the uniformity of film thickness and the deposition characteristics. These include techniques such as:

X-ray diffraction (XRD), which can be used to study the crystalline structure and film thickness

Scanning electron microscope (SEM), which is extensively used for imaging to study surface morphology and microscale structures

Transmission electron microscope (TEM), which gives finer details by imaging of the deposited film

Atomic force micrometer (AFM), which can be used to study the film characteristics by probing over the surface

The results obtained from various samples can be compared with the numerical predictions and linked with the operating conditions to choose appropriate conditions for desired results.

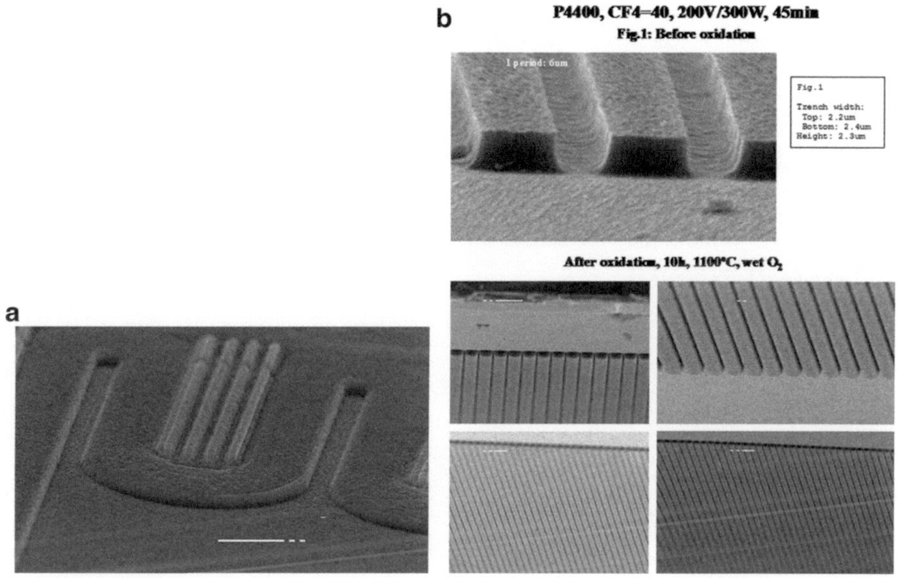

Fig. 7.25 Deposition of silicon carbide to obtain desired surface features and structures

Figure 7.25 shows the results of imaging of silicon carbide deposition. These images can be used to study the important characteristics of the deposition process, including uniformity, local variations, and microscale structures that may be needed in the product. The dimensions can be determined by including an item of known dimensions in the scan. Such images are important in determining if the desired quality and characteristics are achieved and how changes in operating conditions affect the results obtained.

Figure 7.26 shows the results on uniformity by taking samples at different susceptor locations in the CVD reactor. The average as well as the standard deviation of the deposition achieved may be determined in order to see if the deposition variation with the location of the wafer in the reactor is acceptable. If not, adjustments can be made in the inlet flow distribution and surface temperature to achieve better uniformity. Locations, such as those near the center, generally do not yield the desired uniformity and are avoided. Many of these aspects are considered in detail for practical reactors by Wong (2017). Several samples were taken on GaN deposition on a vertical, impinging, rotating susceptor reactor. Numerical predictions and experimental results were compared. Figure 7.27 shows one such result. The experimental average thickness was obtained as 6.399 μm, and the numerical predictions yielded 6.0 μm, indicating good agreement. Similarly, several other samples were taken, and numerical predictions were found to be close to experimental observations. With validation of the models for the flow and heat transfer and of the chemistry, one can proceed to a detailed study to determine the effects of design parameters and operating conditions, leading to optimization to obtain the high-quality films at low cost.

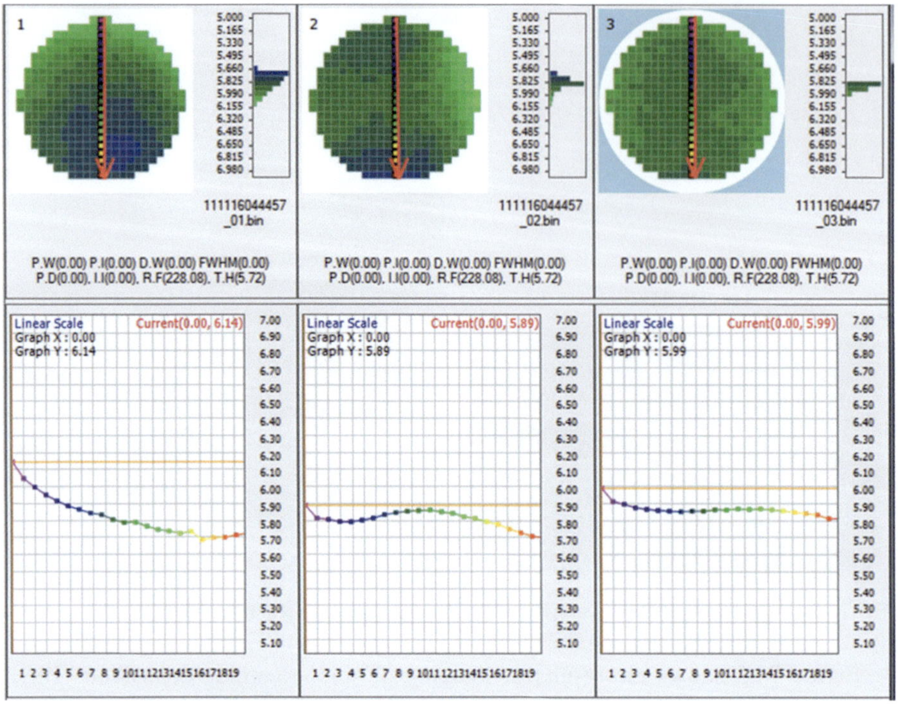

Fig. 7.26 Measurement of uniformity in GaN deposition by taking samples at different locations

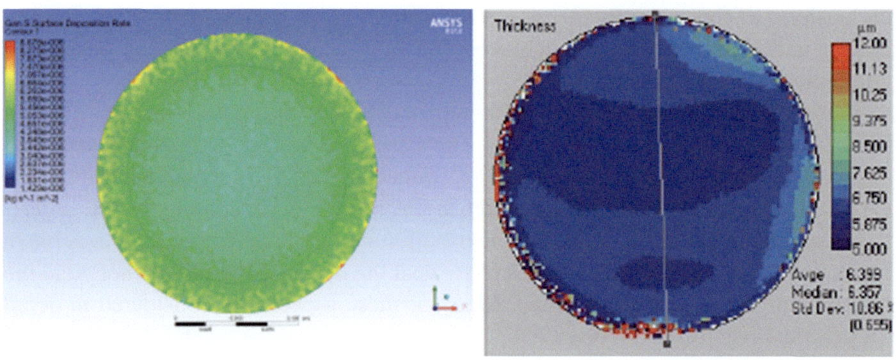

Fig. 7.27 GaN deposition contours and thickness contours obtained experimentally for a sample obtained from a vertical, impingement, rotating susceptor reactor

7.10 Summary

This chapter presents the basic considerations in micro/nanoscale fabrication. The discussion is focused on the chemical vapor deposition process for the fabrication of thin films for materials such as silicon, titanium nitride, and gallium nitride. The different types of reactors, modeling of the flow, heat and mass transfer, and chemical reactions are considered in detail. Numerical and experimental results are presented for a variety of conditions and geometries. Of particular interest are the deposition rate and the quality of the deposited film.

The chemistry is obviously a critical component in these processes, and, though the chemical pathways, chemical kinetics, and properties are known to a fair degree of accuracy for some important materials, the information available for other materials is often quite limited. Similarly, the modeling of the transport processes without the chemical reactions is relatively straight forward, but the inclusion of chemical reactions generally makes the simulation much more complicated and difficult. Experimental data are available for materials like silicon due to their extensive use, but the data for the wide variety of materials of current and future interest are somewhat limited.

Though the focus is on CVD systems, similar considerations arise in other fabrication techniques, such as flame synthesis of various materials (Memon et al. 2011). The basic considerations of quality versus production rate are obviously common to all these processes. The link between the product characteristics and the operating conditions is an important one, and future effort is very much needed in this direction. The approaches presented here may be extended, with suitable modifications, to other processes and applications.

References

A. Acharya, Group III – Nitride semiconductors: Preeminent materials for modern electronic and optoelectronic applications. Himal. Phys. **5**, 22–26 (2014)

H. Amano, M. Kito, K. Hiramatsu, I. Akasaki, P-type conduction in Mg-doped GaN treated with low energy electron beam irradiation (LEEBI). Jpn. J. Appl. Phys. **28**, L2112–L2114 (1989)

M. Amorim, R. Savio, M. Massi, H. Santiago, Applications of SiC-based thin films in electronic and MEMs devices. Phys. Tech. SiC Dev. **1**, 313–336 (2012)

M.J. Buiting, A.F. Otterloo, A.H. Montree, Kinetic aspects of the LPCVD of titanium nitride from titanium tetrachloride and ammonia. J. Electrochem. Soc. **138**, 500–505 (1991)

M.W. Chase, *NIST-JANAF Thermochemical Tables*, 4th edn. (National Institute of Standards and Technology, Gaithersburg, 1998)

S. Chattopadhyay, A. Ganguly, K. Chen, L. Chen, One-dimensional group III-nitrides: Growth, properties, and applications in nanosensing and nano-optoelectronics. Crit. Rev. Solid State Mat. Sci. **34**, 224–279 (2009)

C.H. Chen, H. Liu, D. Steigerwald, W. Imler, C.P. Kuo, M.G. Craford, M. Ludowise, S. Lester, J. Amano, A study of parasitic reactions between NH_3 and TMGa or TMAl. J. Electr. Mat. **25**, 1004–1008 (1996)

W.K.S. Chiu, Y. Jaluria, Continuous chemical vapor deposition processing with a moving finite thickness susceptor. J. Mater. Res. **15**(2000), 317–328 (2000)

W.K.S. Chiu, C.J. Richards, Y. Jaluria, Experimental and numerical study of conjugate heat transfer in a horizontal channel heated from below. J. Heat Transf. **123**, 688–697 (2001)

W.K.S. Chiu, Y. Jaluria, N.G. Glumac, Control of thin film growth in chemical vapor deposition manufacturing system. J. Mfg. Sci. Eng. **124**, 715–724 (2002)

W.A.P. Claasen, J. Bloem, W.G.J.N. Valkenburg, C.H.J. van den Brekel, The deposition of silicon from silane in a low-pressure hot-wall system. J. Cryst. Growth. **57**, 259–266 (1982)

F.C. Eversteyn, P.J.W. Severin, C.H.J. Brekel, H.L. Peek, A stagnant layer model for the epitaxial growth of silicon from silane in a horizontal reactor. J. Electrochem. Soc. **117**, 925–931 (1970)

D.I. Fotiadis, M. Boekholt, K.F. Jensen, W. Richter, Flow and heat transfer in CVD reactors: Comparison of Raman temperature measurements and finite element model predictions. J. Cryst. Growth. **100**, 577–599 (1990)

P. George, Y. Jaluria, H.C. Gea, Optimization of the chemical vapor deposition process for the fabrication of thin films, in *Proceedings of the 13th International Heat Transfer Conference*, Sydney, 2006

P. George, J. Meng, Y. Jaluria, Optimization of gallium nitride metalorganic chemical vapor deposition process. J. Heat Transf. **137**, 061007–1–061007–8 (2015)

C.K. Hu, C.J. Chen, T.C. Wei, T.T. Li, C.C. Wang, C.Y. Huang, Y.J. Lin, A simplified and universal mechanism model for prediction of gallium nitride thin film growth through numerical analysis. Int. J. New Tech. Res. **2**, 7–15 (2016)

C.K. Hu, C.J. Chen, T.C. Wei, T.T. Li, C.C. Wang, C.Y. Huang, Investigation of a simplified mechanism model for prediction of gallium nitride thin film growth through numerical analysis. Coatings **7**, 43 (2017)

K.F. Jensen, D.B. Graves, Modeling and analysis of low pressure CVD reactors. J. Electrochem. Soc. **130**, 1950–1957 (1983)

K.F. Jensen, E.O. Einset, D.I. Fotiadis, Flow phenomena in chemical vapor deposition of thin films. Ann. Rev. Fluid Mech. **23**, 197–232 (1991)

J. Justice, A. Kadiyala, J. Dawson, D. Korakakis, Group III-nitride based electronic and optoelectronic integrated circuits for smart lighting applications. MRS Proc. **1492**, 123–128 (2013)

L. Kadinski, V. Merai, A. Parekh, J. Ramer, E.A. Armour, R. Stall, A. Gurary, A. Galyukov, Y. Makarov, Computational analysis of GaN/InGaN deposition in MOCVD vertical rotating disk reactors. J. Cryst. Growth **261**, 175–181 (2004)

K.C. Karki, P.S. Sathyamurthy, S.V. Patankar, Three-dimensional mixed convection in a chemical vapor deposition reactor. J. Heat Transf. **115**, 803–806 (1993)

P.T. Lin, H.C. Gea, Y. Jaluria, Parametric modeling and optimization of chemical vapor deposition process. J. Manuf. Sci. Eng **131**, 011011–1–011011–7 (2009)

P.T. Lin, H.C. Gea, Y. Jaluria, Systematic strategy for modeling and optimization of thermal systems with design uncertainties, Front. Heat Mass Transf. **1**, 013003–1–20 (2010)

R.L. Mahajan, Transport phenomena in chemical vapor-deposition systems. Adv. Heat Transf. **28**, 339–425 (1996)

R.L. Mahajan, C. Wei, Buoyancy, Soret, Dufour and variable property effects in silicon epitaxy. J. Heat Transf. **113**, 688–695 (1991)

S. Mazumder, S.A. Lowry, The importance of predicting rate limited growth for accurate modeling of commercial MOCVD reactors. J. Cryst. Growth **224**, 165–174 (2001)

N.K. Memon, S.D. Tse, J. Al-Sharab, H. Yamaguchi, A.B. Goncalves, B.H. Kear, Y. Jaluria, E.Y. Andrei, M. Chhowalla, Flame synthesis of graphene films in open-environments. Carbon **49**, 5064–5070 (2011)

J. Meng, Y. Jaluria, Numerical simulation of GaN growth in a metalorganic chemical vapor deposition process. J. Manu. Sci. Eng **135**, 061013–1–061013–7 (2013)

J. Meng, Y. Jaluria, Transient behavior of thin film deposition: coupling micro and macroscale transport, Numer. Heat Transf. **68**, 355–368 (2015)

J. Meng, S. Wong, Y. Jaluria, Fabrication of gallium nitride films in a chemical vapor deposition reactor. J. Ther. Sci. Eng. Appl. **7**, 021003–1–021003–9 (2015)

S. Mokkapati, C. Jagadish, III-V compound SC for optoelectronic devices. Mater. Today **12**, 22–32 (2009)

D. Sengupta, Does the ring compound $[(CH_3)_2GaNH_2]_3$ form during MOVPE of gallium nitride? Investigations via density functional and reaction rate theories. J. Phys. Chem. B **107**, 291–297 (2003)

D. Sengupta, S. Mazumder, W. Kuykendall, S.A. Lowry, Combined ab initio quantum chemistry and computational fluid dynamic calculations for prediction of gallium nitride growth. J. Cryst. Growth **279**, 369–382 (2005)

D. Steigerwald, S. Rudaz, H. Liu, R.S. Kern, W. Götz, R. Fletcher, III–V nitride semiconductors for high-performance blue and green light-emitting devices. JOM **49**, 18–23 (1997)

C. Theodorpoulos, T.J. Mountziaris, H.K. Moffat, J. Han, Design of gas inlets for the growth of gallium nitride by metalorganic vapor phase epitaxy. J. Cryst. Growth **217**, 65–81 (2000)

A. Thon, T.F. Kuech, High temperature adduct formation of trimethylgallium and ammonia. Appl. Phys. Lett. **69**, 55–57 (1996)

R. Wang, R. Ma, M. Zupan, Modeling of chemical vapor deposition of large-area silicon carbide thin film. J. Crys. Growth Des. **6**, 2592–2597 (2006)

S. Wong, Experimental and computational study of GaN chemical vapor deposition reactor. Ph.D. Thesis, Rutgers University, New Brunswick, 2017

S. Wong, Y. Jaluria, Numerical simulation of a practical chemical vapor deposition reactor. Numer. Heat Transf. **70**, 1057–1071 (2016)

B. Wu, R. Ma, H. Zhang, Epitaxy growth kinetics of GaN films. J. Cryst. Growth **250**, 14–21 (2003)

H. Yoo, Y. Jaluria, Thermal aspects in the continuous chemical vapor deposition of silicon. J. Heat Transf. **124**, 938–946 (2002)

Chapter 8
Manufacture of Optical Fibers: Drawing and Coating Processes

This chapter focuses on the manufacture of optical fibers, mainly because of the extensive use of optical fibers in a wide range of applications, from fiber-optic communications to sensors, lasers, and diagnostics. The fabrication of optical fibers involves many processes that are of interest in other manufacturing systems. These processes include fiber drawing, cooling, and coating, which are of interest in processes such as wire drawing, polymer fiber spinning, polymer coating, glass and ceramic processing, continuous processing, and cable manufacture. Also, a variety of optical fibers, such as hollow, doped, and other specialty fibers, have been developed for different applications. The fabrication of such fibers brings in several additional considerations that are similar to those in other manufacturing processes with respect to geometry, surface tension effects, and multilayer and multiscale considerations. Defects generated during the draw process and during coating are also of interest in other processes. The treatment of glass for the draw process provides insight into other processes in the manufacture of glass and other ceramic products. Thus, this chapter, though mainly directed at the fabrication of optical fibers, provides important basic information for a much wider range of manufacturing processes.

8.1 Optical Fiber Systems

Optical fibers, which have broad bandwidth and low attenuation, are extensively used in a wide range of applications, the most common being telecommunications, data transmission, medicine, military, and automotive industry. Recent years have seen an explosive growth in the development and application of fiber-optic networks and systems, allowing high-speed transmission of large amounts of data over short and large distances, with transmission loss as low as 0.2 dB/km and even smaller using doped silica (silicon dioxide SiO_2) glasses. Due to different applications and the need to improve performance in the face of global competition, a range of

Y. Jaluria, *Advanced Materials Processing and Manufacturing*, Mechanical Engineering Series, https://doi.org/10.1007/978-3-319-76983-7_8

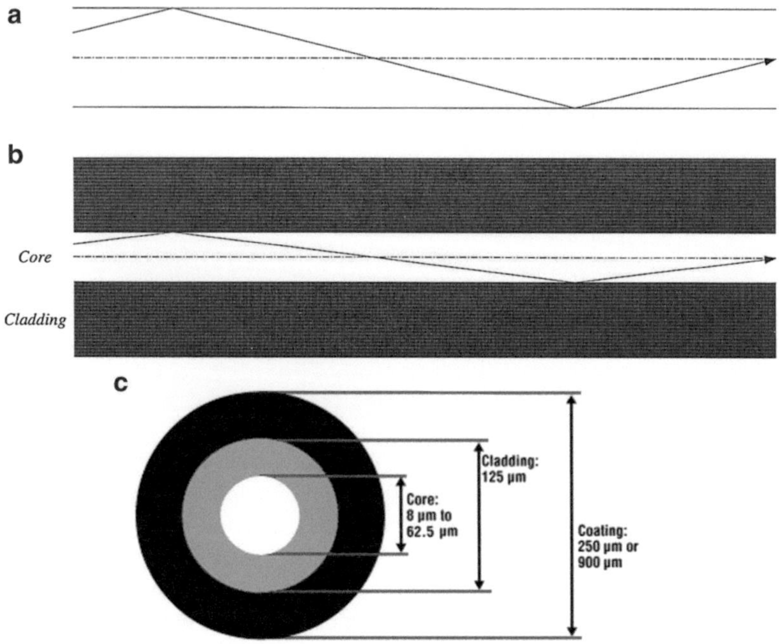

Fig. 8.1 Sketches of optical fibers: (**a**) "unclad" fiber, (**b**) weakly guiding fiber with core and cladding, (**c**) fiber with coating

materials, dopants, fabrication techniques, and fiber configurations have been developed (Izawa and Sudo 1987; Murata 1996; Hecht 2002). Silica glass is the most common material for optical fibers due to its high purity and consequent low loss due to absorption and scattering. However, other glasses, such as fluoride glass and phosphate glass, as well as plastics, are also used to fabricate optical fibers for different applications. Though much of the existing literature deals with the characterization and implementation of optical fibers, the information on the fabrication is somewhat limited (Li 1985; Paek 1986, 1999).

A single-layer optical fiber, as shown in Fig. 8.1a, is called an *unclad* fiber. It is not used for telecommunication purposes because of the excessive optical loss. In practice, an optical fiber usually consists of three components: core, cladding, and coating, as shown in Fig. 8.1b, c. In this multilayer structure, a dielectric layer, known as the *cladding*, surrounds the light-guiding central portion, known as the *core*. In order to confine light to the core by total internal reflection, the core has a higher refractive index than that of the cladding. For weakly guiding fibers, the relative refractive index difference between the core and the cladding is less than 1%. The profile of the refractive index in the core can be uniform, called *step index*, or gradual, called *graded index*, as shown in Fig. 8.2

Optical fibers used for telecommunications are usually made from pure or doped silica glass because of its low transmission loss. The refractive index profiles in the core and cladding can be changed by adding small amounts of chemicals called

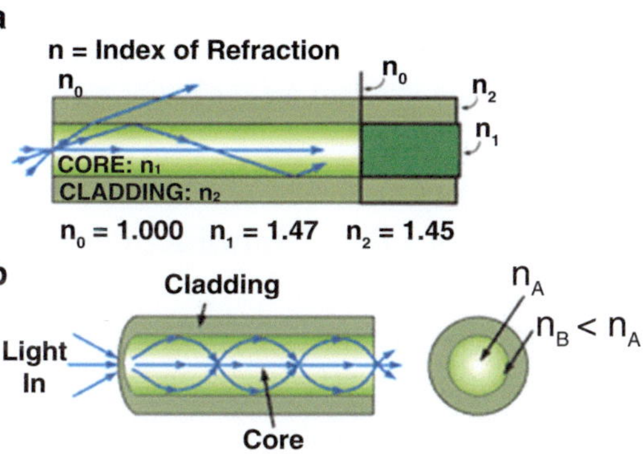

Fig. 8.2 Structure of a typical optical fiber and the transmission of light in a fiber with (**a**) step change in refractive index, (**b**) gradual variation in refractive index

Table 8.1 Typical dimensions of optical fibers

Solid optical fiber core
Single mode: 5–10 μm
Multimode: 50–100 μm
Cladding
Diameter: 125 μm
Coating thickness
Primary coating: 30–50 μm
Secondary coating: 50–100 μm
Hollow fibers
Hole diameter: 10–50 μm
Multi-structured: Hole diameter 5–10 μm

dopants. For example, GeO_2, P_2O_5, TiO_2, and Al_2O_3 are often used to increase the refractive index, whereas F and B_2O_3 are used to reduce the refractive index (Murata 1996). Figure 8.1c shows the typical sizes of the core, the cladding, and the coating. Thus, core sizes of silica fibers vary from 8.3 to 62.5 μm. The single-mode fiber has a small core of diameter around 8.3 μm, which supports propagation of only one mode of light at one time. The multimode fiber can transmit more than one mode of light simultaneously and has a much larger core diameter such as 50 or 62.5 μm. The diameter of the cladding surrounding each of these cores is around 125 μm. Thus, this is the diameter of a typical uncoated fiber. The third portion of a practical optical fiber is the outer protective coating. This coating is typically an ultraviolet (UV)-cured acrylate applied during the manufacturing process. It provides physical and environmental protection to the fiber. During the installation process in a fiber-optic system, this coating is stripped away from the cladding to allow proper termination to an optical transmission system. The typical diameter of coated fibers is from 250 μm to around 500 μm. Other coating materials, including metals, are also used for different applications, such as those in harsh environments. Table 8.1 shows some of the typical dimensions used. In some automobile and medical applications, plastics, particularly polymethyl methacrylate (PMMA), are the material used for optical fibers (Hecht 2002). Compared to silica optical fibers, plastic optical fibers have a much higher transmission attenuation, being above 150 dB/km. However, plastic optical fibers can be used when the transmission distance is short and the ease of handling and low cost are the major requirements.

8.2 Optical Fiber Manufacturing

The optical fiber fabrication system involves several steps or processes, the main ones being:

1. Fabrication of the preform
2. Drawing
3. Cooling
4. Liquid coating
5. Curing

The focus here is on the drawing process, along with the cooling and coating processes. Figure 8.3 shows a sketch of the process, including these three stages that are considered separately and in detail. However, before proceeding to fiber drawing, an important initial process is the fabrication of an appropriate preform, as outlined below. Once the preform is obtained, the optical fiber is drawn in a draw tower such as those shown in Fig. 8.3 for solid and hollow fibers. The preform is fed into a high-temperature furnace and heated above the glass-softening temperature. The optical fiber is pulled or drawn from the heated end of the preform by applying

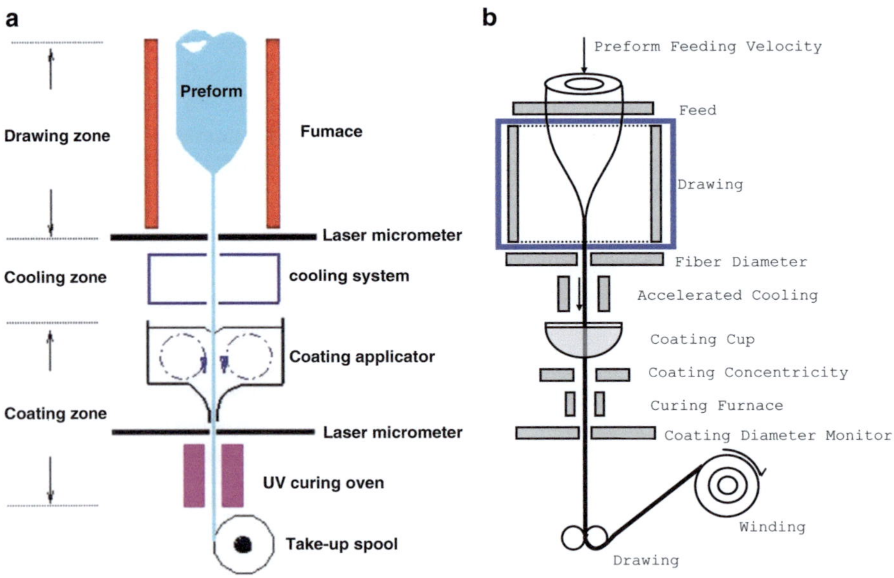

Fig. 8.3 Sketch of the optical fiber drawing process for (**a**) solid-core and (**b**) hollow fibers

axial tension to obtain optical fibers of desired diameter. After the furnace, the fiber is cooled by natural convection in air or by forced convection in an inert environment and then coated in a coating cup. The liquid coating is then solidified by curing using ultraviolet radiation. At the bottom of the draw tower, the fiber is wound on spools.

It may be noted that the drawing, cooling, and coating processes mentioned here with respect to optical fibers are encountered in many other manufacturing processes, such as wire drawing, extrusion, and cable manufacture. Several such processes are discussed elsewhere in the book. The basic aspects and concerns are often quite similar, though the requirements on quality are generally much more stringent for optical fibers.

8.3 Preform Fabrication

A brief description of the preform fabrication process is given here before discussing the different processing stages in the draw tower. Optical fibers are usually made by drawing from a cylindrical preform, which has a similar core-cladding structure as the desired fiber. As mentioned earlier, pure silica, SiO_2, is generally the main component of the preform, and dopants are added to control the refractive index profile. Three common methods for preform fabrication are modified chemical vapor deposition (MCVD), the outside vapor deposition (OVD), and the vapor-phase axial deposition (VAD) (Li 1985; Izawa and Sudo 1987). The most commonly used method is MCVD in which a hollow silica tube is placed horizontally and rotated on a glass-working lathe. Mixed vapors of metal halides along with oxygen are injected into the tube, and an oxyhydrogen flame placed outside the tube is moved up and down along the axis, heating the tube to 1500–1650 °C, see Fig. 8.4. Oxidizing reactions occur in the gas and produce numerous glass particles, which are deposited on the inner surface of the tube and sintered. The tube is eventually filled with glass layers and is collapsed into a preform. The deposited glass layers become the core and cladding of the preform, with the tube becoming the outer layer. A desired refractive index profile in the preform can be achieved by changing the composition of metal halide vapors. Thus low-loss silica fibers can be obtained from preforms fabricated by the MCVD process since it avoids water vapor, resulting in a core that is essentially free of OH molecules that cause attenuation.

OVD was the first successful process to fabricate high-silica fibers. Figure 8.5 shows a sketch of the OVD process. The glass particles are produced by flame hydrolysis, in which a vapor mixture of $SiCl_4$ and $GeCl_4$ react in an oxyhydrogen flame (Cho et al. 1998; Tandon 2005a). The glass particles are then deposited on to a rotating target rod. OVD has a high effective glass deposition rate. As in MCVD, the radial refractive index profile can be controlled layer by layer. When the deposition is complete, the target rod is removed from the porous preform, which is dehydrated and sintered into a preform rod. Doping of silica can be carried out during the sintering process as well (Tandon 2005b).

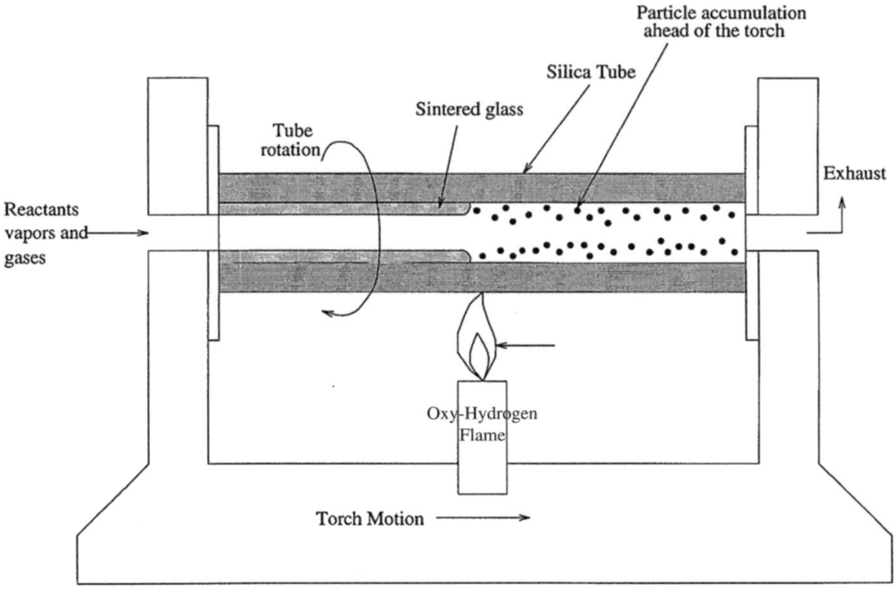

Fig. 8.4 Schematic diagram of the modified chemical vapor deposition (MCVD) process (Li 1985)

The VAD process shares many features with the OVD process. The glass particles are formed by flame hydrolysis, and the porous preform is turned into a preform rod by using consolidation and dehydration. The basic difference in the VAD process is that the glass particles are deposited axially on a seed rod. Therefore, this process has the possibility of continuously producing the porous preform, which the other methods do not. However, it is more difficult to precisely control the refractive index profile across the preform, as compared to the other methods.

8.4 Drawing of Optical Fibers

In a draw tower, the preform is traversed vertically through a high-temperature cylindrical furnace. There are two types of furnaces that are commonly used for optical fiber drawing. These are graphite resistance furnace and zirconia induction furnace. The glass is heated largely by radiation beyond its softening point T_{melt} of about 1900 K for silica and is drawn into a fiber of diameter around 125 μm by applying axial tension. Figure 8.6 shows a sketch of a graphite resistance furnace, along with inflow of inert gases like nitrogen and argon to maintain an inert environment and avoid oxidation of the graphite heating element. The preform diameter is on the order of 10 cm or larger. Thus, during the drawing process, the diameter changes substantially from the preform diameter to the optical fiber of diameter around 125 μm in a fairly short distance of about 0.3 m, which is typically

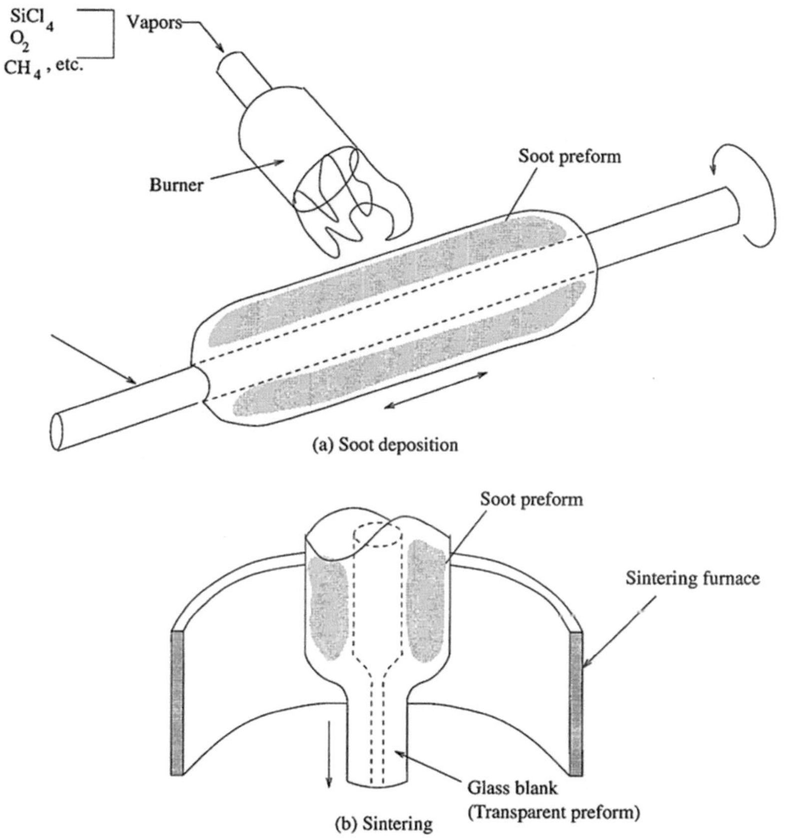

Fig. 8.5 Schematic diagram of the outside vapor deposition (OVD) process (Li 1985)

the furnace height. The region over which the preform develops into a fiber is called *neck-down* region. The neck-down profile of the preform/fiber is determined by the operating conditions such as furnace temperature, preform diameter, applied tension, and drawing speed, as well as the material and furnace design parameters.

Numerical and experimental studies on pure, single-layer, solid, silica optical fiber drawing have been carried out by a number of researchers. Initially, simple one-dimensional modeling was used to determine the profile and temperature distribution within the neck-down region (Paek and Runk 1978; Myers 1989). More comprehensive models were also developed to investigate the flow and the thermal transport in detail, including the convective and radiative heat transfer in the furnace, glass, and the inert gases (Lee and Jaluria 1997; Roy Choudhury and Jaluria 1998a; Cheng and Jaluria 2005). The fiber diameter was taken as 125 μm to represent the core and cladding of the fiber, and different preform diameters were considered. The effects of fiber draw speed, inert gas velocity, furnace dimensions, and furnace temperature on the flow, temperature distribution, and tension were

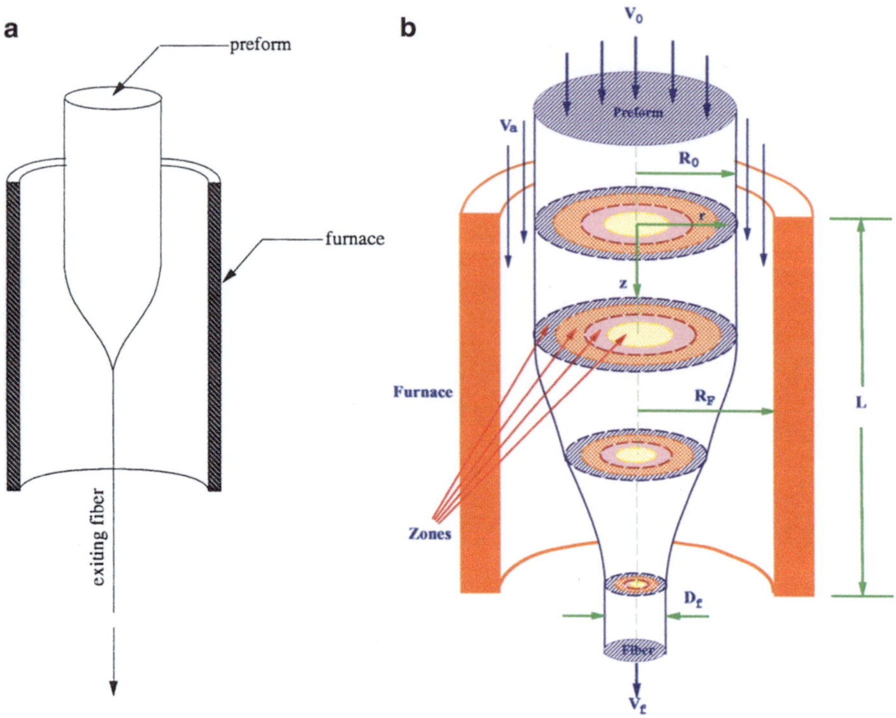

Fig. 8.6 (**a**) Sketch of an optical drawing furnace; (**b**) axisymmetric finite volume zones for the calculation of radiation in the glass preform and fiber

obtained. In initial studies, the preform was assumed to be optically thick, which is not valid, particularly in the lower neck-down region. The radiation transport was later modeled with other models such as the zonal method and the discrete ordinate method (Yin and Jaluria 1998; Wei et al. 2004). High-speed optical fiber drawing, up to around 20 m/s and higher, with preform diameters of 10 cm and larger was investigated. Interest lies in obtaining high-quality optical fibers, as indicated by low concentration of process-induced defects, desired variation of refractive index, low tension, high mechanical strength, and other important measures. A high production rate, as determined by the draw speed, is also desirable, leading to an optimization problem, as discussed later.

8.4.1 Basic Equations

The equations that may be used to model the thermal transport in optical fiber drawing are based on the equations for fluid flow and heat transfer obtained from the conservation of mass and energy and the force-momentum balance (Burmeister 1993; Jaluria 2013). These were given in Chap. 2 for a general three-dimensional, time-dependent process with variable properties. If the transport processes in the glass and in the purge gas in a cylindrical furnace are assumed to be axisymmetric, the governing equations for the glass and the gas are obtained as

$$\frac{\partial v}{\partial z}+\frac{1}{r}\frac{\partial (ru)}{\partial r}=0 \tag{8.1}$$

$$\frac{\partial v}{\partial t}+u\frac{\partial v}{\partial r}+v\frac{\partial v}{\partial z}=-\frac{1}{\rho}\frac{\partial p}{\partial z}+\frac{1}{r}\frac{\partial}{\partial r}\left[r\upsilon\left(\frac{\partial v}{\partial r}+\frac{\partial u}{\partial z}\right)\right]+2\frac{\partial}{\partial z}\left(\upsilon\frac{\partial v}{\partial z}\right) \tag{8.2}$$

$$\frac{\partial u}{\partial t}+u\frac{\partial u}{\partial r}+v\frac{\partial u}{\partial z}=-\frac{1}{\rho}\frac{\partial p}{\partial r}+\frac{2}{r}\frac{\partial}{\partial r}\left(r\upsilon\frac{\partial u}{\partial r}\right)+\frac{\partial}{\partial z}\left[\upsilon\left(\frac{\partial v}{\partial r}+\frac{\partial u}{\partial z}\right)\right]-\frac{2\upsilon u}{r^2} \tag{8.3}$$

$$\rho C_{\mathrm{p}}\left(\frac{\partial T}{\partial t}+u\frac{\partial T}{\partial r}+v\frac{\partial T}{\partial z}\right)=\frac{1}{r}\frac{\partial}{\partial r}\left(rk\frac{\partial T}{\partial r}\right)+\frac{\partial}{\partial z}\left(k\frac{\partial T}{\partial z}\right)+\mu\Phi+S_{\mathrm{r}} \tag{8.4}$$

where $\mu\Phi$ is the viscous dissipation, S_{r} is the radiation source term, and v and u are the velocity components in the axial and radial directions, respectively. The other variables were defined earlier in Chap. 2.

The radiation source term S_{r} in Eq. (8.4) is non-zero for the glass preform/fiber because glass is a participating medium, which emits, scatters, and absorbs energy. The variation of the absorption coefficient with wavelength λ can often be approximated in terms of bands with constant absorption over each band. A two- or three-band absorption coefficient distribution has been used quite effectively. One of the methods that have been used successfully to model the radiation heat transfer is the zonal model (Yin and Jaluria 1998; Modest 2013). Figure 8.6b gives a sketch of the finite zones that may be used for calculating the radiation transport by the zonal method. The radiation transport involves absorption, scattering, and emission in the glass and between the surfaces of the furnace. Much of the effort goes into determining the view factors between surface elements, as well as direct exchange areas for transport between surface and volume elements and that between volume and volume elements. For multilayered systems and for hollow fibers, the calculations become more complicated, as outlined later.

The material properties of glass vary with temperature, composition, and microstructure. The radiation properties and the refractive index are particularly affected by the dopants. Glass is a supercooled liquid at room temperature. The viscosity varies very strongly with temperature. Even a change of a few degrees in temperature in the vicinity of the softening point, T_{melt} can cause substantial changes in viscosity and thus in the flow field and the transport. An equation based on the curve fit of available data for kinematic viscosity υ is given for silica, in SI units, as (Fleming1964; Myers 1989)

$$\upsilon=4545.45\exp\left[32\left(\frac{T_{\mathrm{melt}}}{T}-1\right)\right. \tag{8.5}$$

This equation indicates the strong, exponential variation of υ with temperature T. Variations in all the other relevant properties of glass must also be considered though the variation with temperature is not as strong. Also, the momentum and energy equations are coupled, and, due to large viscosity, viscous dissipation must be included. However, glass flow may be treated as Newtonian at typical draw speeds. Similarly, the properties of the purge gas may be considered to be variable. But the variations are small over the temperature ranges encountered, and average values may be used without incurring significant error.

The neck-down region is characterized by the free surface that arises and that has to be determined in order to study the flow and heat transfer in the process. Many studies in the literature have considered free surface flows in jets, wakes, and fiber formation from the melt, under isothermal conditions (Goren and Wronski 1966; Gifford 1982). In optical fiber drawing, the free surface flow of glass has to be modeled under large temperature differences and large changes in viscosity and in the cross-sectional area. Because of the small diameter of the fiber, several studies assumed uniform temperature across the fiber and in the lower portions of the neck-down region. However, because of the high temperature dependence of viscosity, this assumption does not yield accurate results. Generally, a large number of grid points, typically around 50, are needed even across the fiber radius of around 62.5 μm to capture changes in temperature and the consequent effects on properties, flow, viscous dissipation, thermally induced defects, and dopant movement. A larger number of grid points are needed at larger diameters in the neck-down region. Overall, solution of the flow and heat transfer in fiber drawing is quite a challenging problem because of the large change in diameter over a relatively short distance and strong changes in material properties. Transformations are generally used to simplify the computational domains into cylindrical regions, and nonuniform grids are used in the various regions to obtain the flow and temperature distributions.

8.4.2 Characteristic Results

Typical numerical results in the neck-down region are shown in Fig. 8.7 in terms of the flow, temperature, and viscous dissipation for a specified neck-down profile. It is seen that the flow is smooth and well layered because of the high fluid viscosity. Typical temperature differences that arise across the fiber are on the order of 50–100 °C. Viscous dissipation is quite small compared to the total energy transport in the process. But it is mainly concentrated near the end of the neck-down region and, thus, plays an important role in maintaining the temperature of the fluid above the softening point. Therefore, the calculations must include viscous dissipation effects, as well as property changes with temperature.

The problem is considerably simplified if the neck-down profile is taken as known. However, the actual drawing process involves modeling the free surface flow of glass, and the profile must be obtained as part of the solution. A solution of the transport equations, along with a surface force balance, is needed for the determination of the neck-down profile. The normal force balance and the vertical momentum equations are used to obtain a correction scheme for the neck-down

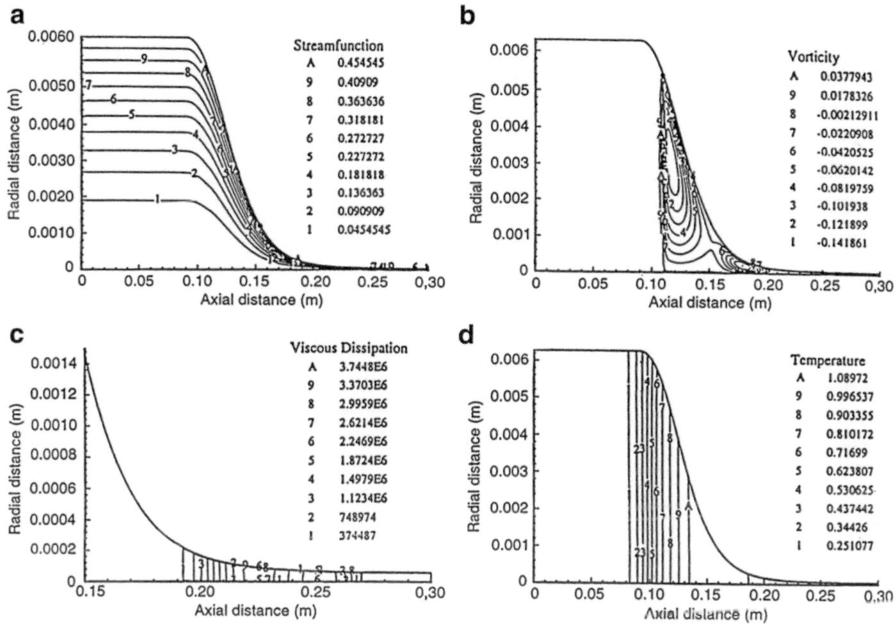

Fig. 8.7 Calculated (**a**) stream function, (**b**) vorticity, (**c**) viscous dissipation, and (**d**) temperature contours in the glass during drawing of a single-layer optical fiber for typical drawing conditions

profile. The vertical force balance equation in the neck-down region for steady flow can be written as

$$\frac{d}{dz}(\dot{m}\,v) = \frac{d}{dz}(\sigma_{zz}\pi R^2) + \rho g \pi R^2 + \frac{d}{dz}(2\pi R\xi \cos\,\theta) - 2(\pi R\tau_a \cos\,\theta) + \frac{d}{dz}(p_a \pi R^2) \tag{8.6}$$

where $\dot{m}$ is the mass flow rate, σ_{zz} the normal stress, τ the shear stress, ξ the surface tension, θ the angle at a surface location with the vertical, and R the radius of the preform/fiber.

Radially lumped axial velocity, normal force balance, and vertical momentum equation are used to obtain a correction scheme. Starting with an initial guessed profile, the profile is corrected using an iteration scheme based on the force imbalance. After a new corrected profile is obtained at each step, the full governing equations are solved for the flow and heat transfer, considering both radiation and convection transport. This process is continued until convergence is achieved, as indicated by the neck-down profile not changing significantly from one iteration to the next. Figure 8.8a shows the iterations in the numerical generation of the neck-down profile with a starting cosinusoidal shape. It is seen that, during the first few iterations, the neck-down profile is quite unrealistic, with a flat region and an abrupt change in radius near the end of the starting polynomial profile. But the shape soon

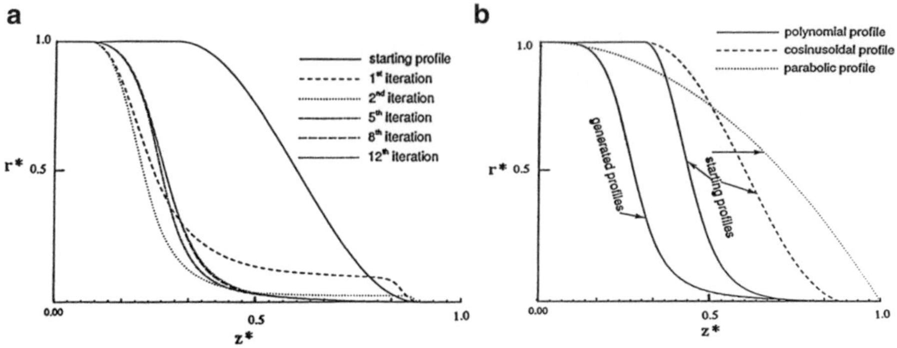

Fig. 8.8 (**a**) Iterative convergence of the neck-down profile in optical fiber drawing. Here, $r^* = r/R$ and $z^* = z/H$, where R is the preform radius and H the furnace height, and (**b**) converged profile for different starting profiles

becomes smooth and monotonically decreasing, eventually reaching a steady, converged profile. The radius approaches the dimensionless fiber radius as the axial distance approaches the furnace exit. For convergent cases, perturbations to the initial profile and different starting shapes lead to the converged neck-down profile, as seen in Fig. 8.8b, indicating the robustness of the scheme and constancy of the drawing process. The viscous and gravitational forces were found to dominate in the determination of the profile. Surface tension effects were found to be small, for the dimensions considered here, as were the external shear and inertial effects. The force balance conditions are also closely satisfied if the iterations converge (Roy Choudhury et al. 1999). However, convergence does not occur in every case for different draw speeds, furnace temperatures, etc. This implies the existence of a finite domain in which the process is feasible, as discussed later.

Experimental work on the transport in optical fiber drawing is complicated because of high temperatures, high draw speeds, and difficulty accessing the inner region of the furnace (Roy Choudhury and Jaluria 1998b). The measurements have largely focused on the characteristics of the fiber for different operating conditions. Experimental results on the draw-down shape of the preform are fairly difficult and are obtained by stopping the draw process and studying the remaining preform. Therefore, only a few results are available that may be used for comparisons with numerical predictions. A comparison with the profile experimentally obtained by Paek and Runk (1978) has been carried out, using the conditions of the experiment and the heat transfer coefficient distribution employed in the experiment. A parabolic furnace temperature profile, with the experimentally obtained maximum temperature of 3000 K and minimum temperature of 2300 K, was used for obtaining the neck-down profiles. The draw tension was also obtained from the forces acting on the fiber. A comparison of the computed results with the experimental data of Paek and Runk (1978) and Paek et al. (1988) shows good agreement, as seen in Fig. 8.9. The agreement is quite good, even though the furnace temperature profile, except for the maximum value, is guessed, and all properties for fused silica are taken from the

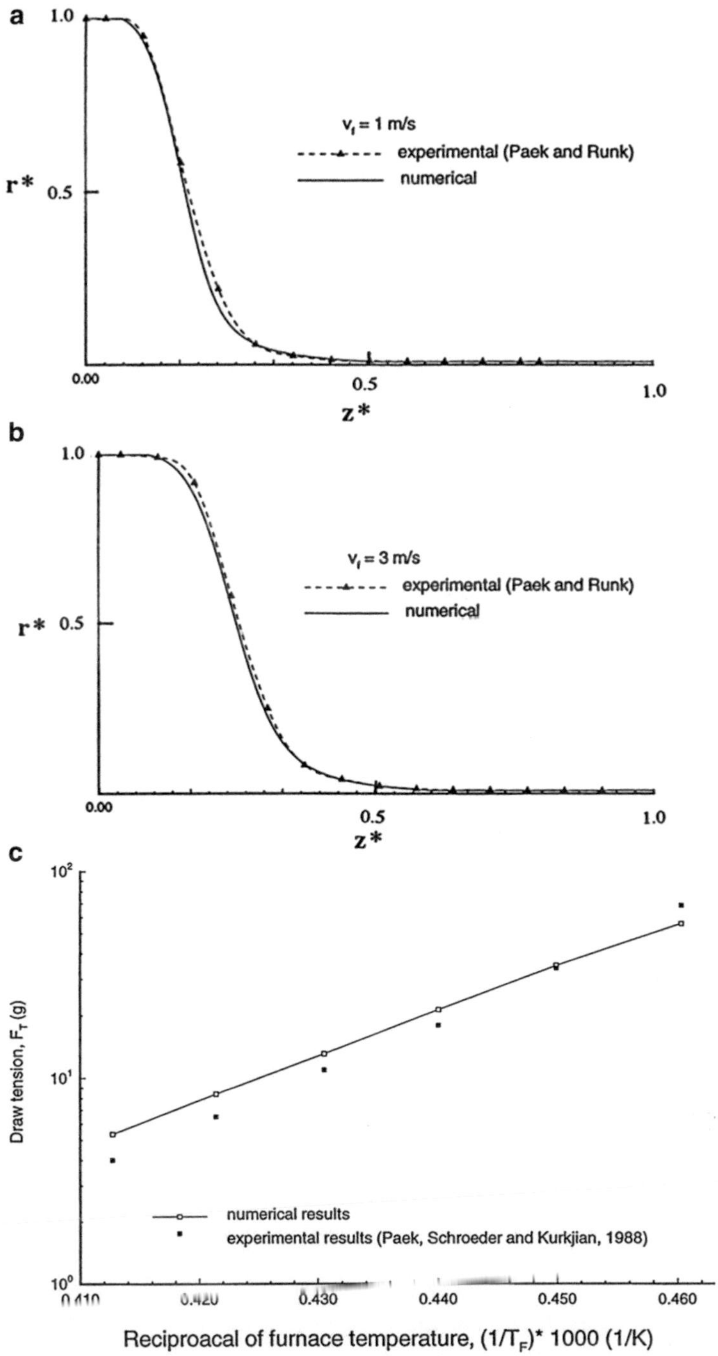

Fig. 8.9 Comparison of (**a, b**) the numerical predictions of neck-down profile and (**c**) draw tension with experimental results of Paek and Runk (1978) and Paek et al. (1988)

literature. These comparisons with experimental results lend support to the approach outlined here for determining the neck-down profile.

Figure 8.10 shows the results on the flow and temperature distributions, in terms of streamlines and isotherms, respectively, at much larger draw speeds and different furnace heights. Again, the smooth, well-layered flow is seen, with small temperature differences across the fiber. The temperature differences across the preform are obviously much higher and depend on the preform diameter as well as the operating conditions. Many such results have been obtained and presented in the literature for different inert gas flow rates and configurations, furnace dimensions, preform diameters, and other operating conditions and design parameters.

8.4.3 Multilayer Fibers and Dopants

It has been mentioned earlier that, for telecommunication purposes, the optical fiber usually consists of a core and cladding, with common core sizes in use being 8.3 μm, 50 μm, and 62.5 μm, depending on whether it is a single-mode or multimode fiber. The diameter of the cladding surrounding each of these cores is around 125 μm. The core and the cladding differ slightly in properties, particularly radiation characteristics, with the cladding at smaller refractive index and may be treated as different materials with different properties. Because of the complex shape of the computational domains, separate body-fitted grids may be applied to different layers and to the inert gas region (Jaluria 2006). Coordinate transformations, such as Landau's transformation discussed earlier, may be used to simplify the domains by converting these to cylindrical regions, as shown in Fig. 8.11 (Landau 1950). Then, each separate layer in the glass is assumed to have a uniform refractive index, bounded by diffuse surfaces. The zonal method is applied to calculate the radiation transfer inside the three enclosures (Chen and Jaluria 2007). The two-band model presented by Myers (1989) and given below may be used for the absorption coefficient a of pure silica. Similarly, different properties and radiation models are used for different glass layers.

$$a = 0, \text{for } \lambda < 3.0\,\mu m \tag{8.7}$$
$$a = 400.0\,\text{m}^{-1}, \text{for } 3.0\,\mu m \leq \lambda < 4.8\,\mu m \tag{8.8}$$
$$a = 15000\,\text{m}^{-1}, \text{for } 4.8\,\mu m \leq \lambda < 8.0\,\mu m \tag{8.9}$$

Figure 8.12a shows a typical mesh for the computational study of the drawing of a two-layer optical fiber. Care has to be exercised in choosing the grid near the bottom of the draw-down region to avoid highly distorted grids due to the small diameter of the fiber. The optimal grid can be obtained by considering the numerical results with grid refinement. In the figure shown, the core is taken as half the diameter of the cladding. However, it is much smaller in diameter for single-mode telecommunication applications, being as small as 8 μm. It is also not accurate to assume uniform temperature and velocity across the glass regions, despite the small diameters, because of the large changes in viscosity due to temperature differences. The

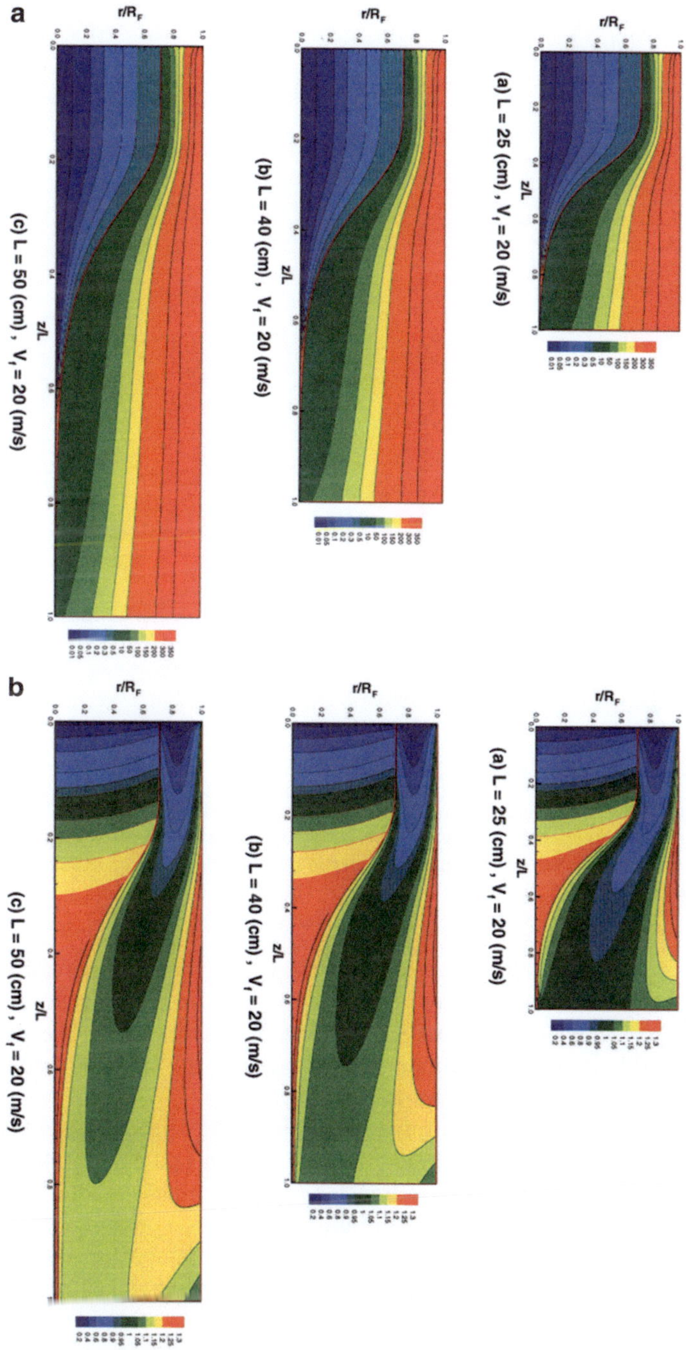

Fig. 8.10 Calculated (**a**) velocity and (**b**) temperature fields in the glass and inert gases during drawing of a single-layer optical fiber for three furnace lengths

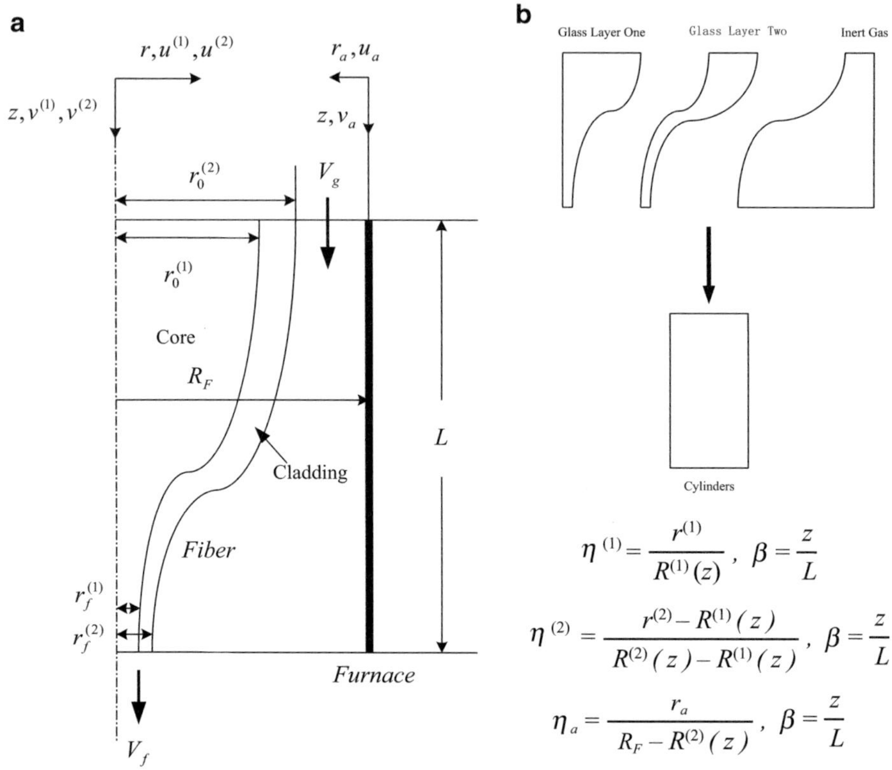

Fig. 8.11 (**a**) Sketch of neck-down profiles for the core and cladding, (**b**) Landau's transformation to simplify the computational domains by converting axisymmetric complex shapes to cylindrical ones. The dimensions given in the equations refer to hollow fiber drawing sketched in Fig. 8.16b and discussed later

boundary conditions in terms of shear, pressure, velocity, and temperature continuity have to be satisfied at the interface between the two layers. A typical example of the numerically calculated neck-down profile for a two-layer optical fiber is shown in Fig. 8.12b at different refractive indexes. As expected, the radiation calculations are much more involved than those for a single-layer fiber.

The refractive index and absorption coefficients are affected when the core and the cladding are doped with various dopants (Izawa and Sudo 1987; Chen and Jaluria 2009). Since radiation heat transfer is the dominant mode of transport in the heating process, a nonuniform distribution of radiation properties in the preform due to doping will cause significant local effect on the temperature and thus on the draw process. Assumed magnitudes of change in the refractive index and absorption coefficients were used to capture the generic effects of variation in these properties. GeO_2 is taken as the dopant for the core to increase the refractive index. This also increases the ultraviolet radiation absorption and lowers the viscosity. Figure 8.13 shows the relevant properties considered here. Three-band models are used for radiation transport for different doping circumstances. Several of these are given

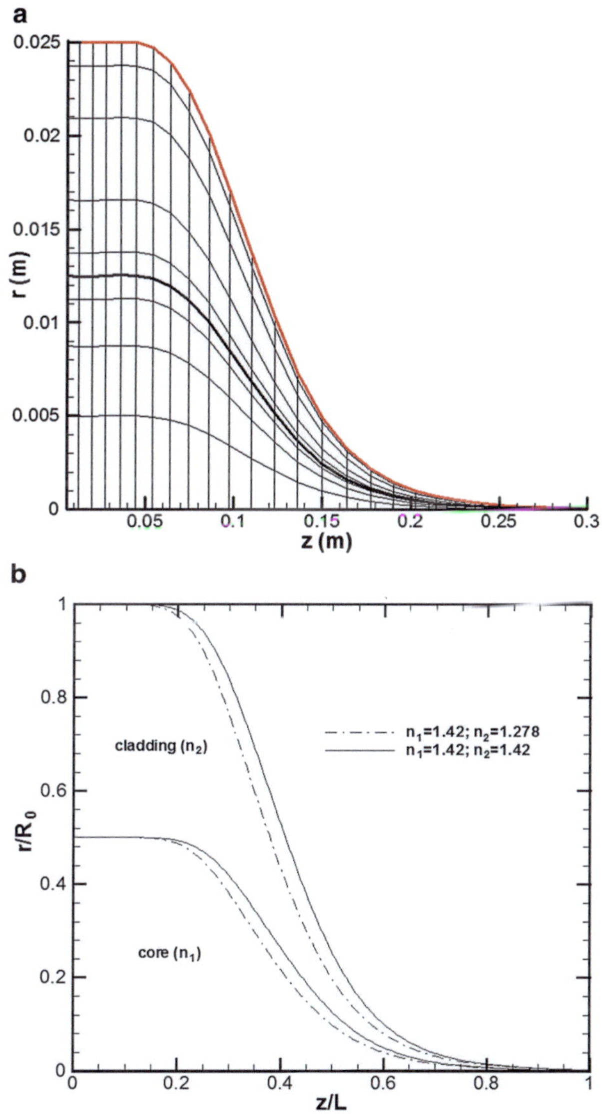

Fig. 8.12 (**a**) Numerical grid and (**b**) calculated neck-down profiles for a preform with a core-cladding structure at two values of the refractive index of the outer layer

by Chen and Jaluria (2009). For instance, for GeO_2 doped silica, the band absorption model used is

$$a = 0, \text{for } \lambda < 0.15\,\mu m \tag{8.10}$$

$$a = 0.0443\,\text{cm}^{-1}, \text{for } 0.15\,\mu m \leq \lambda < 3.0\,\mu m \tag{8.11}$$

$$a = 0.8455\,\text{cm}^{-1}, \text{for } 3.0\,\mu m \leq \lambda < 4.8\,\mu m \tag{8.12}$$

$$a = 345.0\,\text{cm}^{-1}, \text{for } 4.8\,\mu m \leq \lambda < 8.0\,\mu m \tag{8.13}$$

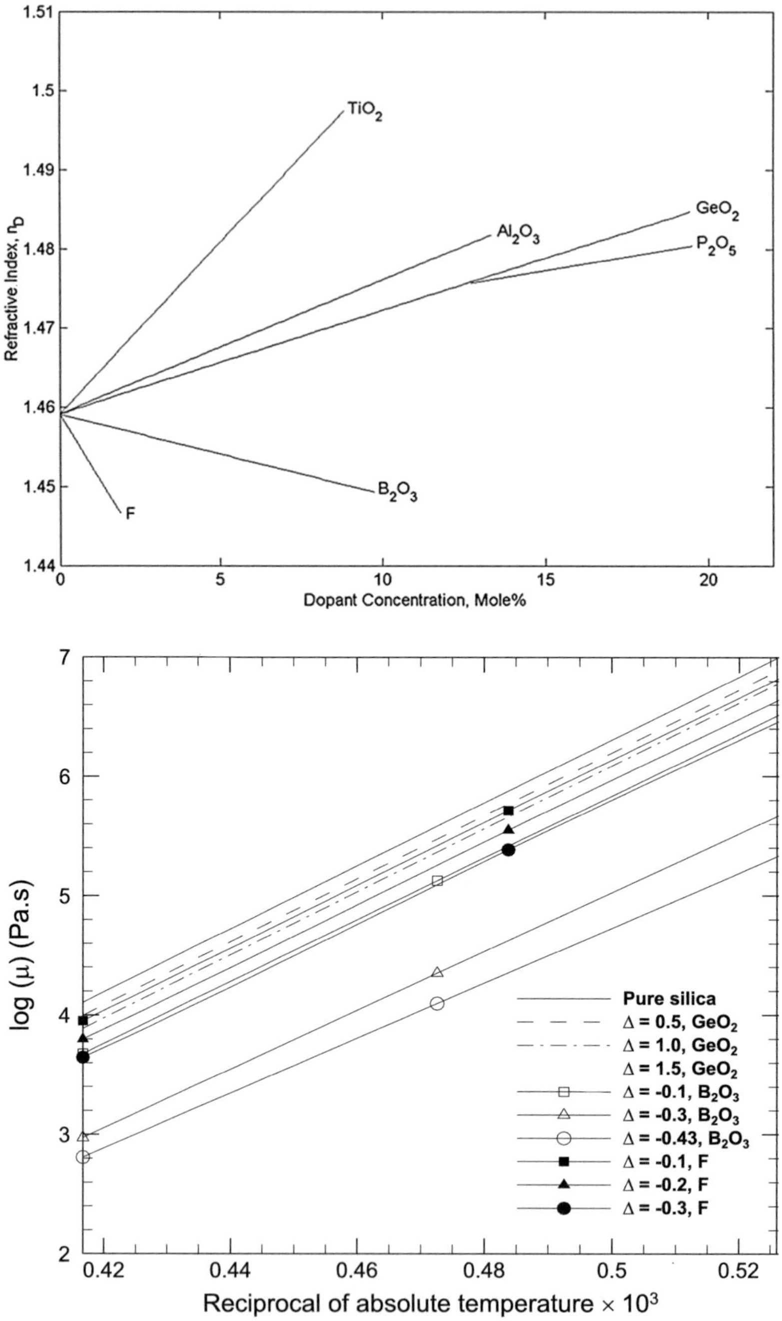

Fig. 8.13 Variation of silica glass properties with dopant concentration and temperature

Figure 8.14 shows experimental measurements on the absorption coefficients from Izawa et al. (1977) and Izawa and Sudo (1987) and the three-band model that may be used to fit the data. However, most of the data are available at room temperatures and not at the elevated temperatures at which the drawing process occurs. Also, there is generally a lack of accurate property data that may be used in the simulation.

Results are shown here for the core doped with 5.5 mole%, 11.1 mole%, and 16.6 mole% GeO_2, to obtain a relative refractive index difference of 0.5%, 1.0%, and

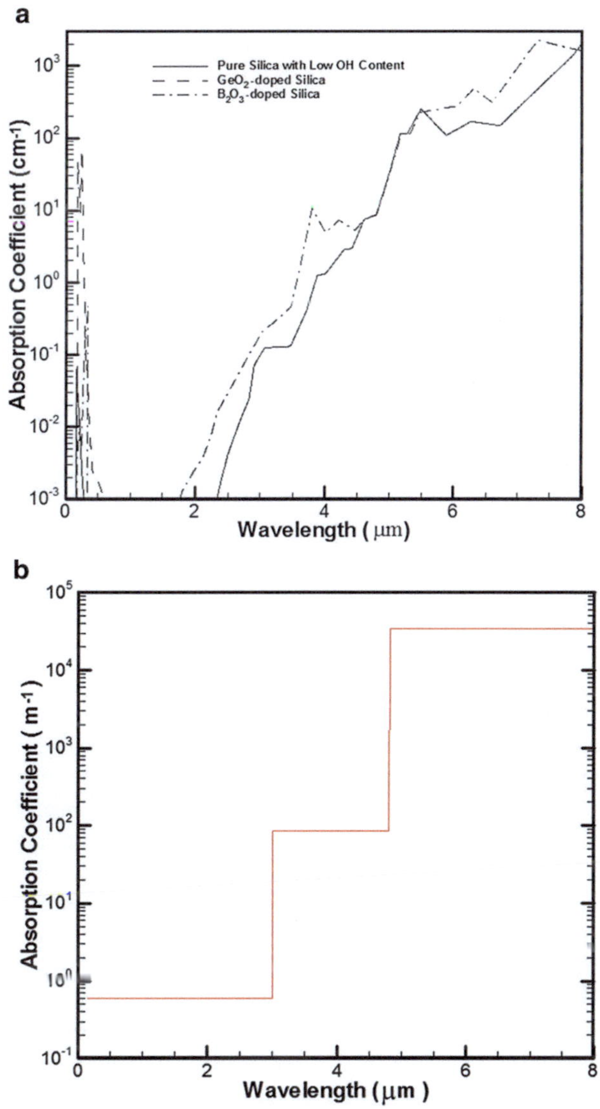

Fig. 8.14 (**a**) Measured absorption coefficients of pure fused silica with low OH content, GeO_2-doped silica, and B_2O_3-doped silica (Izawa and Sudo 1987; Izawa et al. 1977); (**b**) a typical three-band radiation model to fit the experimental data

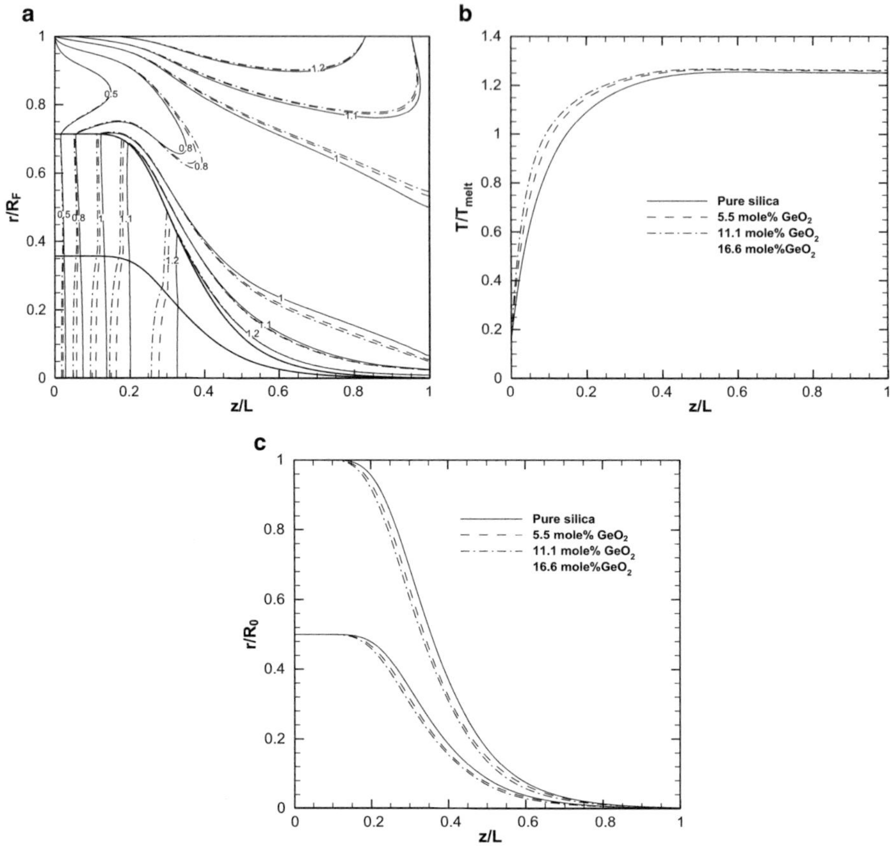

Fig. 8.15 (**a**) Isotherms for various GeO_2 concentrations (solid line, pure silica; dashed line, 5.5 mol % GeO_2; dashed dotted line, 11.1 mol % GeO_2; dotted line, 16.6 mol % GeO_2), (**b**) temperature variation along the centerline for various GeO_2 concentrations, (**c**) neck-down profiles

1.5%, respectively, in the fiber. In Fig. 8.15a, the isotherms for various GeO_2 concentrations are shown, with a reference profile to indicate the variations in temperature levels in the axial direction. It is seen that, when the core is doped with GeO_2, the core is heated up to a given temperature level in a shorter distance than the cladding. The distance to reach a certain temperature level for the whole preform also decreases with increasing GeO_2 concentration. This is due to the fact that GeO_2-doped silica core has a larger radiation coefficient in the ultraviolet region and the transmissivity of pure silica cladding is very high in this region. Thus, the GeO_2-doped silica core absorbs more energy by radiation than pure silica core. Figure 8.15b shows the temperature of the preform along the centerline for pure silica and GeO_2-doped silica core with a pure silica cladding. As expected, a significant increase in the centerline temperature is observed in the upper neck-down region with greater GeO_2 doping. Beyond that, the fiber with higher GeO_2

concentration maintains a slightly higher temperature along the centerline. Figure 8.15c shows the effect of the dopant concentration on the neck-down profile.

The preceding discussion was focused on the furnace drawing of single and multilayer fibers, with dopants added to enhance the transmission characteristics. This modeling can be extended to include more than two layers and other designs of optical fibers. A continuous variation in the refractive index may be modeled as step changes in different layers. Nano-engineered optical fibers often involve additional layers such as depressed cladding and trench fiber designs that reduce bending losses (Matsuo et al. 2007; Tandon et al. 2013). Such additional features in the fiber design are incorporated to improve the performance, particularly for reducing the bending losses. The preform fabrication by methods such as OVD is where these design features are obtained, followed by the furnace drawing of the multilayered fiber. Also, though only steady axisymmetric flow results are presented here, transient solutions are of interest in the consideration of the stability of the process, particularly the stability of the free surface. Three-dimensional solutions are needed to check on the axisymmetry of the fiber, as well as of the process.

8.4.4 *Hollow Fibers*

In recent years, an important new class of optical fiber, hollow or microstructured fiber, has emerged. The fiber is built of one material, which is usually silica, and guiding of light is achieved through the presence of air holes in the region surrounding the solid core (van Eijkelenborg 2003; Harrington 2004). Different periodic and nonperiodic patterns for the arrangement of holes have been used. Photonic crystal fibers, where propagation occurs within a photonic defect or due to photonic bandgap effect, are a subgroup of microstructured fibers. Among these fibers, *holey* fibers use air holes in their cross-sections, hole-assisted fibers guide light by a higher-index core modified by the presence of air holes, and Bragg fibers have concentric rings of multilayer film. Figure 8.16 shows sketches of hollow-core and microstructured fibers. Hollow optical fibers have several advantages over the conventional solid-core optical fibers. These include high laser power thresholds, no end reflection, low insertion loss, ruggedness, and small beam divergence. Microstructured fibers achieve low attenuation through structural design rather than high transparency material selection and can be used to provide infrared laser beam and high-power delivery. Microstructured optical fibers are widely used in medical applications, sensor technology, diagnostics, and laser delivery, besides traditional telecommunications.

Let us first consider the drawing of a hollow-core fiber, as shown in Fig. 8.3b, where the hollow core is of diameter 10–80 μm. The multilayer fiber modeling discussed earlier may be used for modeling the process. The neck-down region consists of two neck-down profiles that determine the geometry of the final hollow fiber (Fitt et al. 2001). The neck-down profiles strongly depend on the fiber drawing conditions, material properties, and preform dimensions. Some typical results are

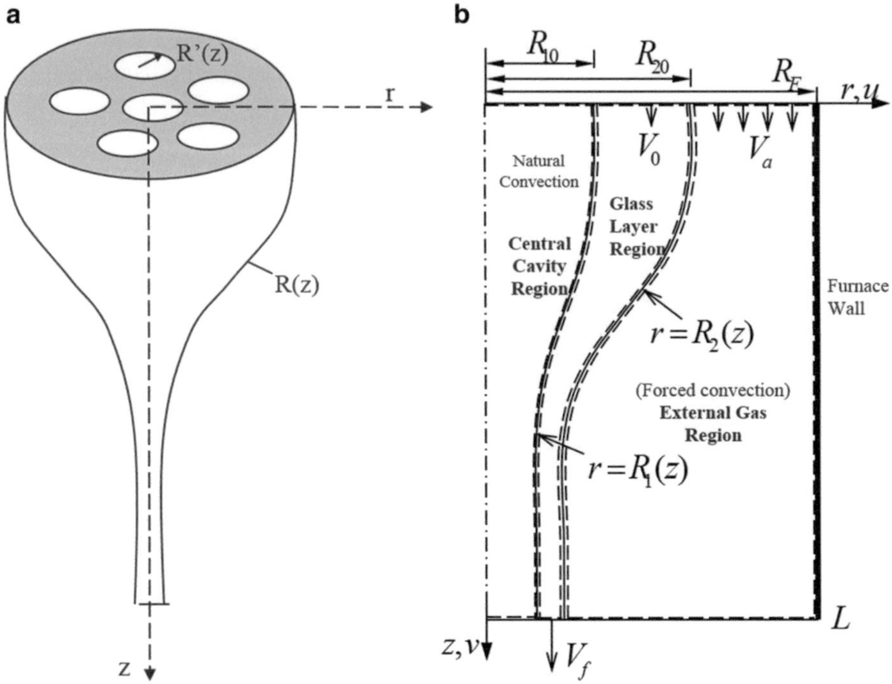

Fig. 8.16 (**a**) Sketch of a microstructured optical fiber, (**b**) neck-down profiles in central-core hollow fiber drawing

presented here. The diameter of the cylindrical furnace is taken as 7 cm and its length as 30 cm. The inner and outer diameters of the preform are taken as 2.5 cm and 5 cm, respectively. The furnace wall temperature is assumed to be a parabolic profile with a maximum of 2500 K in the middle and a minimum of 2000 K at the two ends. The fiber drawing speed is 10 m/s and the velocity of the purge gas at the inlet is 0.1 m/s. The results are shown in Fig. 8.17 in terms of streamlines and isotherms (Yang and Jaluria 2009a, b). The thicker black lines indicate the two neck-down profiles obtained from the numerical solution. The isotherms are given in terms of the nondimensional temperature, with the softening temperature taken as the characteristic temperature. It is seen that the flow and temperature distributions in the fiber and in the external gas in hollow fiber drawing are similar to those in solid-core fiber drawing. Because of the small diameter of the central core, the air in the core is dragged by the moving fiber. Thus, a slug flow approximation is made for this airflow to simplify the analysis. Thermal diffusion dominates in the central cavity since the natural convection flow is found to be quite weak.

An important concern is the collapse or closing down of the central core. A collapse ratio C may be defined in terms of the dimensions shown in Fig. 8.16b to quantify the collapse process as

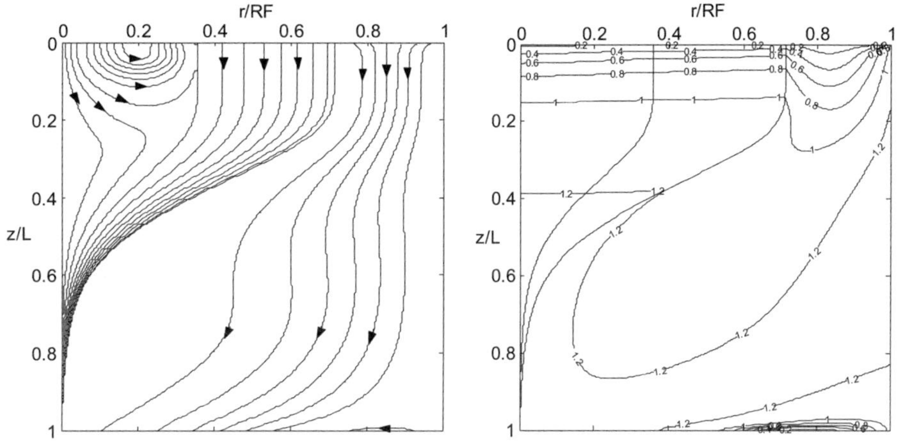

Fig. 8.17 Streamlines and isotherms in the glass and the furnace for a typical case of hollow fiber drawing with a parabolic furnace temperature distribution at a drawing speed of 10 m/s

$$C(z) = 1 - \frac{R_1(z)/R_2(z)}{R_{10}/R_{20}} \tag{8.14}$$

Thus, when the radius ratio of the final fiber equals the initial radius ratio, $C = 0$. When the central cavity is closed, $C = 1$. The effects of the preform feeding and fiber drawing speeds and of the furnace temperature on the collapse ratio are studied. The variation of the collapse ratio C with furnace temperature for different drawing speeds is shown in Fig. 8.18a. It is seen that the collapse ratio increases with a decrease in drawing speed. It was found that the effect of preform feed rate on the collapse ratio was much larger than that of the drawing speed. The collapse ratio is found to increase with the furnace temperature. All these trends are similar to the results in the literature (Fitt et al. 2001; Chakravarthy and Chiu 2005).

The observed trends can be explained by considering the *collapse time* of preform/fiber. When the drawing or feeding speed decreases, the total time spent in the furnace by a point in the preform/fiber increases. The total time in the furnace is comprised of the time to heat up the preform to the softening temperature and the time for the preform/fiber to collapse. The collapse time increases since the time to heat the preform is nearly the same for different drawing or feeding speeds. Therefore, the collapse ratio increases due to an increase in the collapse time when the drawing or feeding speed decreases. Since the drawing speed is much larger than the feeding speed, the collapse time is more sensitive to the feeding speed. Also the time to heat the preform is shorter at higher furnace temperature, which implies that the collapse time increases at the same feeding and drawing speeds. Hence, the collapse ratio increases at higher furnace temperature. In order to avoid collapse of the central core, drawing and feeding speeds may be increased, furnace temperature decreased, or preform radius ratio increased. The collapse ratio is also influenced by the pressure difference between the purge gas and the central cavity. It was found that,

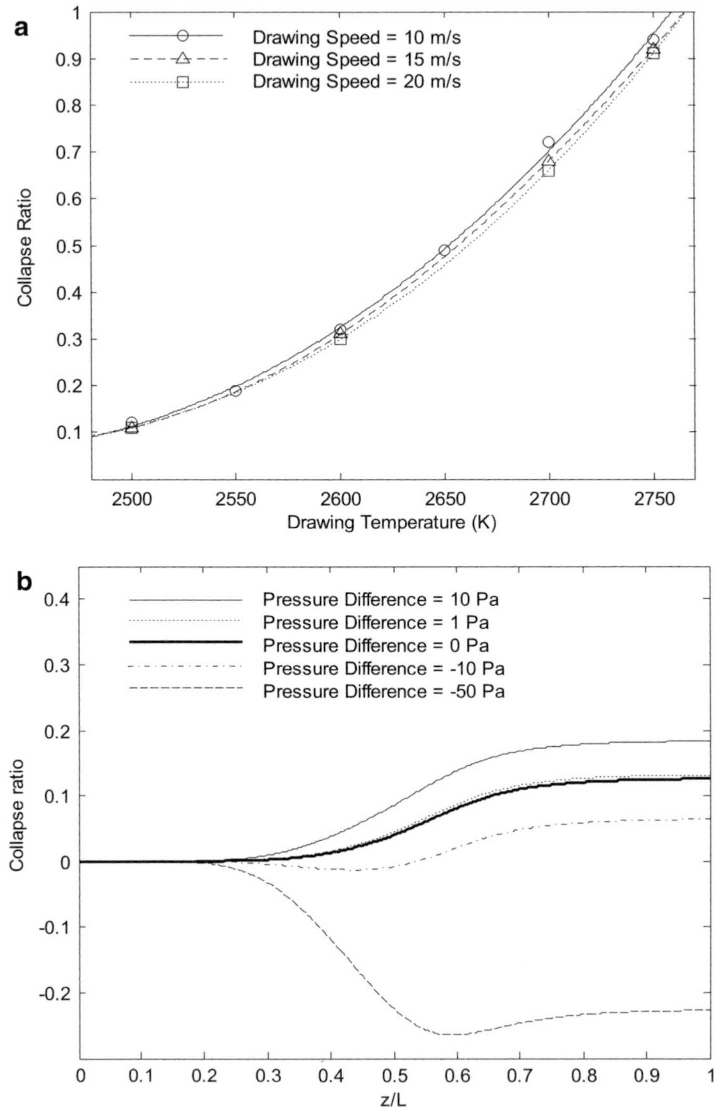

Fig. 8.18 (**a**) Variation of the collapse ratio C with the drawing temperature for different drawing speeds, (**b**) variation of C along the axis with different pressurizations in the core

as expected, higher pressure in the central cavity tends to prevent collapse of the central cavity. It was also seen that a high pressure may cause the central cavity to enlarge during the drawing process; see Fig. 8.18b. This is called *explosion* in the drawing process. Because of the size of the core, surface tension is an important parameter and plays a significant role in the collapse.

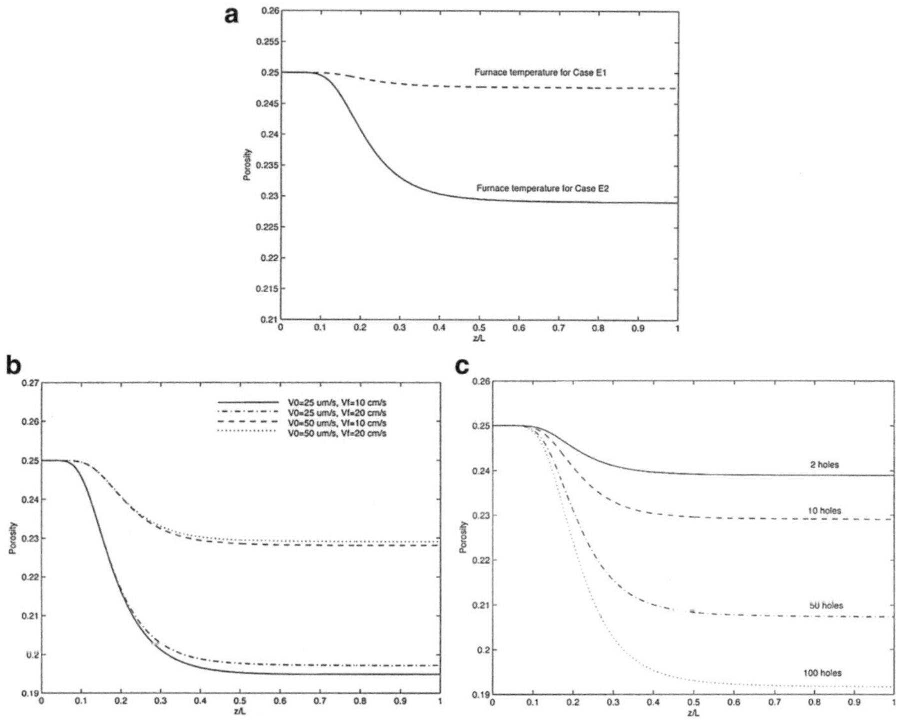

Fig. 8.19 Effect of operating conditions on the porosity and thus the resulting geometry of a microstructured polymer optical fiber during the drawing process. (**a**) Effect of furnace temperature, with $E2$ being at higher temperature than $E1$. (**b**) Effect of feeding speed V_o in μm/s and drawing speed V_f in cm/s, (**c**) effect of number of holes in the microstructured polymer fiber

Similarly, the drawing of microstructured optical fibers, with several channels, can be modeled. The problem is obviously complicated due to the geometry. However, a simple model can be developed considering the preform as a porous material, with the porosity being defined as the fraction of the void area in the total cross-sectional area. The neck-down profile is denoted by $R(z)$. The number and radius of circular air holes are denoted by N and $R'(z)$. The porosity is then defined as

$$\phi = \frac{A_{\text{air}}}{A_{\text{total}}} = \frac{N[R'(z)]^2}{[R(z)]^2} \tag{8.15}$$

Therefore, $\phi = 0$ applies for a solid-core optical fiber. The effective properties of the microstructured fiber are obtained by using the usual approximations for porous media, and the flow and heat transfer are investigated as for single-layer optical fibers. Figure 8.19 shows the effect of the main operating conditions on the porosity of a microstructured polymer optical fiber (MPOF) and thus on the changing geometry of the fiber. Here, it is assumed that all the holes are preserved, and, therefore, the resulting microchannel diameters can be obtained with the overall fiber diameter taken as 125 μm. As expected, the porosity decreases at higher furnace temperature, indicating that the diameter of the holes is reduced. Similarly, a larger number of holes, with the same initial porosity, imply smaller starting diameter and larger decrease in diameter during the process. The basic trends are similar to those discussed earlier for hollow fibers. Additional results were obtained on this process (Yang 2008). More on hollow and microstructured fibers can be obtained from the references given here as well as from Cregan et al. (1999) and Monro et al. (2001).

8.4.5 Furnace Wall Temperature Distribution

A study of the flow and thermal transport associated with the draw furnace requires accurate knowledge of the furnace wall temperature distribution. Even though an average, uniform temperature has been used in many of the preceding results, the dependence on the wall temperature distribution has also been investigated by a few investigators; see Cheng and Jaluria (2005). However, the wall temperature distribution is often not known. Due to high temperature, furnace geometry, and inaccessibility, it is not possible to use sensors on the walls. Generally, a single temperature sensor is used for controlling the temperature of the wall and can be used to provide the value at a specific location. However, for accurate modeling of the process, the actual temperature distribution must be specified.

An experimental procedure involving mounting rods of different materials and diameters and feeding them axially within the furnace cavity was employed for this purpose by Issa et al. (1996). Figure 8.20 shows a typical fiber drawing furnace with a centrally located graphite rod. Each rod was instrumented with thermocouples inserted through an axial hole along the centerline. The temperature measurements were used along with the numerical model for the flow and heat transfer in the furnace in order to obtain the furnace wall temperature profile. This is an inverse problem since the centerline temperature in the rod is known, whereas the furnace thermal conditions are not known. The results obtained using graphite rods of different diameters indicated that the furnace temperature was not significantly affected by rod size. Figure 8.21a shows a sketch of the experimental system. The measured temperature distribution along the axis of a graphite rod is shown, along with the wall temperature distribution obtained from the solution of the inverse problem. The dashed lines represent the water-cooled portion of the furnace wall. An optimization procedure is used to narrow the uncertainty in the nonunique solutions and obtain an essentially unique wall temperature distribution. This distribution is used to calculate the graphite rod temperature, and a good agreement with the experimental data is observed. The computed element temperature at the control sensor location was also in good agreement with the measured value, lending support to the model for the flow and thermal transport in the furnace. The small difference between the computed element temperatures for the two rod sizes is in support of the previous statement that there was a small influence of rod size on the furnace temperature distribution. Similar results were obtained for other furnace

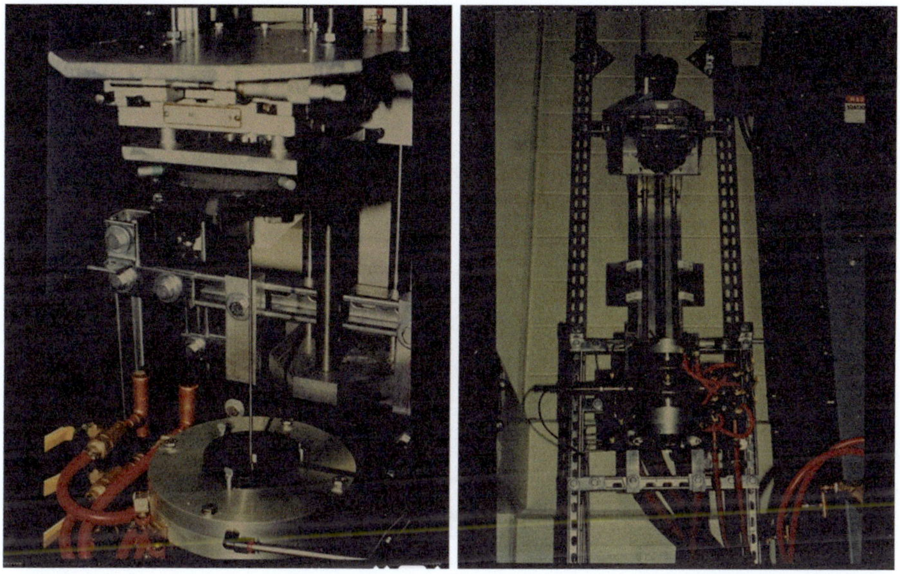

Fig. 8.20 System for measuring the centerline temperature of rods placed at the axis of a fiber drawing furnace

temperatures and for other materials, including silica glass (Liu and Polymeropoulos 1998).

8.5 Process Feasibility

As might be expected, it is possible to draw the fiber only over a range of operating conditions. If the furnace temperature is not high enough, the fiber could break due to lack of material flow, a phenomenon that is known as *viscous rupture*. Numerically, this is indicated by the divergence of the correction scheme used for obtaining the neck-down profile. Physically, it is given by excessive tension in the fiber. Similarly, for a given furnace temperature, there is a limit on the draw speed beyond which drawing is not possible, since this leads to rupture. Therefore, a region in which drawing is feasible can be identified, in terms of the furnace temperature and draw speed, as shown in Fig. 8.22a, which also gives the computed tension in the fiber. Beyond the boundaries of this region, drawing is not possible. Other combinations of physical and process variables, such as inert gas flow rate, flow configuration, furnace wall temperature distribution, furnace dimensions, and preform and fiber diameters, may similarly be considered to determine the feasibility of the process (Cheng and Jaluria 2004).

Similar to the situation with solid-core fibers, the hollow and microstructured fibers cannot be drawn at any arbitrary combination of critical drawing parameters.

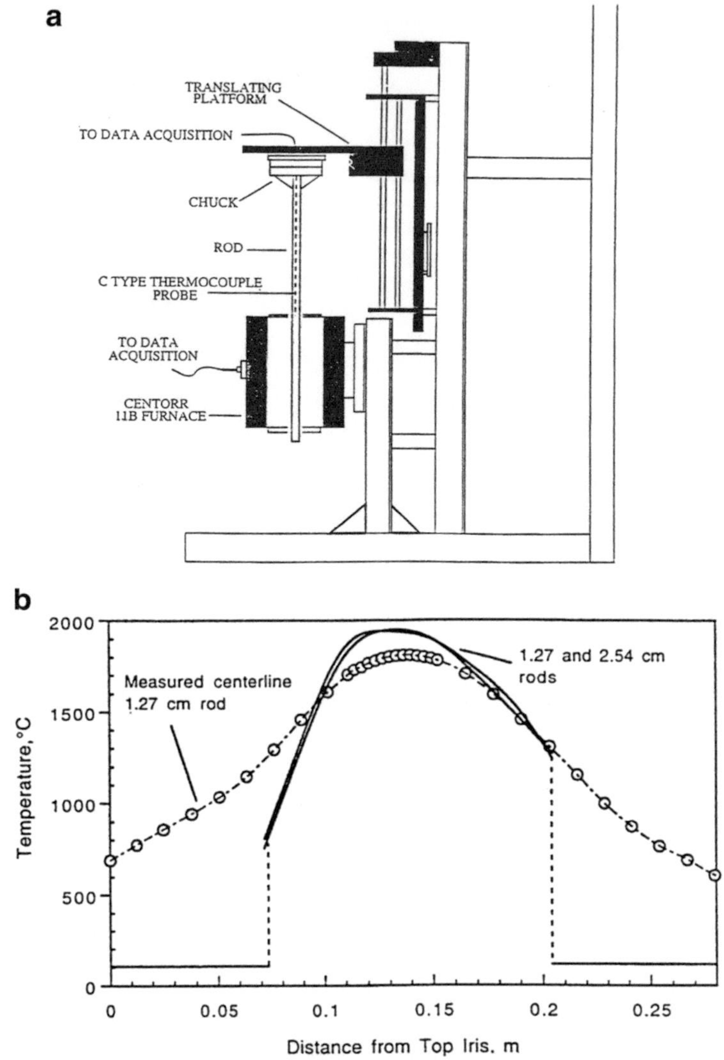

Fig. 8.21 (**a**) Schematic of an experimental system for measuring the temperature distribution in a rod located in an optical fiber drawing furnace; (**b**) computed furnace wall temperature distributions (solid line) from graphite rod data. Experimental points are from the 1.27 cm diameter rod

In the hollow fiber case as well, lack of material flow can lead to the process being infeasible. Numerically, the iterative process to determine the neck-down profiles does not converge, similar to solid-core fiber drawing, and the draw tension exceeds the constraint for viscous rupture. However, there is also the possibility of the collapse of the central core, making the process unacceptable. The collapse process discussed earlier imposes its own constraints on feasibility. Figure 8.22b shows the calculated feasible domain in terms of drawing temperature and drawing speed. The

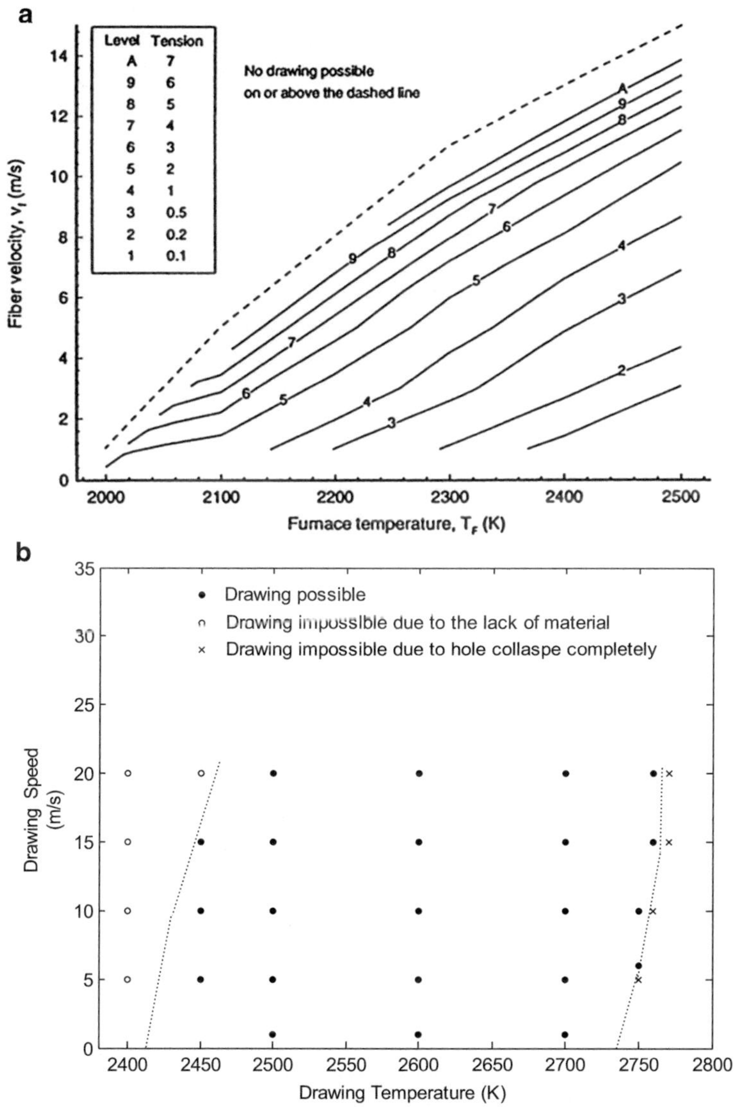

Fig. 8.22 (**a**) Feasibility of the optical fiber drawing process, as indicated by regions where drawing is not possible due to viscous rupture resulting from excessive draw tension, shown in terms of "iso-tension" contours for the feasible range of fiber drawing, (**b**) feasible domain for hollow fiber drawing in terms of the drawing speed and temperature

drawing speed is in the range of 1 m/s to 20 m/s. The drawing process is possible only in the region between the two dashed lines. The left boundary exists because the continuous drawing process fails due to the lack of material flow at low furnace temperature or high drawing speeds, which is initially indicated by the divergence of the numerical correction scheme for the profiles. The right boundary exists because high furnace temperature or low drawing speeds may cause the central air core to collapse. At the right side of this boundary line, the central air core collapses during the drawing process. Therefore, there are two limitations on the furnace temperature for a given drawing speed. These two boundary lines are seen to be nearly vertical since the effect of the drawing speed was found to be relatively weak, as indicated by the earlier parameter study (Yang and Jaluria 2009a, b).

Other furnace wall temperature distributions, such as Gaussian, parabolic, or experimentally determined distribution, obtained from the solution of an inverse problem as discussed earlier, may be considered, depending on the given system or process. Similarly, variations of different design parameters and operating conditions may be considered to determine feasibility of the process and to vary these appropriately to obtain a feasible process. Figure 8.23 shows the results for a particular example, which employs a Gaussian wall temperature distribution, and feasibility is determined in terms of furnace height and maximum wall temperature. Obviously, this is an important consideration in the fiber drawing process and in the design of the draw tower for desired optical fibers.

8.6 Thermally Induced Defects

An important concern in the fabrication of optical fibers is the quality of the product, as measured in terms of transmission loss with distance. Various defects are generated during the fabrication process, and these could adversely affect the performance. The role of bending in transmission loss has been mentioned earlier, along with some design features that help in reducing these losses. Similarly, tension applied during the draw process can cause microstructural defects and should be minimized by varying the temperature level (Li 1985; Oh and Paek 2012). Also important are thermally induced defects that are generated during the fiber drawing process. One such defect is the E' defect, which is a point defect generated at high temperature during drawing and which causes transmission loss and mechanical strength degradation in the fiber. A differential equation for the time dependence of the E' defect generation was formulated by Hanafusa et al. (1985) based on the theory of the thermodynamics of lattice vacancies in crystals. It was assumed that the E' defects are generated through the breaking of the $\mathrm{Si}=\mathrm{O}$ band. A fraction of the defects recombine to form $\mathrm{Si}=\mathrm{O}$ again. Then, the net concentration of the E' defects is the difference between the generation and the recombination. The equation for E' defect concentration was given by Hanafusa et al. (1985) as

$$v\frac{dn_{\mathrm{d}}}{dz} = n_{\mathrm{p}}(0)\nu\exp\left(-\frac{E_{\mathrm{p}}}{KT}\right) - n_{\mathrm{d}}\nu\left[\exp\left(-\frac{E_{\mathrm{p}}}{KT}\right) + \exp\left(-\frac{E_{\mathrm{d}}}{KT}\right)\right] \tag{8.16}$$

where n_{d} and E_{d} represent the concentration and activation energy of the E' defect, n_{p} and E_{p} represent those of the precursors, ν is a frequency factor, and K is Boltzmann

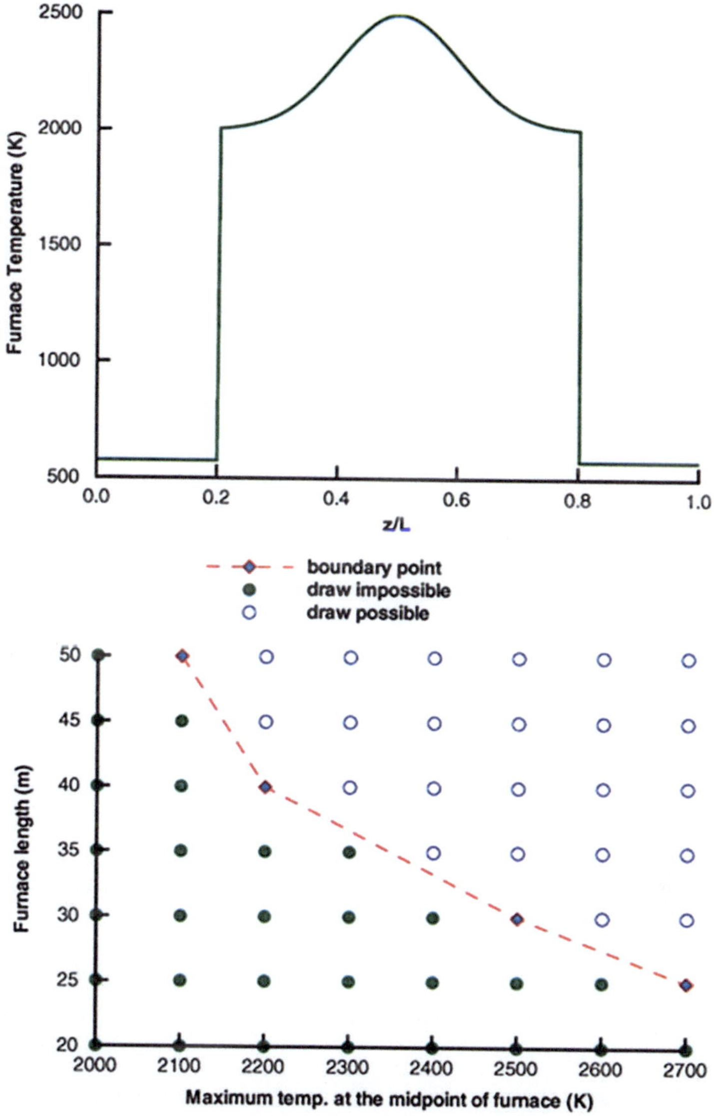

Fig. 8.23 Feasibility of the optical fiber drawing process at a fiber speed of 15 m/s. The specified Gaussian temperature distribution at the wall is shown, along with the feasible domain in terms of the furnace height and maximum wall temperature

constant. The initial values and constants were given as $n_d(0) = 0$, $n_p(0) = 7 \times 10^{22}\mathrm{g}^{-1}$, $E_p = 6.4087 \times 10^{-19}\mathrm{J}$, $E_d = 0.3204 \times 10^{-19}\mathrm{J}$, $\nu = 8 \times 10^{-3}\mathrm{sec}^{-1}$, and $K = 1.380658 \times 10^{-23}\mathrm{J/K}$.

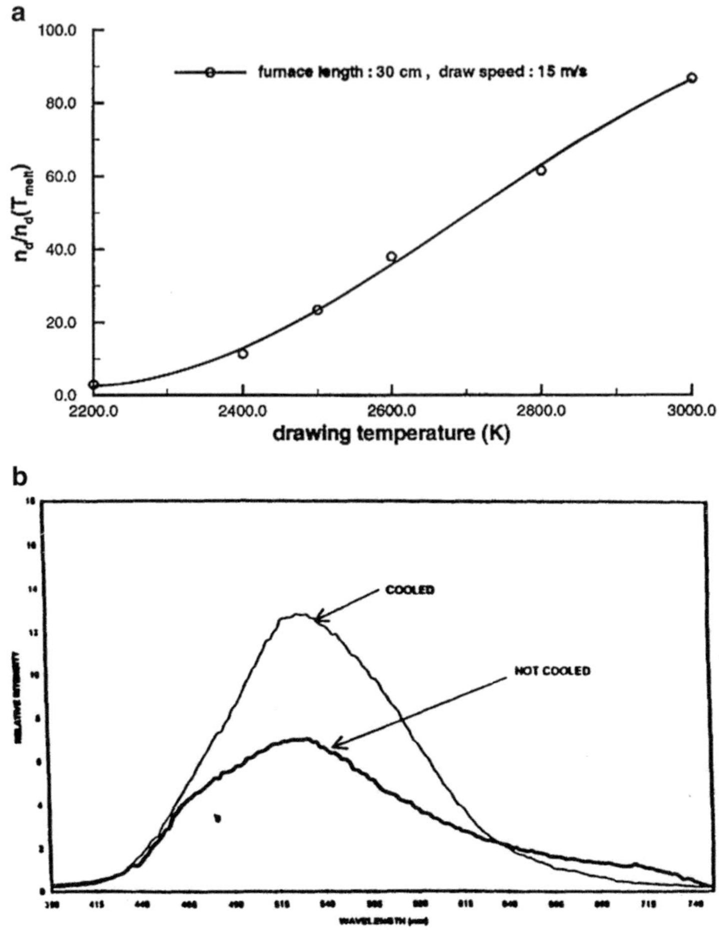

Fig. 8.24 (**a**) Dependence of average concentration of E' defects on furnace wall temperature, (**b**) luminescence versus wavelength of fibers drawn at 2050 °C and 80 m/min, indicating concentration of E' defects for fibers that are cooled or not cooled by forced convection

Figure 8.24a shows the variation of the average concentration of E' defects with the draw temperature, indicating an increase with temperature, as expected from the higher breakage of the Si = O bond. Figure 8.24b shows the final concentration of the defects after the fiber has cooled down, as obtained from the luminescence of the fibers. It is seen that, if the fiber is cooled slowly after the draw furnace, the defects are reduced due to the annealing of the fiber resulting in recombination of the broken bonds. If the fiber is cooled very fast by forced convection, the defects are frozen and yield a higher concentration. This implies that a slow cooling of the fiber, probably by natural convection, is helpful in reducing the final concentration of these defects. Therefore, slow cooling over a large distance is used in typical manufacturing systems for optical fibers. Figure 8.25a shows the variation of E' defect

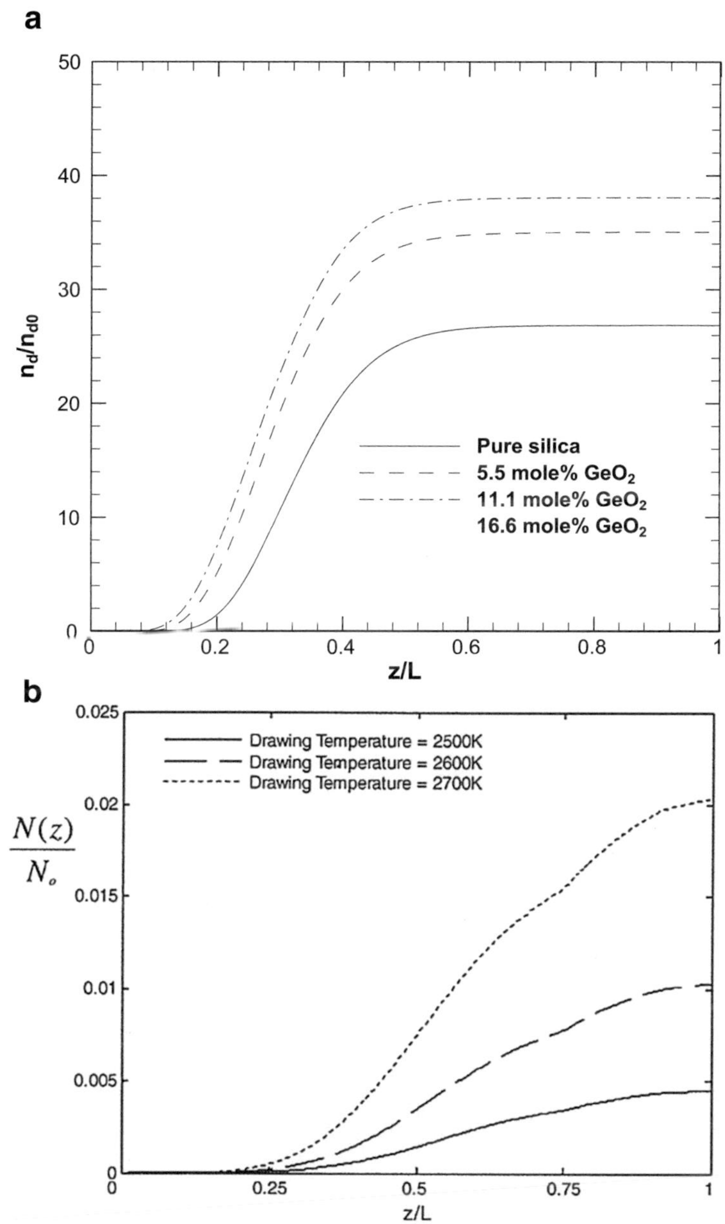

Fig. 8.25 (**a**) Concentration of E' defects along the solid fiber centerline for various GeO_2 concentrations; (**b**) variation of normalized concentration of NBOHC defects with the axial location z for hollow fiber drawing

concentration for a doped fiber. It was seen earlier that the double-layer preform leads to higher temperatures, which give rise to a higher concentration of E' defects. Thus, it is seen that the fiber quality is degraded with an increase in refractive index difference between the core and the cladding in terms of E' defects. In all cases, the defect concentration increases with the axial location z since additional defects are generated as the fiber moves downstream.

Another important defect is the non-bridging oxygen hole center (NBOHC), which can cause optical absorption and greater transmission loss. The generation of NBOHCs is affected by the shear stress as well as the temperature history undergone by the fiber. The equation to determine NBOHCs has been given by Hibino and Hanafusa (1986). It has been observed that the 630 nm absorption is related to the NBOHCs' concentration in high-OH-content silica glass. In low-OH-content silica glass, the 630 nm absorption is related to the sum of the concentrations of NBOHCs, E' defect centers, and peroxy radicals. Figure 8.25b shows the variation of the NBOHC defect concentration, normalized by a constant N_o given by Hibino and Hanafusa (1986), with the axial location z. As seen earlier, the defect concentration increases downstream and with increasing temperature. Therefore, it is important to consider such drawing-induced defect centers. Obviously, a low value of the defect concentration is desirable for a higher quality of the fiber. This aspect can be used for the optimization of the process to obtain the highest possible quality of optical fiber, as outlined later.

8.7 Fiber Cooling

After the drawing process in the furnace, the optical fiber has to be cooled in order to reach appropriate temperature levels before liquid coating is applied (Jochem and Ligt 1986). Rapid cooling is desirable to reduce the overall height of the draw tower. However, as mentioned earlier, slow cooling helps in annealing the hot fiber so that the defects and residual stresses are reduced. Thus, a combination of natural and forced convective cooling may be used to obtain the best results. Accelerated cooling involves passing the fiber through a cooling section located downstream of the draw furnace. Various flow configurations may be used, such as aiding and opposing annular flow where the fiber is passed through a cooled tube with cooling gas flowing in a direction aiding or opposing the fiber motion, as well as peripheral flow entering and leaving on the side of the cooling section.

Substantial work has been done on the flow in the forced cooling process, considering laminar and turbulent flow in the cooling section. An analysis of the cooling of the moving fiber in forced flow has been carried out to determine the resulting temperature decay downstream, the temperature variation in the fiber, and the heat transfer rates. Different gases, such as nitrogen and helium, different flow configurations, and different designs of the cooling region were considered to determine the conditions needed to obtain the desired decrease in the fiber temperature. Numerical models were developed to include the effects of fiber motion,

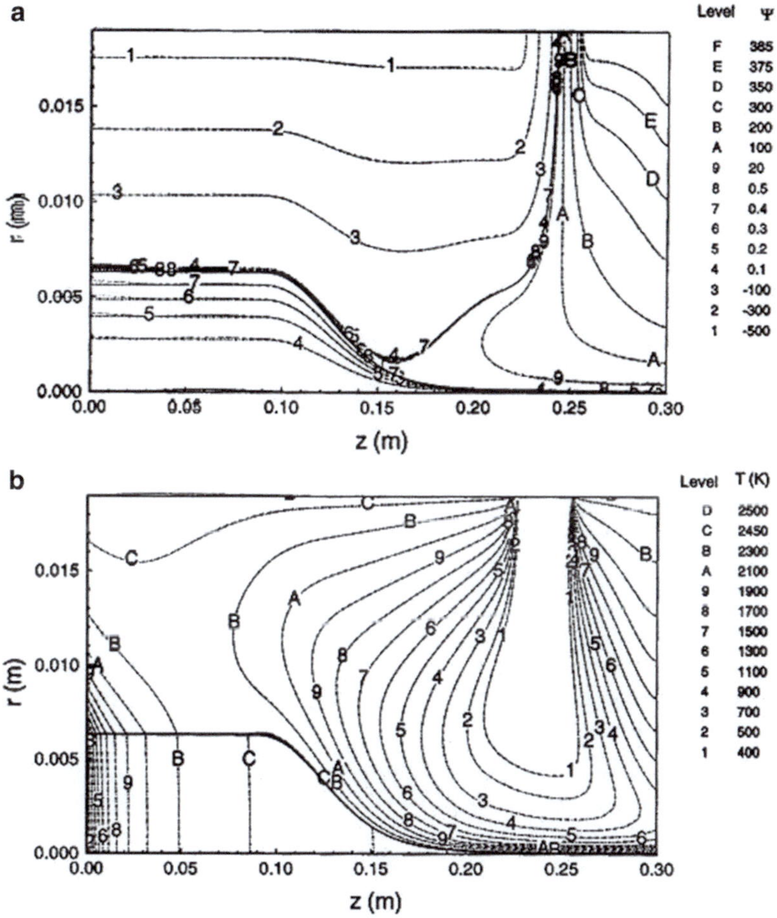

Fig. 8.26 Inert gas flow and temperature distributions in an optical fiber drawing furnace for side entry (**a**) Streamlines, (**b**) Isotherms

forced flow, geometry, variable properties, and dimensions (Vaskopoulos et al.1993). The experimental work employed a specially designed cooling section, which was installed, tested, and used on the full-scale fiber draw tower facility at the Fiber Optic Materials Research Center at Rutgers University (Roy Choudhury et al.1994). The fiber was around 125 μm in diameter, and the diameter of the cooling jacket adjacent to the fiber varied over the range of 300–1000 μm.

The thermal transport is a strong function of the geometry, particularly the locations of the inlet/outlet channels and of the flow rate. The flow can affect local transport rates and thus cause local hot or cold spots. Figure 8.26 shows the calculated streamlines and isotherms for peripheral flow configuration. The side inlet, though more convenient to design and operate, directs the cold fluid at the fiber, causing possible local cooling and breakage. Similarly, flows that aid or

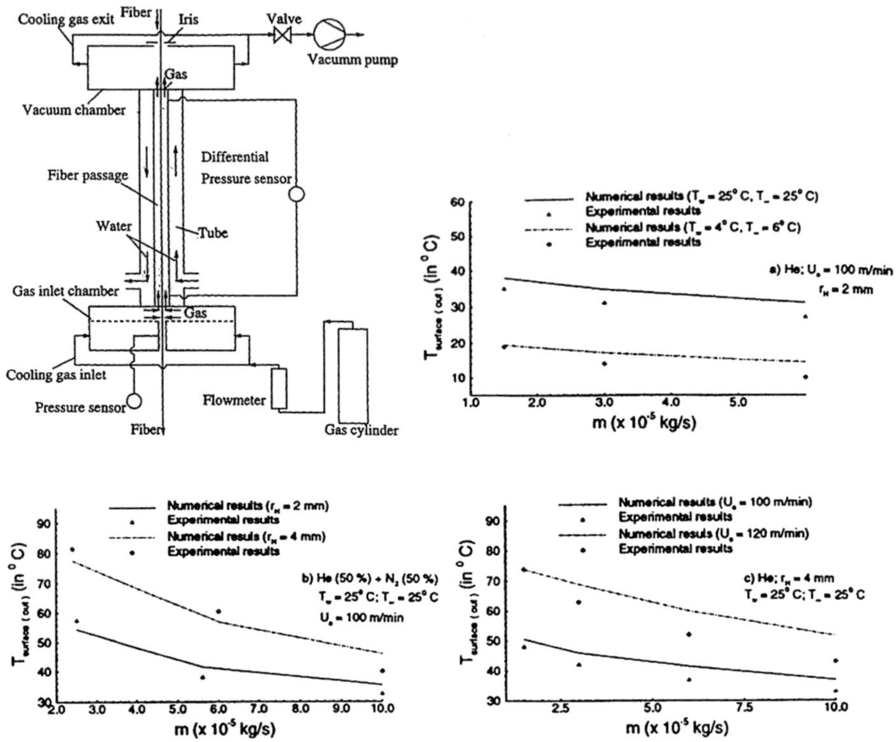

Fig. 8.27 Comparison of numerical results with experimental temperature data at the exit of a forced convection cooling channel with peripheral flow entrance for different conditions in fiber cooling (Roy Choudhury et al. 1994)

oppose fiber motion, respectively, increase or decrease the heat transfer, significantly affecting the process. Figure 8.27 shows the experimental arrangement and comparisons between numerical and experimental results for several circumstances, indicating good agreement. The dependence of the temperature decay on the operating conditions and on the flow configuration was determined. These results were also employed to generate guidelines for the design of accelerated cooling sections that can be used in production draw towers.

8.8 Fiber Coating

After the furnace drawing and cooling, optical fibers are coated with a jacketing material for protection against abrasion, to increase strength and to reduce stress-induced microbending losses. Typical primary coating thicknesses are on the order of 40–50 μm and are applied to the uncoated fiber, with a secondary coating of thickness in the range 50–100 μm generally applied to the coated fiber. In most cases, the basic coating process involves drawing the fiber of diameter around 125 μm through a reservoir of coating liquid, with inlet and outlet dies. Thus, a liquid coating is picked up by the fiber. This is immediately followed by a curing process of the polymer coating material to obtain a solidified coating. Figure 8.28a shows a schematic of a typical coating applicator and die system. The chamber, which contains the coating liquid, is typically 1–2 cm in diameter, and the entrance and exit dies are channels with appropriate diameters to yield gap thickness between the moving fiber and the die wall of 50–100 μm. Curable acrylates are commonly used for the coating material. Viscous shear due to the moving fiber results in a recirculation within the fluid. A balance between the various forces, such as surface tension, viscous, gravitational, and pressure forces, results in an upstream meniscus at the cup entrance; see Fig. 8.28b. Similarly, a downstream meniscus at the die exit results from a balance of forces, as shown in Fig. 8.28c. Figure 8.28d shows an instability, known as the *saw-tooth* instability, which frequently arises and leads to a breakdown of the upper meniscus, resulting in air entrapment in the coating. Centering forces within the tapered die contribute to the positioning of the fiber at the center of the die exit. Interest lies in obtaining coatings that are concentric, of uniform thickness, and free of particle inclusions or bubbles (Blyler and DiMarcello 1980; Paek 1986; Quere 1999; Simpkins and Kuck 2000).

Under ideal conditions at the entrance die, the coating liquid, which forms an upstream meniscus between the air, liquid, and the fiber with a dynamic contact angle, replaces the air entrained by the moving fiber. However, at high speeds, the meniscus breaks down, with saw-tooth instability, and air is entrained into the coating. The use of high draw rates requires consideration of alternate pressurized applicator designs, where pressure-induced motion of the coating material is used to reduce the shear at the fiber surface and help in the establishment of a stable free surface flow. Such pressurized dies have been found to minimize gas bubbles entrained at the coating cup entrance and then trapped within the coating layer. Many other designs of coating applicators have been developed over the years, including using vacuum above the fiber entrance and a microchannel at the entrance, to reduce the entrainment of gas bubbles.

The coating material is drawn out with the fiber at the die exit, forming a downstream meniscus. The shape and dimensions of this meniscus are primarily determined by a balance of viscous, gravitational, and inertia forces. It is critically important to control the characteristics of the coating, particularly its thickness, concentricity, uniformity, and bubble inclusions. These considerations have become particularly important as the coating speeds have been increased to values beyond 20 m/s to increase productivity and as interest in specialty fibers and fibers of different materials, including polymer fibers, has grown. The physical properties of the polymer coating materials, particularly the viscosity, are important in the coating process. As expected, surface tension has a significant effect on the free surface, which is the interface between liquid and gas, with respect to the shape, stability, and other characteristics. Surface tension affects the force balance at the free surface and thus affects the equilibrium shape of the interface. For typical optical

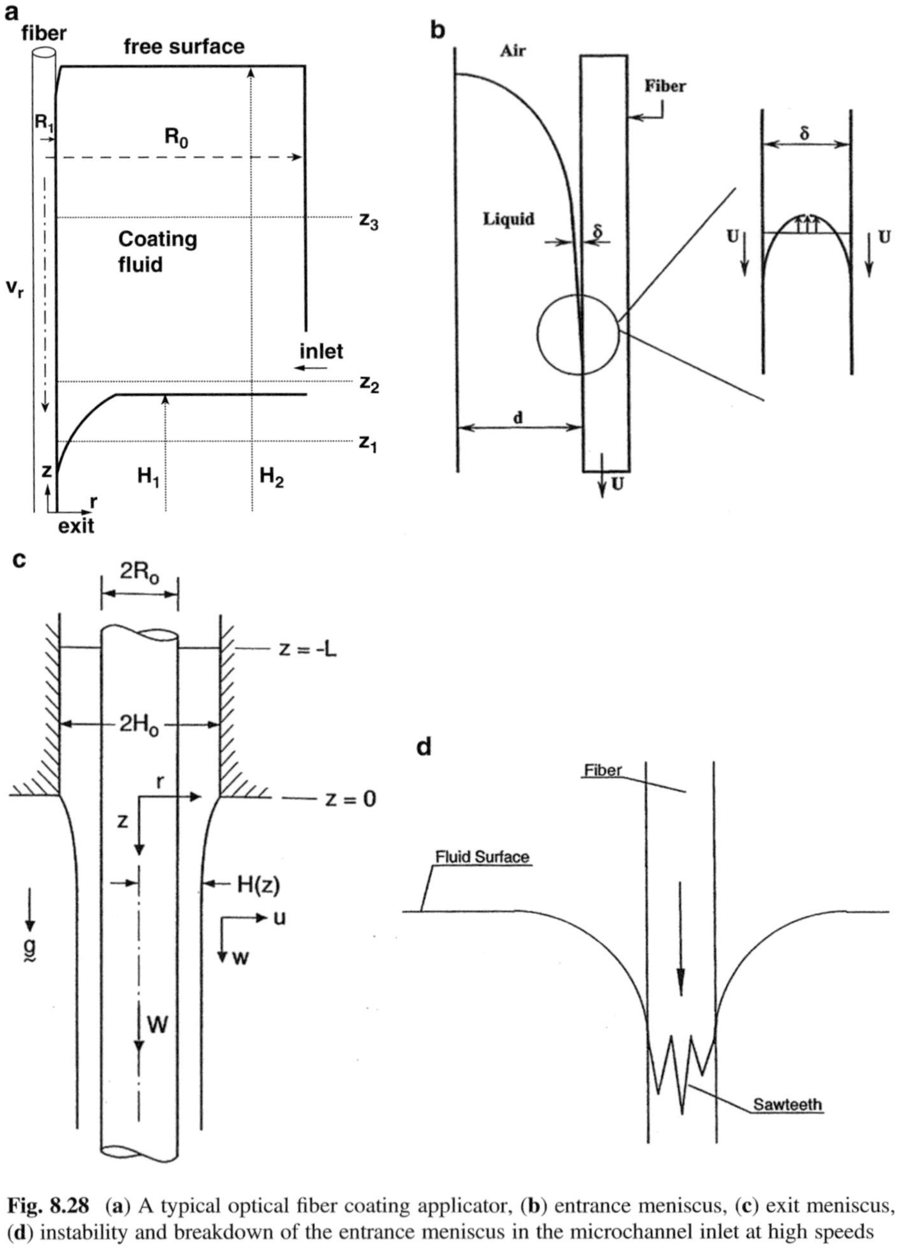

Fig. 8.28 (**a**) A typical optical fiber coating applicator, (**b**) entrance meniscus, (**c**) exit meniscus, (**d**) instability and breakdown of the entrance meniscus in the microchannel inlet at high speeds

fiber dimensions, surface tension effects are often large, and gravitational effects are small.

A few typical results from various studies are discussed here (Ravinutala et al. 2001; Ravinutala and Polymeropoulos 2002; Yoo and Jaluria 2007). The

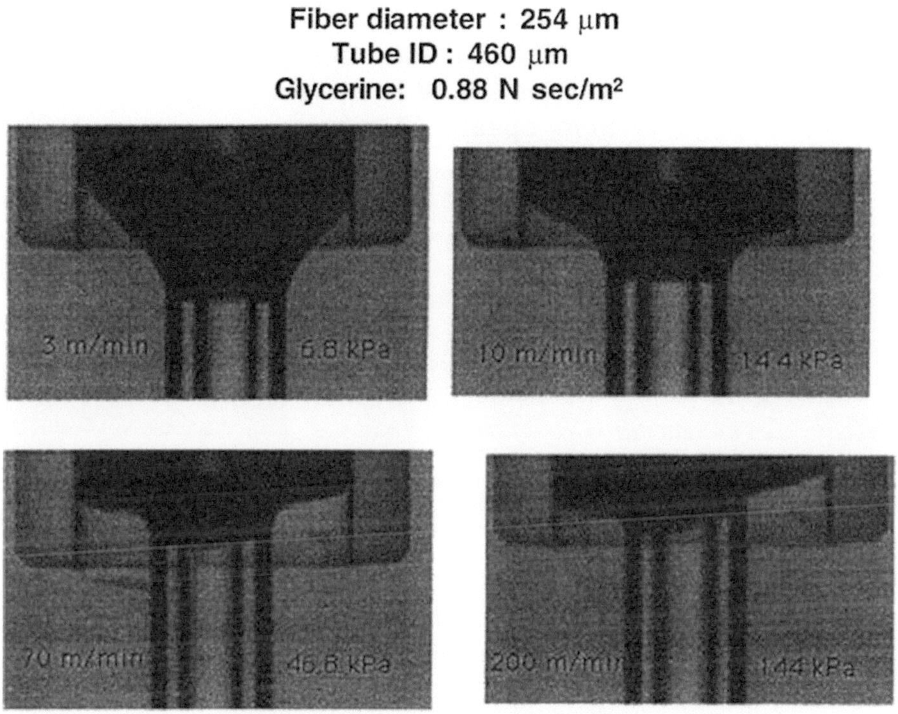

Fig. 8.29 Entrance flow in the microchannel inlet of annular gap thickness 103 μm, representing an inlet die, at different pressures and draw speeds (Ravinutala and Polymeropoulos 2002)

experimental work was carried out on a full-scale fiber coating facility to study the upstream and downstream menisci, with glycerin/water solutions taken as the test fluid. Glycerin is convenient for such studies because its viscosity at 20 °C is in the range of 1 N.s/m^2, which is similar to that used in coating applications. The inlet and exit dies can be changed so that different geometries and diameters can be investigated. Figures 8.29 and 8.30 show photographic images of the meniscus formed with the fiber moving through a micropipette inlet tube into the applicator. The fiber speed ranges from around 3 to 200 m/min. Different diameters and imposed pressures were employed. Figure 8.30c clearly shows the breakdown of the meniscus into saw-tooth patterns and tip streaming, as mentioned earlier. At higher imposed pressures, the meniscus images appeared to be smooth, suggesting suppression of large-scale breakdown at the same fiber speed. The unpressurized meniscus was found to generate a large number of relatively large air bubbles compared to the pressurized case. A comparison of the shape of the menisci showed that the effect of the imposed pressure was to flatten the meniscus and to increase the slope of the liquid-air interface near the fiber, compared to that for an unpressurized meniscus. This results in a smaller air volume available for entrainment, accounting for the difference

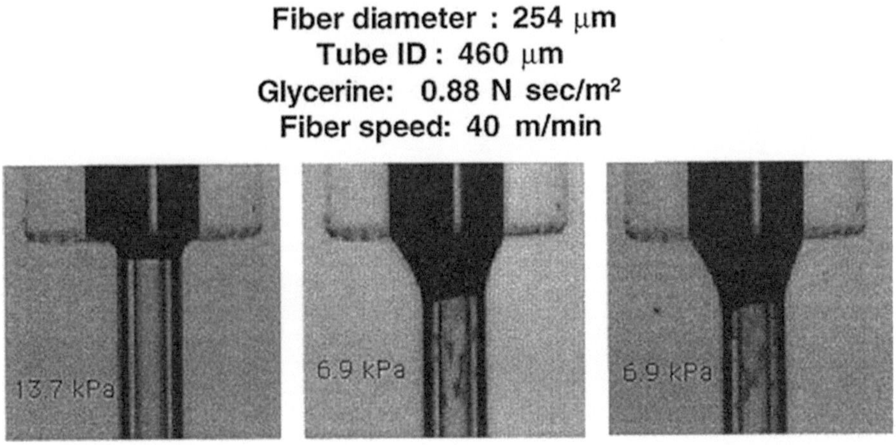

Fig. 8.30 Entrance flow in a microchannel with an annular gap thickness of 103 μm, representing an inlet die, at different pressures, showing breakdown at low pressure (Ravinutala and Polymeropoulos 2002)

between the figures. The effect of the increased imposed pressure is seen as both flattening of the meniscus shape and upward movement of the meniscus.

The observed trends are summarized in Fig. 8.31. The first part gives the fiber speed at which breakdown occurs as a function of the applicator pressure, indicating higher speeds for breakdown at larger pressure. Figure 8.31b shows the location of the meniscus in the microchannel, which forms the inlet die, as a function of the imposed pressure for a given fiber speed. Clearly, the pressure flattens and moves the meniscus upward, imparting greater stability and reducing the chances of a breakdown.

An important consideration in the coating process is the nature, characteristics, and shape of the exit meniscus, which represents the profile of the free surface as the fluid exits from the die due to the viscous drag from the moving fiber (Blyler and DiMarcello 1980; Paek 1999). An approach similar to that mentioned earlier with respect to the optical fiber drawing process can be used to determine the exit meniscus. Thus, the governing equations are solved to obtain the temperature and flow distributions, from which the shear at the free surface is determined. The additional forces due to gravity, surface tension, and external shear due to air are included to determine the overall force balance. The force imbalance is used to iterate, starting with a guessed profile, until the force balance is satisfied and a converged meniscus is obtained (Yoo and Jaluria 2008). Figure 8.32 shows the numerical results obtained, along with experimental data from Raj (2004). A fairly good agreement between the two is observed, indicating the validity and accuracy of the modeling.

The fluid flow and heat transfer in the coating applicator and die system have also been investigated. The problem may be simplified by specifying the shape of the meniscus on the basis of experimental data and assuming axisymmetric transport.

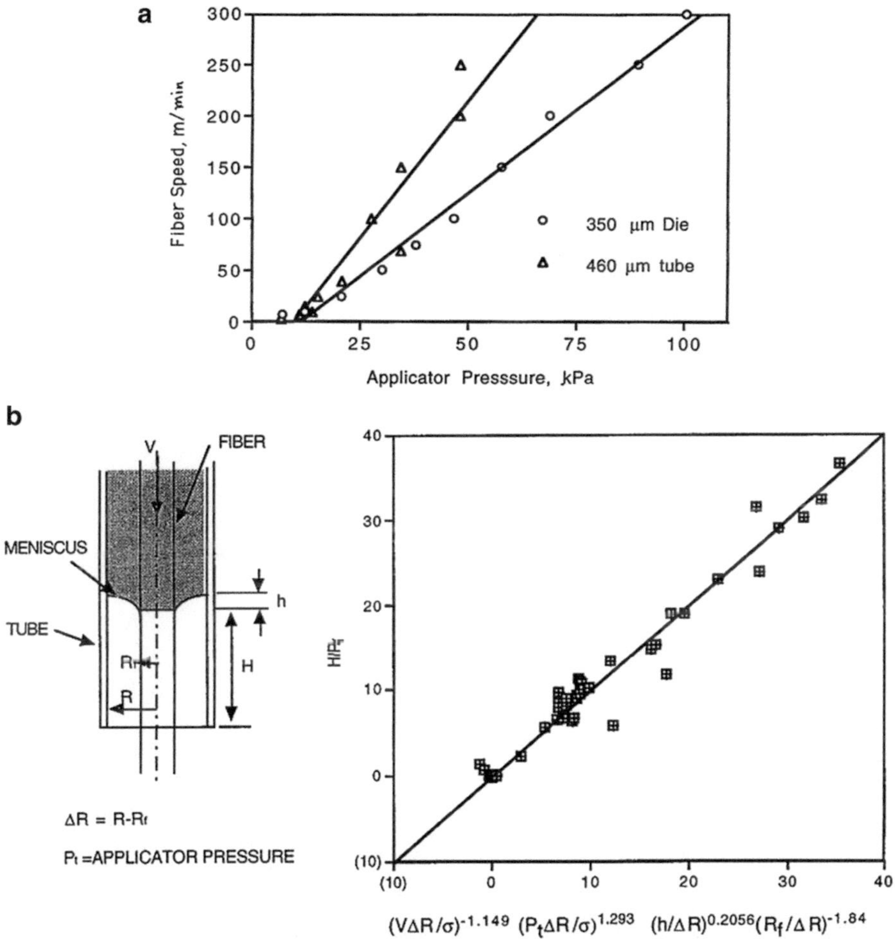

Fig. 8.31 (**a**) Dependence of fiber speed for meniscus breakdown, for a fiber entering a microchannel inlet, on the imposed pressure, (**b**) location of the meniscus in the microchannel as function of the pressure (Ravinutala and Polymeropoulos 2002)

The typical height of the meniscus varies from around 10 μm to 100 μm. Typical numerical results are shown in Fig. 8.33, in terms of the computed velocity field in the applicator, with a prescribed upper meniscus. Even though the flow near the meniscus changes significantly with a variation in the meniscus, the flow far away remains largely unaffected. A nanosecond double-pulse laser source was used to map and probe the flow field in the applicator, particularly in the vicinity of the moving optical fiber, using particle image velocimetry. Figure 8.34 shows a comparison between the numerical and experimental results in terms of the velocity profiles. Several other such measurements and computations were carried out. The agreement between the two provided strong support to the model.

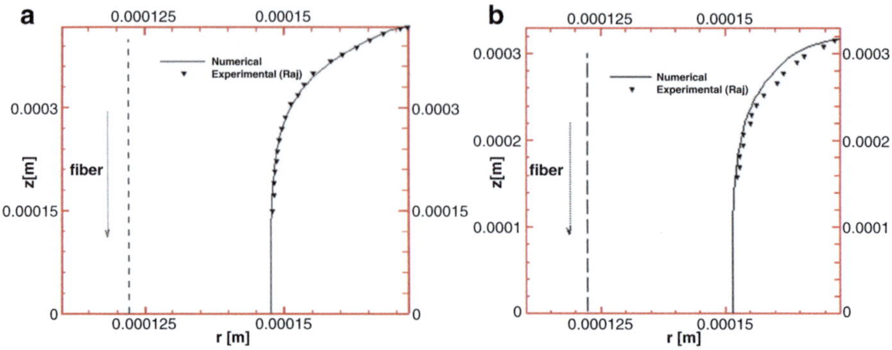

Fig. 8.32 Calculated meniscus at the exit of the coating die, along with experimental measurements of the profile, for glycerin at fiber speeds of (**a**) 20 m/min and (**b**) 75 m/min

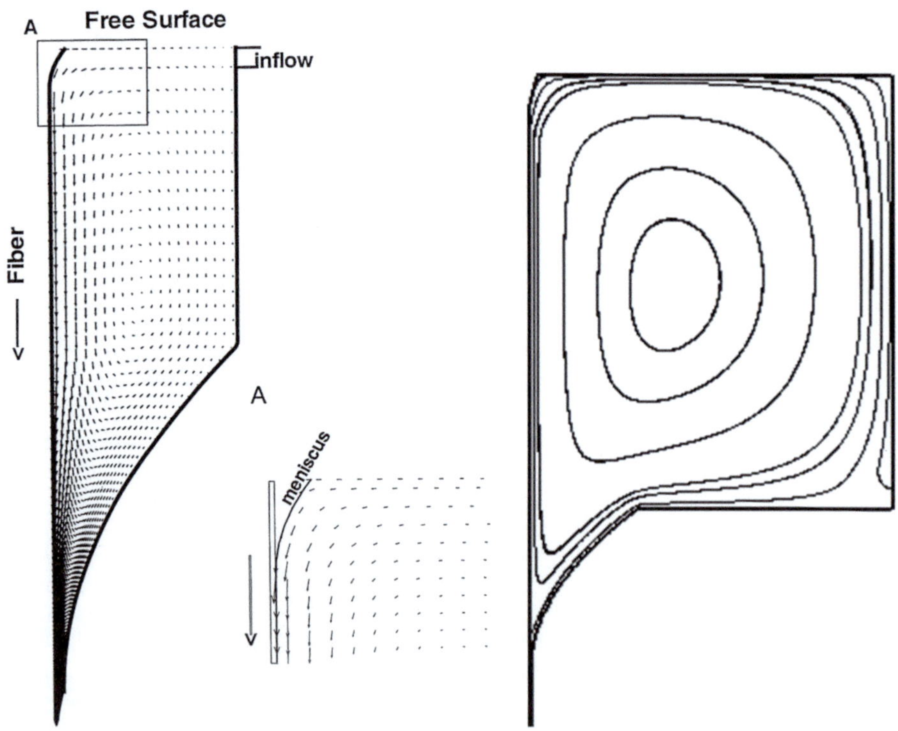

Fig. 8.33 Calculated flow field in the applicator and exit die, with a prescribed meniscus of height around 100 μm obtained experimentally

The exit die is a microchannel of varying cross-section. In most cases, it is a converging channel with the exit being the desired final diameter of the coating. However, other geometries are also used in practice for stability and better control

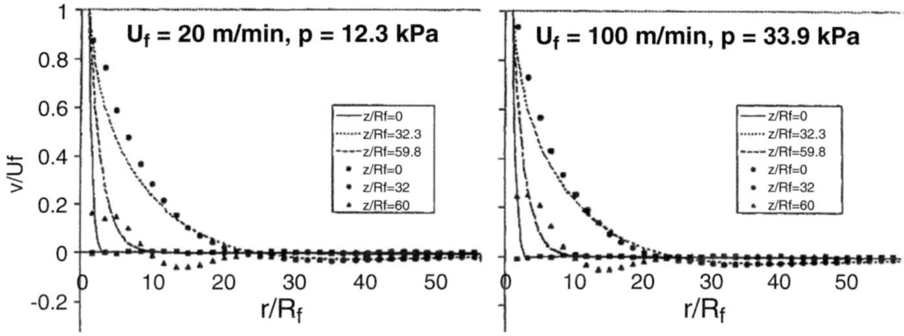

Fig. 8.34 Calculated and measured velocity distributions near the moving optical fiber

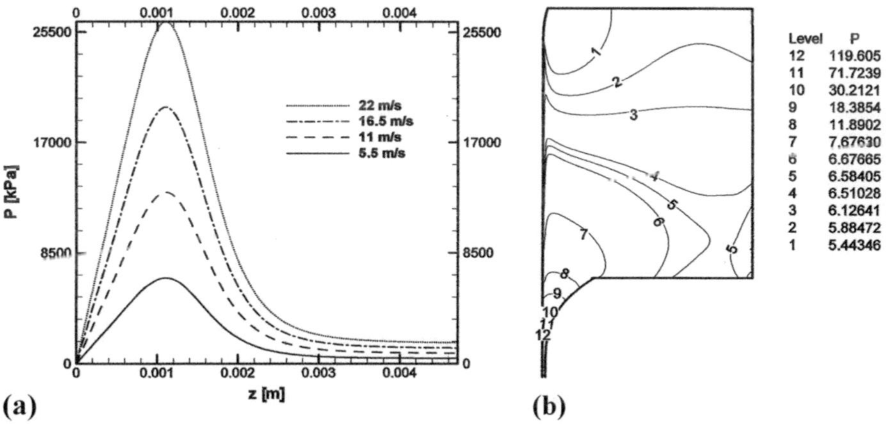

Fig. 8.35 Pressure distribution in the chamber and the exit die, which consists of a converging microchannel, for polymer coating of a moving fiber (**a**) under isothermal conditions and (**b**) when thermal effects are included

over coating characteristics. Figure 8.35a shows typical results obtained in terms of the pressure distribution in the applicator and the exit die for isothermal conditions. It is seen that a maximum arises somewhere near the midsection of the die. This pressure was found to be quite large due to the small diameter of the microchannel and to be much larger than the typical pressure in the applicator. Thus, the exit conditions are largely determined by the die shape and fiber speed, rather than the pressure in the applicator chamber. A properly designed die can help in controlling the thickness and quality of the coating. Thermal effects can also have a strong effect on the flow due to the dependence of fluid properties, particularly viscosity, on temperature. The corresponding pressure distribution, when thermal effects, due to heat transfer from the fiber, viscous dissipation, and heat loss to the surroundings are included, is shown in Fig. 8.35b. The pressures were found to decrease as the temperatures increase due to the reduction in fluid viscosity. The flow was also found to be more stable, leading to greater uniformity of the coating, as the temperatures were increased. Thus, the thermal conditions at the boundaries can be used to improve the quality of the coating.

8.9 Multiscale Modeling

It is clear from the preceding that large variations in length scales, as well as in time scales, occur in the optical fiber manufacturing process. The fiber and core diameters, as well as the coating thicknesses, are in the micrometer range, whereas the furnace and coating applicator are in centimeters, and the overall system is at engineering scales of several meters. The characteristics and quality of the material being processed are determined by the transport processes that occur at the micro- or nanometer scale in the material, for instance, in the formation of defects, collapse of the hollow core, and transport of dopants. The engineering aspects, such as the operating conditions, are concerned with the macroscale, involving practical physical dimensions, systems, and appropriate boundaries. Therefore, in order to model the fabrication process, it is crucial to link the modeling or experimentation at the micro- or nanoscale with those for the system (Jaluria and Yang 2011; Jaluria 2009). The different length scales lead to different dominant mechanisms and thus different governing equations and boundary conditions. Many of the results presented in the preceding sections were obtained with multiscale modeling for substantially different length scales.

As seen earlier, the profile of the neck-down region in the optical fiber drawing process is governed by the viscous flow of molten glass, which is in turn determined by the thermal field in the glass. The properties of the material, particularly the radiative properties, are affected by microscale dopants added to the material in order to enhance or modify the optical behavior of the fiber. These properties affect the heating of the preform and fiber, making it important to accurately determine the properties as functions of dopant concentrations. The transport of the dopants in the material flow during the process is also important in the determination of the characteristics of the final fiber obtained. Thermally induced defects arise due to the breaking of the $Si = O$ bonds, micro-cracks result from the tension, and other defects are caused by local instabilities. All these occur at small length scales, and their generation kinetics need to be determined by modeling or experimentation. Similarly, the behavior of the core diameter in hollow fibers and of the holes in microstructured fibers, particularly their collapse, needs to be determined as a function of the processing parameters. The governing equations are significantly different in the region close to the collapse and in regions far away, mainly due to the dominance of surface tension effects in the collapse. Again, the operating conditions are imposed at the system level. In coating as well, the meniscus is typically a few microns in height, whereas the applicator is at macroscale. In all these cases, the modeling at different length scales may be carried out separately and then coupled to obtain the overall simulation. If the properties are assumed to be independent of structural changes, the problem becomes decoupled, so that the transport processes in the system can be modeled and then the microscale changes can be determined from the local conditions using the available kinetics and other models. An understanding of the microscale mechanisms that determine material characteristics is important, but these must be linked with the boundary conditions that are usually imposed at the macroscale level in the system. A wide range of multiscale problems have been considered in the literature, along with approaches to model the different scales and link the different simulations (Joshi 2012; He and Tao 2012).

As discussed in detail here, analytical/numerical modeling and experimentation have resulted in a detailed, quantitative understanding of the optical fiber manufacturing process and of the basic thermal transport mechanisms that arise. The effects of various operating conditions and design parameters are determined, in order to predict the transport process undergone by the material and the transformations that occur. However, one of the major goals of such a detailed basic study is to employ the results obtained for designing and optimizing an appropriate thermal system. The feasibility of the drawing process was considered and this aspect may be used to obtain the acceptable design domain that meets the given requirements without violating constraints arising from temperature, pressure, and tension limitations, as well as those from size, cost, energy, and other considerations. A feasible or acceptable design is generally not the best or optimal design, as judged on the basis of cost, efficiency, quality, production cost, or other such measures. Thus, an important parameter in optimization is the quantity or objective function U, which is to be minimized or maximized (Jaluria 2008). Because of the complexity and nonlinearity of these processes, search methods are among the most important optimization strategies for materials processing. A number of designs are systematically generated and the best among these is selected. More is said on the optimization of manufacturing systems later in the book.

8.10 Summary

This chapter discusses the fabrication of optical fibers, focusing on the drawing, cooling, and coating of fibers. The basic transport mechanisms that arise are discussed, along with results from analytical, numerical, and experimental studies. Starting with the fabrication of the preform, the paper discusses the thermal transport in the draw furnace and the flow in silica glass. Of particular interest are the neck-down profile, defects generated during the process, feasibility of the process, and possible optimization of the process. The consideration is also extended to hollow, microstructured, and doped fibers that are of interest in different applications. The cooling and the liquid polymer coating processes are similarly considered, to discuss the underlying phenomena and present characteristics results. In all these considerations, a particular aspect that is brought up is the quality of the fiber, as well as of the coating, in terms of defects, imperfections, entrapment of bubbles in the coating, dopant distribution, viscous rupture, and so on. Also considered is the production rate, as indicated by the drawing speed and the ranges over which operating conditions may be varied to obtain a high quality coated fiber with low transmission loss.

Clearly, many important aspects need further study. Of particular concern is the availability of accurate property data at the applicable temperature levels. Radiation properties and viscosity for various doped fibers are of particular interest because of the current lack of data at drawing temperatures. Experimental data on the various steps during the fabrication process are critically needed for physical insight and for validation, improvement, and development of analytical/numerical models. The transport processes at the micro- and nanometer scales, where most of the changes in the material occur, need further detailed investigation, along with linking with the operating conditions that are imposed at the system level. The effect of the changes at the small length scales on the characteristics of the final optical fiber needs to be accurately determined. Further work is also needed on feasibility, process control, design, and optimization, particularly for multiobjective optimization that is of interest in this process.

As has been pointed out at a few places in this chapter, even though the presentation is directed at the fabrication of optical fibers, many of the basic aspects, methods for analysis, concerns, and trends are similar to those applicable to other manufacturing processes. For instance, cable manufacture involves wire drawing, which is similar to fiber drawing; coating of the wires, which is similar to the coating process discussed here; a fairly similar cooling process; and many similar concerns with productivity and quality. There are similar drawing, cooling, and coating processes that arise in other systems. The approaches used to determine feasibility of the process and to optimize the system are often quite similar. Some of these similarities are discussed in other chapters as well in order to link the basic aspects of different manufacturing processes and thus impact on new and existing processes and systems.

References

L.L. Blyler, F.V. DiMarcello, Fiber drawing, coating and jacketing. Proc. IEEE **68**, 1194–1198 (1980)

L.C. Burmeister, *Convective Heat Transfer*, 2nd edn. (Wiley Interscience, New York, 1993)

S.S. Chakravarthy, W.K.S. Chiu, Collapse prediction during hollow optical fiber fabrication, in *Proceedings of the 2005 ASME Summer Heat Transfer Conference*, San Francisco, Paper No. HT2005–72508, 2005

C. Chen, Y. Jaluria, Numerical simulation of transport in optical fiber drawing with core-cladding structure. J. Heat Transf. **129**, 559–567 (2007)

C. Chen, Y. Jaluria, Effects of doping on the optical fiber drawing process. Int. J. Heat Mass Transf. **52**, 4812–4822 (2009)

X. Cheng, Y. Jaluria, Feasibility of high speed furnace drawing of optical fibers. J. Heat Transf. **126**, 852–857 (2004)

X. Cheng, Y. Jaluria, Effect of furnace thermal configuration on optical fiber heating and drawing. Num. Heat Transf. **48**, 507–528 (2005)

J. Cho, J. Kim, M. Choi, An experimental study of heat transfer and particle deposition during the outside vapor deposition process. Int. J. Heat Mass Transf. **41**, 435–445 (1998)

R.F. Cregan, B.J. Mangan, J.C. Knight, T.A. Birks, P.S.J. Russell, P.J. Roberts, D.C. Allan, Single mode photonic band gap guidance of light in air. Science **285**, 1537–1539 (1999)

A.D. Fitt, K. Furusawa, T.M. Monro, C.P. Please, Modeling the fabrication of hollow fibers: Capillary drawing. J. Lightwave Technol. **19**, 1924–1931 (2001)

J.D. Fleming, Fused Silica Manual, Final Report for the U. S. Atomic energy commission, Oak Ridge, TN, Project B-153, 1964

W.A. Gifford, A finite element analysis of isothermal fiber formation. Phys. Fluids **25**, 219–225 (1982)

S.L. Goren, S. Wronski, The shape of low-speed jets of Newtonian liquids. J. Fluid Mech. **25**, 185–198 (1966)

H. Hanafusa, Y. Hibino, F. Yamamoto, Formation mechanism of drawing-induced E′ centers in silica optical fibers, J. Appl. Phys. **58**, 1356–1361 (1985)

J.A. Harrington, *Infrared Fibers and their Applications* (SPIE Pub, Bellingham, 2004)

Y.L. He, W.Q. Tao, Multiscale simulations of heat transfer and fluid flow problems. J. Heat Transf. **134**, 031018–1–031018–13 (2012)

J. Hecht, *Understanding Fiber Optics* (Prentice Hall, Upper Saddle River, 2002)

Y. Hibino, H. Hanafusa, Defect structure and formation mechanism of drawing- induced absorption at 630 nm in silica optical fibers. J. Appl. Phys. **60**, 1797–1801 (1986)

J. Issa, Z. Yin, C.E. Polymeropoulos, Y. Jaluria, Temperature distribution in an optical fiber draw tower furnace. J. Mater. Proc. Mfg. Sci. **4**, 221–232 (1996)

T. Izawa, S. Sudo, *Optical Fibers: Materials and Fabrication* (KTK Scientific Publishers, Tokyo, 1987)

T. Izawa, N. Shibata, A. Takeda, Optical attenuation in pure and doped fused silica in the IR wavelength region. Appl. Phys. Lett. **31**, 33–35 (1977)

Y. Jaluria, Numerical modeling of manufacturing processes, W.J. Minkowycz, E.M. Sparrow, J. Murthy, (eds.), in *Handbook of Numerical Heat Transfer* (Wiley, New York, 2006) p. 729–784

Y. Jaluria, *Design and Optimization of Thermal Systems*, 2nd edn. (CRC Press, Boca Raton, 2008)

Y. Jaluria, Microscale transport phenomena in materials processing. J. Heat Transf. **131**, 033111–1–033111–17 (2009)

Y. Jaluria, Thermal issues in materials processing. J. Heat Transf. **135**, 061701–061714 (2013)

Y. Jaluria, J. Yang, A review of microscale transport in the thermal processing of new and emerging advanced materials. J. Heat Transf. **133**, 060906–060914 (2011)

M.G. Jochem, W.C. Ligt, Cooling and bubble-free coating of optical fibers at high drawing rate. Lightwave Tech. **4**, 739–742 (1986)

Y. Joshi, Reduced order thermal models of multiscale microsystems. J. Heat Transf. **134**, 031008–1–031008–11 (2012)

H.G. Landau, Heat conduction in a melting solid. Q. Appl. Math. **8**, 81–94 (1950)

S.H.-K. Lee, Y. Jaluria, Simulation of the transport process in the neck-down region of a furnace drawn optical fiber. Int. J. Heat Mass Transf. **40**, 843–856 (1997)

T. Li (ed.), *Optical Fiber Communications, Vol. 1: Fiber Fabrication* (Academic Press, New York, 1985)

Y. Liu, C.E. Polymeropoulos, Measurement and prediction of thermocouple probe temperatures within glass rods subjected to radiative heating, in *Proceedings of the 11th International Heat Transfer Conference*, Kyongju, Korea, Taylor and Francis, Philadelphia, 4, pp. 27–32, 1998

S. Matsuo, T. Nunome, T. Yoshita, T. Hamada, K. Himeno, Design optimization of trench index profile for the same dispersion characteristics with SMF, OFC/NFOEC'2007, Paper JWA2, 2007

M. Modest, *Radiative Heat Transfer*, 3rd edn. (Academic Press, New York, 2013)

T.M. Monro, W. Belardi, K. Furusawa, J.C. Baggett, N.G.R. Broderick, D.J. Richardson, Sensing with microstructured optical fibres. Meas. Sci. Technol. **12**, 854–858 (2001)

H. Murata, *Handbook of Optical Fibers and Cables*, 2nd edn. (Marcel Dekker, New York, 1996)

M.R. Myers, A model for unsteady analysis of preform drawing. AICHE J. **35**, 592–602 (1989)
K. Oh, U.C. Paek, *Silica Optical Fiber Technology for Devices and Components: Design, Fabrication and International Standards* (Wiley, New York, 2012)
U.C. Paek, High-speed high-strength fiber drawing. J. Lightwave Tech., LT **4**, 1048–1060 (1986)
U.C. Paek, Free drawing and polymer coating of silica glass optical fibers. J. Heat Transf. **121**, 774–788 (1999)
U.C. Paek, R.B. Runk, Physical behavior of the neck-down region during furnace drawing of silica fibers. J. Appl. Physiol. **49**, 4417–4442 (1978)
U.C. Paek, C.M. Schroeder, C.R. Kurkjian, Determination of the viscosity of high silica glasses during fibre drawing. Glass Technol. **29**, 263–266 (1988)
D. Quere, Fluid coating on a fiber. Annu. Rev. Fluid Mech. **31**, 347–384 (1999)
A. Raj, Pressure effects on the exit meniscus of an optical fiber coating applicator. MS thesis, Rutgers University, New Brunswick, 2004
S. Ravinulata, C.E. Polymeropoulos, Entrance meniscus in a pressurized optical fiber coating applicator. Int. J. Exp. Heat Transfer Fluid Mech **26**, 573–580 (2002)
S. Ravinutala, C.E. Polymeropoulos, Y. Jaluria, G. Elliott, G. Finberg, Experimental and calculated velocity distributions within a pressurized optical fiber coating applicator, in *Proceeding of the 50th International Wire Cable Symposium*, Orlando, 2001
S. Roy Choudhury, Y. Jaluria, Thermal transport due to material and gas flow in a furnace for drawing an optical fiber. J. Mater. Res. **13**, 494–503 (1998a)
S. Roy Choudhury, Y. Jaluria, Practical aspects in the drawing of an optical fiber. J. Mater. Res. **13**, 483–493 (1998b)
S. Roy Choudhury, Y. Jaluria, T. Vaskopoulos, C.E. Polymeropoulos, Forced convective cooling of optical fiber during drawing process. J. Heat Transf. **116**, 790–794 (1994)
S. Roy Choudhury, Y. Jaluria, S.H.-K. Lee, A computational method for generating the free-surface neck-down profile for glass flow in optical fiber drawing. Num. Heat Transf. **35**, 1–24 (1999)
P. Simpkins, V. Kuck, Air entrapment in coatings via tip streaming meniscus. Nature **403**, 641–643 (2000)
P. Tandon, Fundamental understanding of processes involved in optical fiber manufacturing using outside vapor deposition method. Int. J. Appl. Ceramic. Tech. **2**, 504–513 (2005a)
P. Tandon, Doping of silica during sintering. J. Non-Cryst. Solids **35**, 1466–1472 (2005b)
P. Tandon, M.-J. Li, D.C. Bookbinder, S.L. Logunov, E.J. Fewkes, Nano-engineered optical fibers and applications. Nanophotonics **2**, 383–392 (2013)
M.A. van Eijkelenborg, A. Argyros, G. Barton, I.M. Bassett, M. Fellew, G. Henry, N.A. Issa, M.C.J. Large, S. Manos, W. Padden, L. Poladian, J. Zagari, J., recent progress in microstructured polymer optical fibre fabrication and characterisation. Opt. Fiber Technol. **9**, 199–209 (2003)
T. Vaskopoulos, C.E. Polymeropoulos, A. Zebib, Heat transfer from optical fiber during the draw process. J. Mater. Process. Manuf. Sci. **1**, 261–271 (1993)
Z. Wei, K. Lee, S.W. Tchikanda, Z. Zhou, S. Hong, Free surface flow in high speed fiber drawing with large-diameter glass preforms. J. Heat Transf. **26**, 713–722 (2004)
J. Yang, Numerical modeling of hollow optical fiber drawing process, Ph.D. Thesis, Rutgers University, Piscataway, 2008
J. Yang, Y. Jaluria, Transport processes governing the drawing of a hollow optical fiber. J. Heat Transf. **131**, 072102–1–072102–9 (2009a)
J. Yang, Y. Jaluria, Feasibility and optimization of the hollow optical fiber drawing process. Int. J. Heat Mass Transf. **52**, 4108–4116 (2009b)
Z. Yin, Y. Jaluria, Thermal transport and flow in high-speed optical fiber drawing. J. Heat Transf. **120**, 916–930 (1998)
S.Y. Yoo, Y. Jaluria, Fluid flow and heat transfer in an optical fiber coating process. Int. J. Heat Mass Transf. **50**, 1176–1185 (2007)
S.Y. Yoo, Y. Jaluria, Numerical simulation of the meniscus in the non-isothermal free surface flow at the exit of a coating die. Num. Heat Transf. **53**, 111–131 (2008)

Chapter 9
Other Manufacturing Processes

In the preceding chapters, several important manufacturing and materials processing techniques were considered, and typical numerical and experimental results were presented. Besides the basic transport processes, some practical aspects were also pointed out. However, materials processing is a vast field, and it is not possible to cover all the different methods that are currently being used, as well as new and emerging techniques. There are obviously many more manufacturing processes in which transport mechanisms are of critical importance and further work on modeling, simulation, design, and optimization is needed. In many of these, the concerns and requirements are quite different, though the basic aspects are similar to the ones outlined earlier and similar approaches may be applied to study them.

Some of the additional processes that may be mentioned are:

Heat treatment
Joining and bonding processes
Fabrication of microchannels and microdevices
Additive manufacturing
Thermal sprays
Laser processing
Powder metallurgy
Machining and material cutting
Microgravity materials processing

These processes involve the basic transport mechanisms that have been discussed in the preceding chapters, and thus their study may be based on what has been presented earlier. This chapter discusses some of major additional considerations in these processes.

Y. Jaluria, *Advanced Materials Processing and Manufacturing*, Mechanical Engineering Series, https://doi.org/10.1007/978-3-319-76983-7_9

9.1 Heat Treatment

Heat treatment is a common and important step in many manufacturing processes. Basically, it involves heating and subsequent cooling of an item in order to make it undergo a given temperature cycle, allowing it to obtain desired physical properties and characteristics, such as becoming harder, tougher, or softer. Though heat treatment is of particular importance and interest in metals and alloys, it is also used for various other materials such as ceramics, glass, and semiconductors. Among the common techniques used are annealing, hardening, normalizing, tempering, and case hardening. Traditionally, many of these processes were directed at different steels because of the overwhelming interest in these materials. However, much of the focus has shifted to other materials and these processes are applied, with appropriate variations in the temperature cycle, to achieve the desired properties. For instance, annealing is used extensively for a wide range of materials. As mentioned earlier and sketched in Fig. 1.2b, annealing is used to soften the material and remove residual stresses to make the material more ductile for further processing. The material to be annealed is heated beyond its recrystallization temperature, at which point the distorted or deformed grain structure of the material is replaced by a strain or defect-free one, maintained or soaked at this temperature level for some time for temperature homogeneity, and then cooled slowly to obtain a structure with reduced stress and more ductile material. Generally, this slow cooling is obtained by cooling in the furnace itself, with the heat turned off.

Similarly, other processes involve heating the material to an elevated temperature; maintaining it at this level for some time, which depends on the size of the item and the material properties; and then cooling, as shown in Fig. 3.15. Depending on the rate of cooling, the material properties obtained are different. For instance, a rapid cooling of steel held beyond the critical temperature, where steel is in a form known as austenite, by quenching in water or oil results in the formation of martensite, a hard form of crystalline steel structure and which imparts hardness to the material. Similarly, normalizing, which removes stresses, refines the grain structure, and improves machinability in steel, is obtained by cooling in air by natural convection, with the rate of cooling somewhat faster than for annealing. Spheroidizing is used to obtain small, rounded, globules and is obtained by heating the item to a temperature below the critical temperature at which austenite starts forming, followed by slow cooling. Tempering involves heating to a temperature below the critical temperature and then quenching to harden the steel and increase its toughness. Figure 9.1 shows a schematic of these different cooling processes. The elevated temperature level needed for a given operation, as well as the cooling rate, is determined by the material, particularly its composition if it is an alloy. The soaking time is determined by the need to obtain essentially uniform temperature in the item undergoing heat treatment, and thus it depends on the thermal conductivity and size of the item. The phase diagram of the material under consideration, such as that for steel shown in Fig. 9.2, is used to obtain the desired temperature cycle (Ghosh and Mallik 1986). Different crystalline structures for steel are also indicated in this

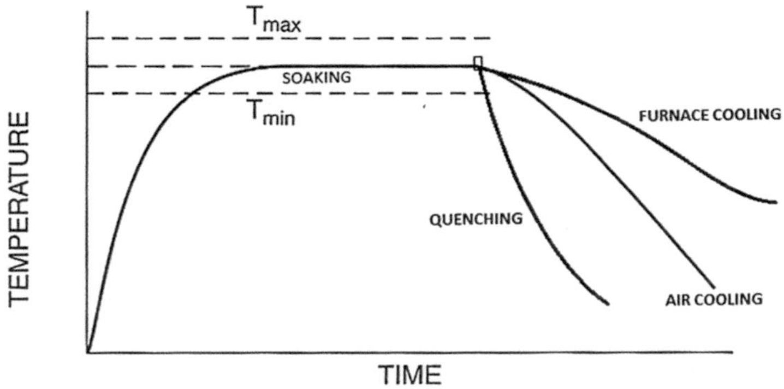

Fig. 9.1 Temperature variation to obtain different material characteristics by varying the rate of cooling from a temperature beyond a critical value, such as the recrystallization temperature

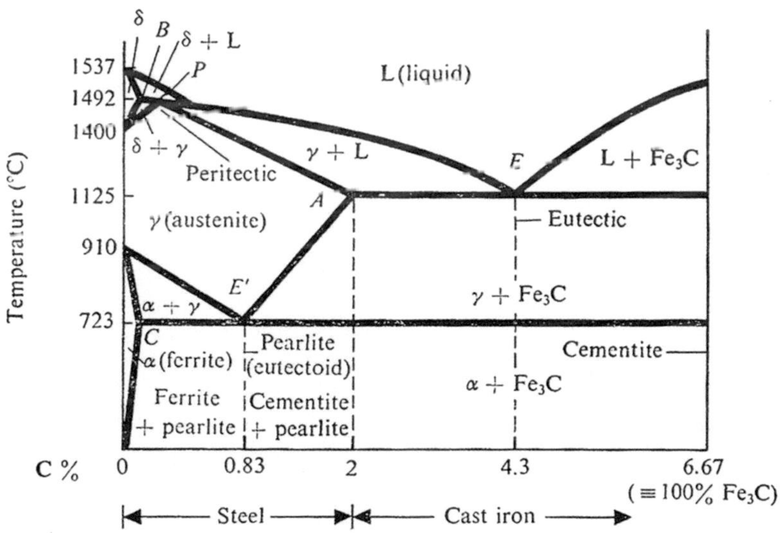

Fig. 9.2 Iron-carbon equilibrium phase diagram for steel (Ghosh and Mallik 1986)

figure. Similarly, phase diagram for other alloys and materials is available in the literature (Black and Kohser 2017).

9.1.1 Basic Aspects

Even though the heat treatment process mainly involves heating, followed by cooling, to obtain a given time-dependent temperature variation in the material, there are several basic issues that have to be addressed for a satisfactory product. The metallurgical and structural changes are coupled with the temperature variation, and consequently, the desired thermal cycle for a given process is strongly dependent on the material. A detailed information on the effect of temperature on the material microstructure, characteristics, and properties is necessary. Once the desired temperature cycle is specified, the boundary conditions needed to achieve the cycle must be determined. This is an inverse problem, as mentioned earlier, and involves finding the conditions that would lead to a desired result. The solution is typically non-unique, and optimization is used to narrow the domain of uncertainty (Orlande et al. 2011; Ozisik 2000; Orlande 2012). Also, the imposition of boundary conditions is linked with the system and would involve its various components, making the problem quite complicated for practical processes.

The temperature distribution in the solid material is determined by the heat transfer mechanisms, as given by the equation

$$\nabla.(k\nabla T) = \rho C\frac{\partial T}{\partial t} \tag{9.1}$$

where the properties are generally temperature dependent and C is the specific heat of the solid. The boundary conditions are typically convection plus radiation during heating and convection during cooling. For soaking, a balance between heating and cooling is maintained so that the material is able to attain essentially uniform temperature. It is desirable to reduce the heating time to increase productivity, but high heat fluxes lead to larger temperature differences in the material and resulting higher thermal stresses. Therefore, constraints on heating rates arise due to material limitations and item size.

These considerations are similar to those discussed in Chap. 6 with respect to plastic thermoforming, regarding heating, soaking, and cooling. Considering a similar practical configuration of a solid rod encased in a polymer insulation, Fig. 9.3 shows the temperature increase with time for a constant heat flux input at the surface. The temperatures increase linearly a short time after the onset of the process, as expected for a constant energy input. Clearly, the rate of temperature increase can be enhanced by increasing the heat flux, but the temperature difference across the material also increases. Also, the heat input must be cut off before the surface temperature exceeds the allowable value at which point the material is damaged. Figure 9.4 shows the situation with pure convective heating. The rates of temperature rise decrease with time, but there is better control on the temperature, since the surface temperature is constrained by the fluid temperature. Therefore, a combination of radiative and convective transport is the best approach, and the inverse problem may be solved to obtain the appropriate conditions. For metals and alloys, the temperatures are generally much higher, and thermal stresses lead to similar constraints on heating rates. The cooling rates are governed by the desired material characteristics such as hardening and annealing.

Figure 9.5 shows the temperature at constant heat flux input, followed by different rates of convective cooling. The heat flux is cut off at the maximum allowable temperature, and no soaking of the material is considered. However,

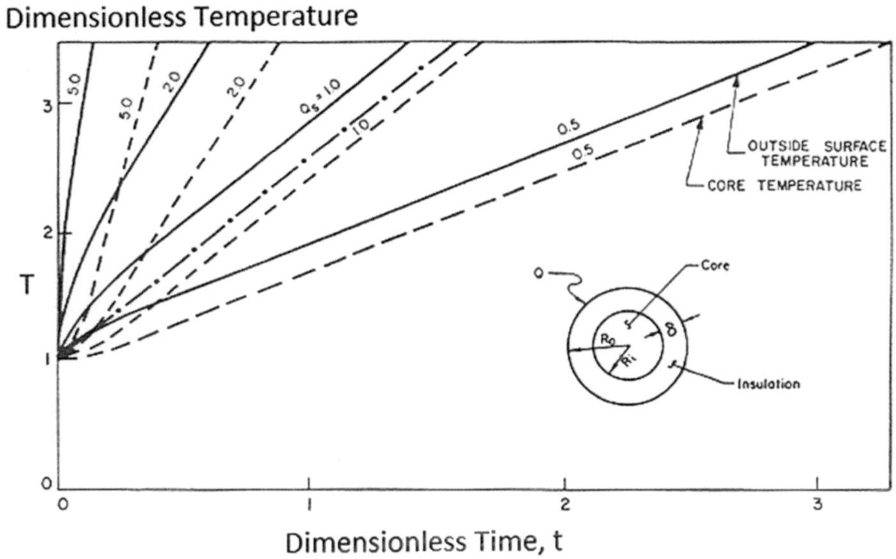

Fig. 9.3 Transient temperature response in an insulated cylindrical rod to a constant heat flux input at the outer surface

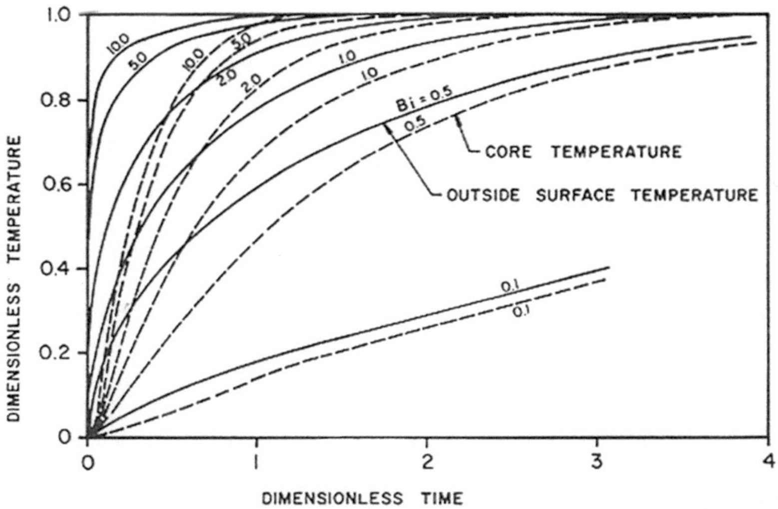

Fig. 9.4 Temperature variation in an insulated cylindrical rod for convective heat input at the outer surface

unless the component is relatively small, with a low thermal capacity, soaking is necessary to ensure essentially uniform temperature in the component to obtain the desired heat treatment throughout the material. For soaking, the heat input may be controlled, using various available control strategies, to keep the surface temperature

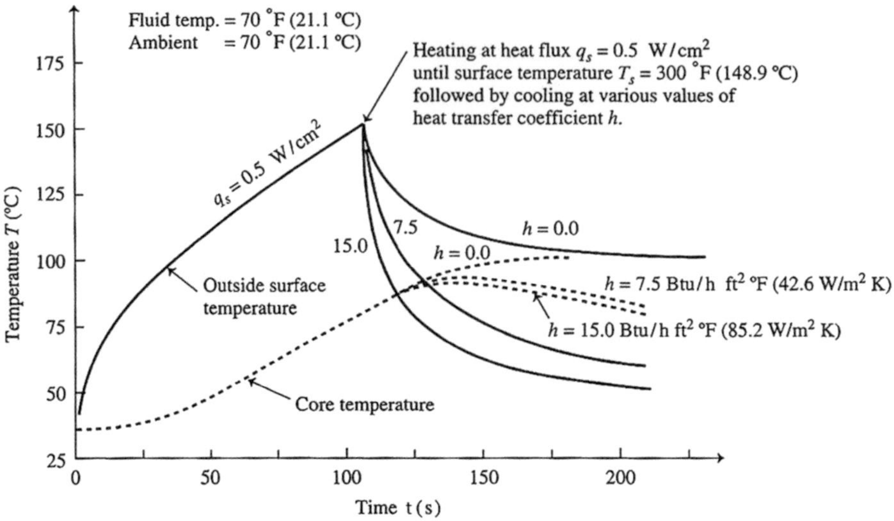

Fig. 9.5 Results for heating at constant heat flux, followed by convective cooling

within the two limits of $T_{\max}$ and $T_{\min}$, while the remaining material heats up to the desired temperature level. However, if interest lies in heat treatment only in a region close to the surface, soaking time may be reduced or eliminated, depending on the component size and material. Thus, this figure illustrates the use of radiation and convection heat transfer to control the temperature cycle to which the material is subjected.

In view of the preceding discussion, we may list the basic aspects in a heat treatment process as:

Material considerations for desired temperature cycle
Constraints and/or requirements on rate of temperature increase and decrease
System considerations for imposed boundary conditions
Solution of inverse problem to obtain the appropriate boundary conditions
Optimization to narrow the domain of non-unique solutions
Control system for the temperature cycle, particularly during soaking
Dependence of material properties and characteristics on temperature
Different heating methods such as electric resistance or induction heating, gas, and laser
Possible inclusion of mass transfer

9.1.2 A Batch Annealing Furnace

An example of a heat treatment system is shown in Fig. 9.6, which gives a sketch of a batch annealing furnace for steel sheets rolled up into cylindrical coils (Harvey 1977; Jaluria 1984, 1988; Stikker 1970). The material has to be heated beyond its

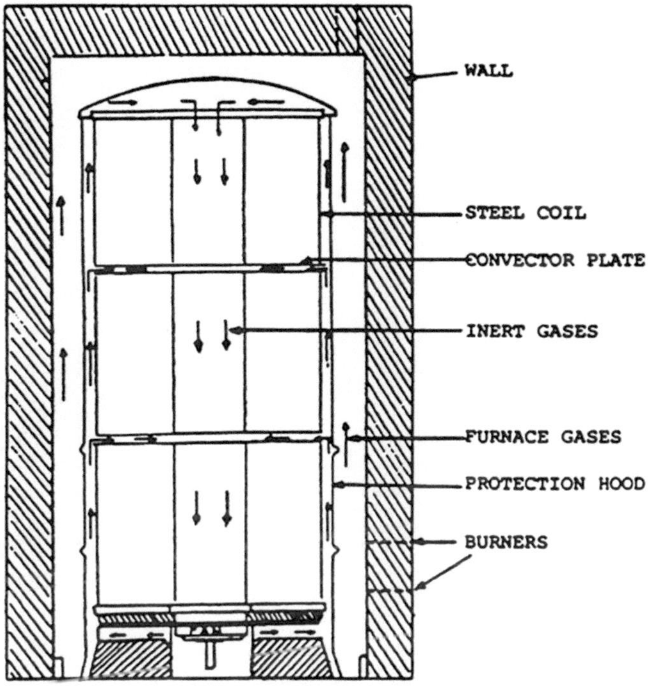

Fig. 9.6 Sketch of a batch annealing furnace for steel

recrystallization temperature of around 723 °C and then cooled very slowly to achieve the desired annealing process, which removes residual stresses and restores ductility. Figure 9.7 shows, qualitatively, the desired temperature cycle, which involves initial slow cooling in the furnace and then faster cooling with the furnace top removed. The faster cooling is undertaken after reaching the temperature by which the grain structure is settled and annealing is completed. The boundary conditions, particularly the time-dependent flow of flue gases from the blast furnace, which provides the heat input into the furnace, need to be determined to obtain the desired temperature variation with time in the coils. The basic considerations mentioned earlier with respect to the heat treatment process are considered in deciding the desired temperature cycle.

This problem has to be numerically modeled and simulated to obtain the dependence of the temperature distributions on the boundary conditions. Different components, or subsystems, of the furnace are first modeled and then coupled to obtain the model for the entire system, as outlined in Chap. 3. Figure 9.8 illustrates this process by showing some of the major components of this practical batch annealing furnace for steel sheets. In the coiled steel sheets, the thermal conductivity k_r in the radial direction is often much smaller than k_z in the axial direction, due to gaps within the coils, and the governing conduction equation may be written taking this effect into account. Simplifications are made wherever possible. For instance, the

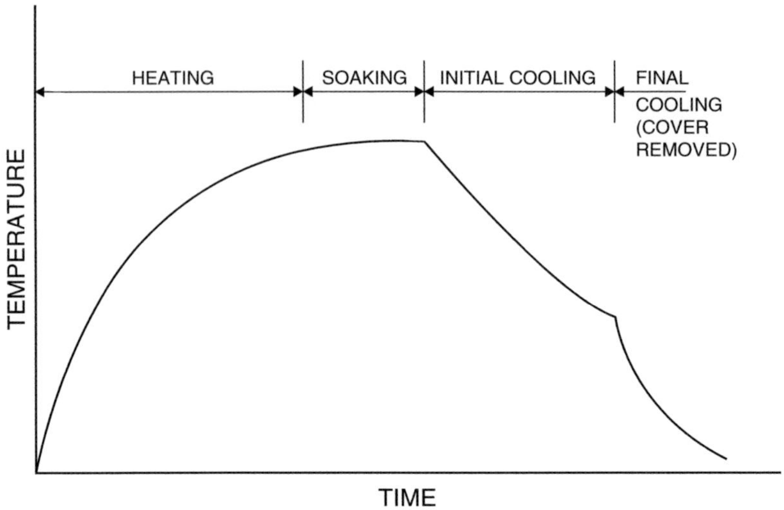

Fig. 9.7 Typical temperature variation at a given location in the steel coils undergoing annealing in the furnace of Fig. 9.6

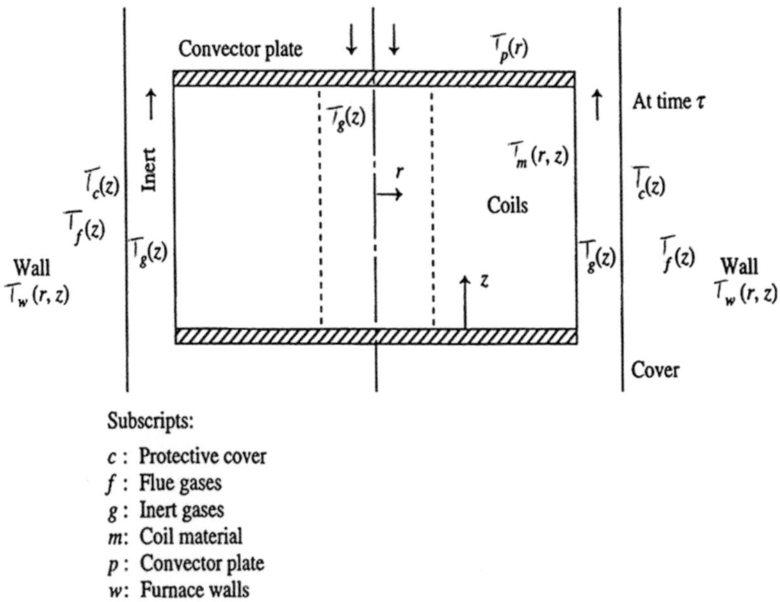

Fig. 9.8 Sketch of some of the components of the furnace used for developing the mathematical model

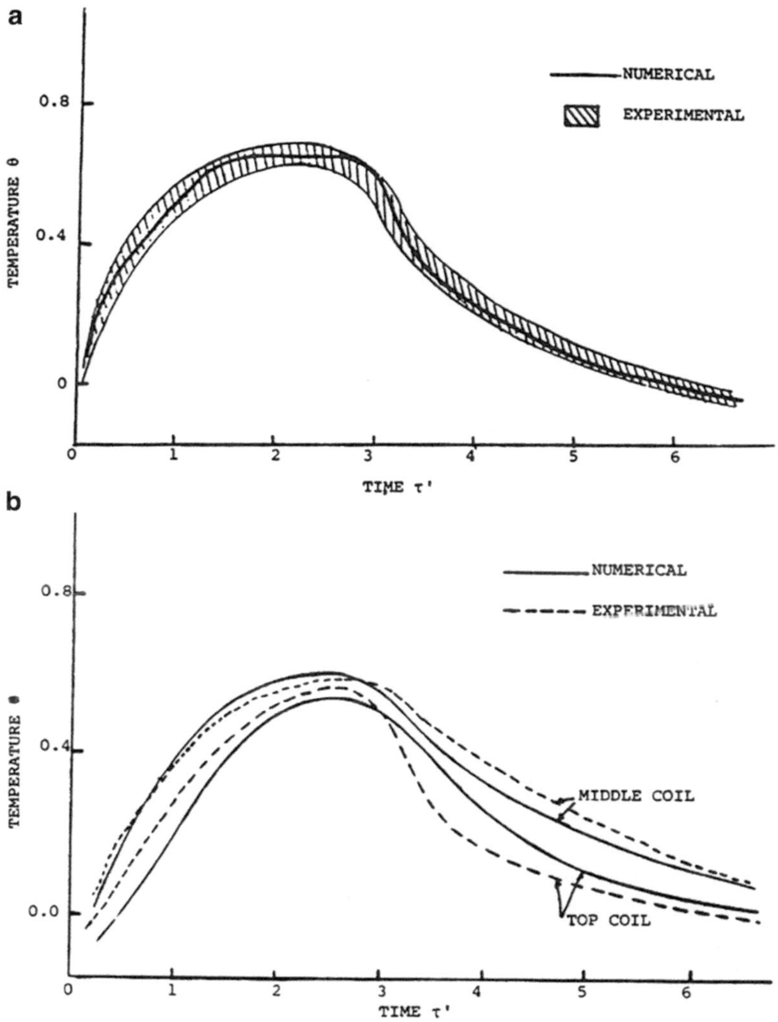

Fig. 9.9 Comparison between the numerical simulation results and experimental measurements in the steel coils. (**a**) General trends, and (**b**) Specific temperature data at chosen points

thin cover, or protective hood, and the convector plate may be assumed to have uniform temperature across the thickness, the walls may be treated as one-dimensional transient cases, and the gases may be taken as well mixed, all these assumptions being based on scale analysis and governing parameters (Jaluria 2008). All the transport mechanisms, radiation, convection, and conduction, are considered, as appropriate, to yield conjugate problems, which are simulated numerically.

Once the simulation of the system is achieved, experimental validation is undertaken to ensure that the results obtained are valid and accurate. Figure 9.9 shows

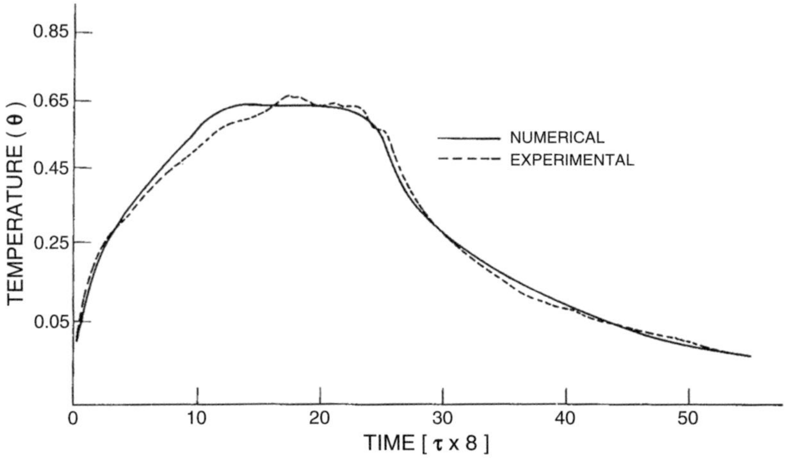

Fig. 9.10 Comparison between the numerical simulation results and experimental measurements of the control thermocouple in the annealing furnace

comparisons between the numerical and experimental results, indicating fairly good agreement and the accuracy of the numerical predictions. Thus, the numerical simulation results may be employed for the design and optimization of the process as well as of the system. The results obtained from the simulation provide the necessary inputs for improving existing designs, developing new designs, improving productivity, and improving the product quality for this manufacturing process. For instance, the dimensions and materials may be varied to reduce costs and improve the product, these changes being based on a thorough numerical simulation. For soaking, different control strategies may be considered, such as on/off systems, and simulated to choose an appropriate approach.

The heat treatment furnace is controlled by means of a temperature sensor located at the base of the furnace and in contact with the bottom surface of the lowest steel coil. The temperatures measured by this sensor are used to control the flow rate of the hot gases in order to achieve the desired annealing process. Using the simulation results, the desired temperature variation at the sensor to achieve the given thermal cycle was determined by solving the resulting inverse problem. Figure 9.10 shows a comparison between the numerical predictions and the experimental measurements from the control sensor. Again, fairly good agreement is observed, lending support to the model and the solution of the inverse problem to determine the boundary conditions, in terms of hot gas inflow rate, needed to achieve the given thermal process. Similar processes are used for the annealing of parts made of glass, ceramics, aluminum, and other materials.

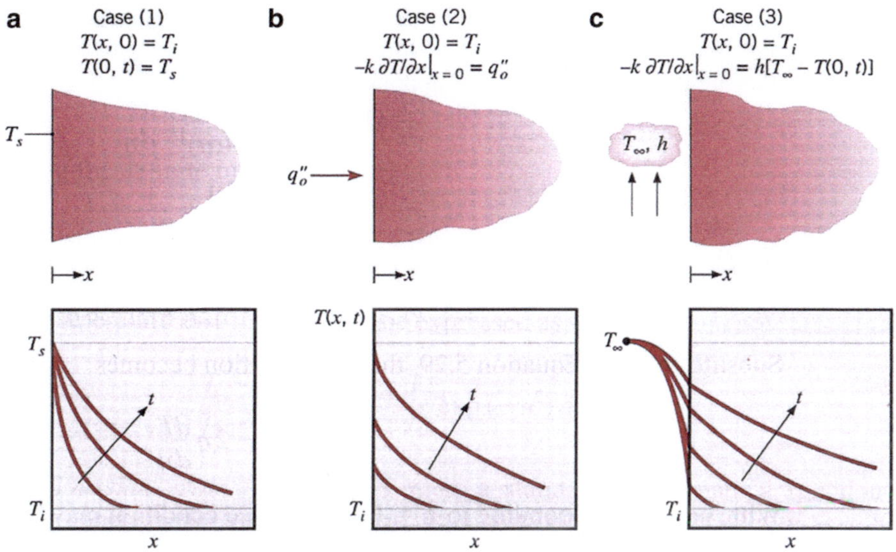

Fig. 9.11 Transient temperature distributions in a semi-infinite solid for three surface conditions. (**a**) Constant surface temperature; (**b**) Constant surface heat flux; (**c**) Constant convective heat transfer coefficient, obtained as step changes from a uniform temperature initial condition (Bergman et al. 2011)

9.1.3 Surface Heat Treatment

In many applications, interest lies in the heat treatment of a region close to the surface, leaving the remaining material untreated. The region near the surface may range from a few microns to a few millimeters, depending on the application. Among the common surface heat treatment processes are hardening to make the surface more wear resistant; surface curing that is used for various polymeric materials; softening to make, for instance, a pressure bond; and forming in order to obtain a desired shape or structure on the surface. The basic approach is similar to the heat treatment processes discussed in the preceding, except that the process is confined to a narrow region near the surface, so that the bulk properties and characteristics of the item are maintained.

Besides the thermal cycle that the material must undergo to obtain a particular heat treatment, the depth to which the material is to be heat treated is an important parameter. The requirements on the temperatures, heating and cooling rates, as well as constraints are similar to those discussed earlier and depend on the material and the desired process. The focus is on the surface and region close to it. The duration of the process is kept small so that the interior of the material is not affected. Thus, the penetration of thermal affects is small, compared to the dimension of the component, since the penetration depth varies as $(\alpha t)^{0.5}$, as shown by Eckert and Drake (1972) for various boundary conditions. Then, the process may be modeled as transport in a semi-infinite region, as shown in Fig. 9.11 (Bergman et al. 2011). Analytical solutions are possible in some cases. However, numerical methods provide greater versatility and convenience for the combined modes and time-dependent boundary conditions that are generally encountered in practice.

The governing energy equation for constant material properties is

$$\frac{\partial T}{\partial t} = \alpha \frac{\partial^2 T}{\partial x^2} \tag{9.2}$$

where x is normal to the surface. The initial temperature may be the result of an earlier process or simply the ambient temperature. The boundary condition at the surface is typically of the form

$$-k\frac{\partial T}{\partial x} = q_{\mathrm{c}} + q_{\mathrm{r}} \tag{9.3}$$

where q_{c} is the conductive and/or convective heat input and q_{r} the radiative energy input. For conjugate transport, the conduction in the solid material is coupled with the convective flow in the ambient medium. For the three cases shown in Fig. 9.11, the solution can be obtained analytically for the three cases, (a), (b), and (c), respectively, as

$$\frac{T - T_{\mathrm{s}}}{T_{\mathrm{i}} - T_{\mathrm{s}}} = \mathrm{erf}(\eta) \quad \text{where } \eta = \frac{x}{2\sqrt{\alpha t}} \tag{9.4}$$

$$T - T_{\mathrm{i}} = \frac{2q_{\mathrm{o}}(\alpha t/\pi)^{1/2}}{k}\exp\left(\frac{-x^2}{2\sqrt{\alpha t}}\right) - \frac{q_{\mathrm{o}}x}{k}\mathrm{erfc}(\eta) \tag{9.5}$$

$$\frac{T - T_{\mathrm{i}}}{T_{\infty} - T_{\mathrm{i}}} = \mathrm{erfc}(\eta) - \left[\exp\left(\frac{hx}{k} + \frac{h^2\alpha t}{k^2}\right]\left[\mathrm{erfc}\left(\eta + \frac{h\sqrt{\alpha t}}{k}\right)\right] \tag{9.6}$$

where erf (η) is the Gaussian error function, erfc $(\eta) = 1 - \mathrm{erf}\ (\eta)$, q_{o} the surface heat flux, T_{s} the surface temperature, T_i the initial temperature, and T_{∞} the ambient temperature. In most practical problems, combined modes are involved, material properties are not constant, and the input conditions vary with time, making numerical solutions necessary.

Multidimensional transient problems are also frequently encountered, since the heat transfer condition at the boundary may not be uniform, as shown in Fig. 9.12a. Explicit finite difference and finite volume methods are often convenient, and the solution is obtained by marching in time. Computation proceeds away from the surface till the change from the initial temperature is less than a specified small quantity. Therefore, the penetration of thermal effects into the region is determined in terms of the numerically obtained isotherms. Figure 9.12 shows these effects qualitatively for localized heating with a stationary or moving source, such as a laser. In many cases, the surface may be taken as infinite in extent. This applies away from the boundaries of the material whose surface is being heat treated. Near the boundaries, the transient 2D or 3D problem must be simulated.

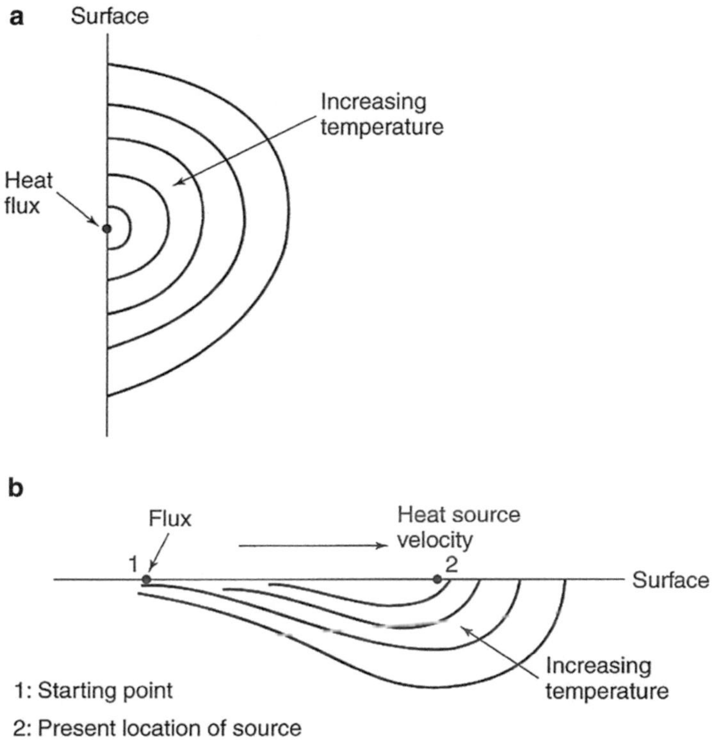

Fig. 9.12 Sketch of isotherms resulting from a localized surface-heating process for a semi-infinite body. (**a**) Stationary source; (**b**) Moving thermal source

To obtain the desired surface heat treatment, the time for heating must be stringently controlled in order to restrict the effect to the chosen depth. High heat flux inputs over short times are commonly used to attain the desired heat treatment temperature, without exceeding the allowable limit, followed by slow or rapid cooling to obtain the desired heat treatment. The heat flux, as well as the convective heat transfer coefficient, may be taken as functions of time, with a combination of the two modes allowing the greatest amount of flexibility. The constant temperature case is also used in some cases by bringing the surface in contact with an isothermally heated body. Then the surface temperature rises rapidly, depending on the contact resistance between the two surfaces, and the depth to which the heat treatment temperature is attained is determined by the contact time. Periodically varying heat input may also be used to heat up a small region close to the surface, without significantly affecting the interior of the material.

9.1.4 Case Hardening

Case hardening is largely used for low carbon steels, in which the low concentration of carbon makes it difficult to harden the material by transforming austenite into martensite by quenching, as can be seen from the phase diagram for steel. Then carbon penetration into the material can be used to raise the percentage of carbon to, typically, 0.8–1.2%. Also, the focus is on hardening the region near the surface. So, generally, the penetration of carbon is restricted to a depth of 0.5–2.0 mm. The component is heated beyond the upper critical temperature to obtain austenite in an atmosphere of carbon. Carbon diffuses into the region near the surface to reach desired concentration levels, and quenching imparts the desired hardened surface layer. Nitrogen can also be used as the hardening agent for various alloy steels. An atmosphere of ammonia is used in this case, the temperatures needed being much lower because of the diffusing species. Similarly, a cyanide atmosphere maybe used to obtain both carburizing and nitriding.

The diffusion process is similar to that discussed earlier for heat transfer in semi-infinite solids. Thus, the concentration C at a depth x and time t is obtained as

$$\frac{C - C_\mathrm{i}}{C_\mathrm{s} - C_\mathrm{i}} = 1 - \mathrm{erf}\left(\frac{x}{2\sqrt{D_\mathrm{cs} t}}\right) \tag{9.7}$$

where D_cs is the diffusion coefficient of carbon into steel, C_s is the surface concentration, and C_i is the initial concentration. Using this equation, the time needed for achieving a desired concentration to a given depth may be calculated. This is often on the order of several hours. This equation applies for constant properties. However, the temperature can significantly affect D_cs, and the corresponding variable-property mass diffusion equation may be written. Numerical solutions are generally needed for practical problems due to complex boundary conditions and variable properties.

Similar processes are used for changing the characteristics of the surface layer in a wide range of materials, such as ceramics, semiconductor materials, and polymers. Diffusion of various dopants into the material is used to obtain desired properties. An example of this is the use of rare-earth elements as dopants in optical fibers to modify the transmission characteristics.

9.2 Joining and Bonding Processes

There are a number of processes that are used to bond or join different items or components. These processes are often an important step in the overall manufacturing process. Among the common joining processes used in manufacturing are:

Soldering and brazing
Welding
Pressure bonding
Chemical bonding
Explosive bonding

In all cases, the important considerations are the strength of the bond, the contact between the surfaces, time taken for the process, and the feasibility of the bonding method for given materials. The material that is used to create the bond may be similar in composition to the base materials or quite different. Though phase change plays a critical role in processes like welding and soldering, pressure and chemical reactions, without a strong thermal effect, are used in other cases.

9.2.1 Soldering

Soldering, brazing, and welding are the most extensively used bonding processes. They involve melting a filler material for joining two components, allowing it to flow to an equilibrium position and then allowing it to solidify to form the bond. In soldering, which is extensively used in the manufacture of electronic circuitry and equipment, two or more items, which are usually made of metal such as copper, are joined together by melting and putting the molten solder into the joint. The work pieces are not melted and solder generally has a lower melting point than the adjacent metal, typically less than 427 °C. Brazing is similar to soldering, except that the filler metal melts at a higher temperature, typically above 427 °C. Copper and its alloys, such as copper-tin, copper-zinc, and copper-silver, are commonly used in brazing, depending on the materials to be joined. Tin-lead alloys were commonly used for solder, with different percentages of tin and lead, respectively. A 63/37 alloy is eutectic, which melts at a single melting point of 183 °C, 60/40 alloy melts over the range 183–190 °C, and 50/50 alloy melts between 183 and 215 °C. Due to environmental concerns, lead-free alloys have been developed and increasingly used. Silver solder contains substantial amounts of silver and melts at much higher temperatures, e.g., 56% silver results in a melting point of around 620 °C. Welding involves melting the pieces to be joined, and the filler material is the same or similar to the material of the pieces being joined. Different energy sources such as electric arc, gas, laser, friction, electron beam, and explosion are employed to melt the materials.

A few examples of these processes have been given in earlier chapters. Figure 1.2c showed the different steps in the formation of a solder joint between a pin or terminal and a circuit board with a plated-through hole. Solid solder, known as solder preform and containing the flux needed to clean the surfaces, is placed around the pin in the form of a doughnut and heated beyond its melting point. The molten solder is driven by surface tension and gravitational forces to form a joint, which solidifies on cooling to give the desired bond between the pin and the copper pads on the board. This process is known as *reflow soldering* in which solid solder or solder paste with flux is heated to melt it, thus allowing it to flow and form the joint. Surface tension affects the force balance on a free surface and plays a critical role in determining the equilibrium shape of the solder joint (Chu 1975). Figure 9.13a shows a typical temperature cycle undergone by a solder joint formed by melting of a solid preform and Fig. 9.13b the solder joint obtained. The time needed for initial heating, melting, flux action and flow, solidification, and cool down depends

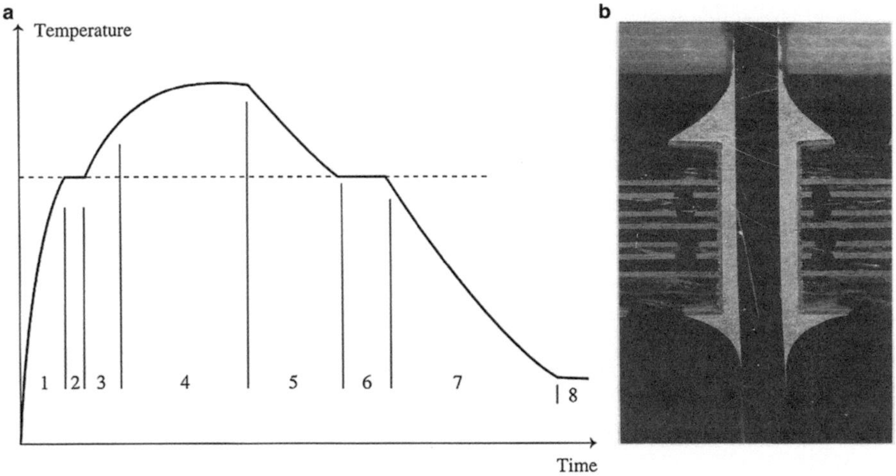

1. Initial heating
2. Melting of solder
3. Further heating and flux action
4. Solder flow and approach to equilibrium
5. Initial cooldown
6. Solidification of solder
7. Further cooling to room temperature
8. Aging

Fig. 9.13 (**a**) Typical temperature cycle undergone by a solder joint formed by melting of a solid preform; (**b**) A typical solder joint resulting from the process sketched in Fig. 1.2c

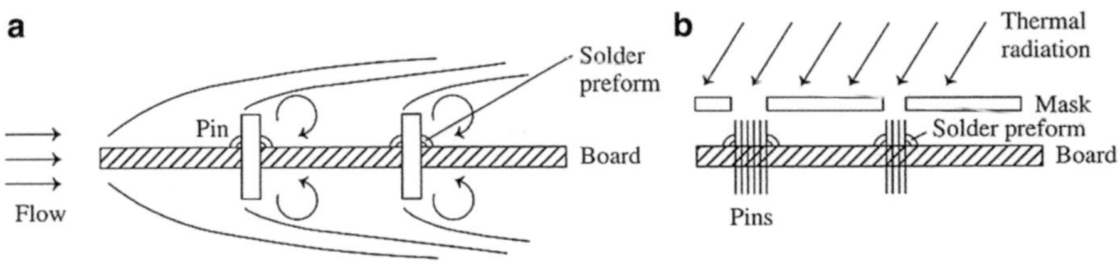

Fig. 9.14 Heating of solid solder preform for reflow soldering by (**a**) forced convection and (**b**) thermal radiation

on the dimensions and materials involved. The amount of solder used also depends on the dimensions of the wetted surfaces and the desired joint size. Though this process is a fairly traditional one, the basic mechanisms of bond formation are illustrated well by this example. It is also seen that the bonding process involves heating, melting, flow, and solidification. Thus, temperature constraints due to thermal stresses in the material, effectiveness of the flux to clean or degrease the surfaces, voids formed during solidification, dewetting of the surfaces, and the capillary action due to the flow channel dimensions are among the important aspects in the process.

The heating of the solder may be done by radiation or by convection, as shown schematically in Fig. 9.14. Excessive and nonuniform heating of the boards is a

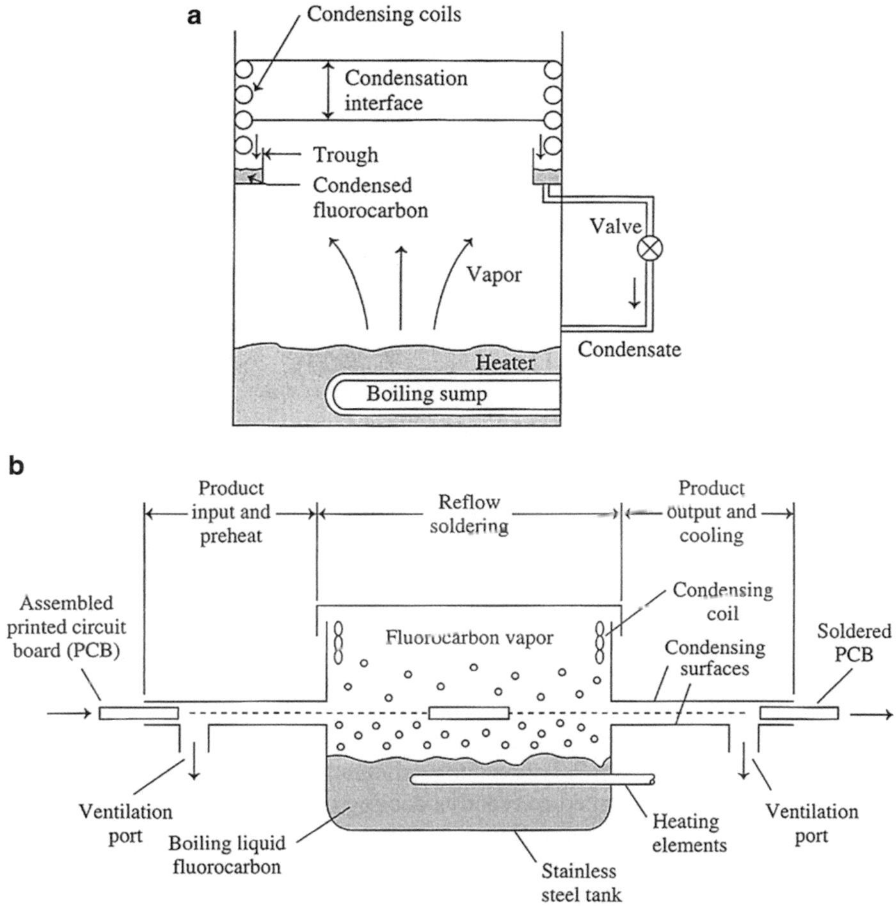

Fig. 9.15 (**a**) Schematic of a condensation soldering facility for electronic circuitry manufacture; (**b**) Another concept for a condensation soldering machine (Dally 1990)

common problem with radiation, and masks are needed to avoid local overheating. Cleaning of the fluid for liquid convection and low heat transfer for air convective heating are concerns with convection. In response to these, a method based on vapor condensation was developed and used for various electronic systems. The basic system to generate a condensation chamber is shown in Fig. 9.15a. A condensation interface is generated by the presence of cooling coils at the top, and the vapor is generated by heat input into the liquid at the bottom. The vapor condenses on an electronic circuit board immersed into the condensing region and thus heats the solid solder placed at various locations, such as doughnuts mentioned above, melts it, and forms the bond as it is cooled. Good uniformity in heating, without overheating, and a clean environment are obtained. Figure 9.15b shows a different concept of a condensation soldering facility. Similarly, other system designs may be developed to obtain a stable condensation region (Dally 1990; Jaluria 1979).

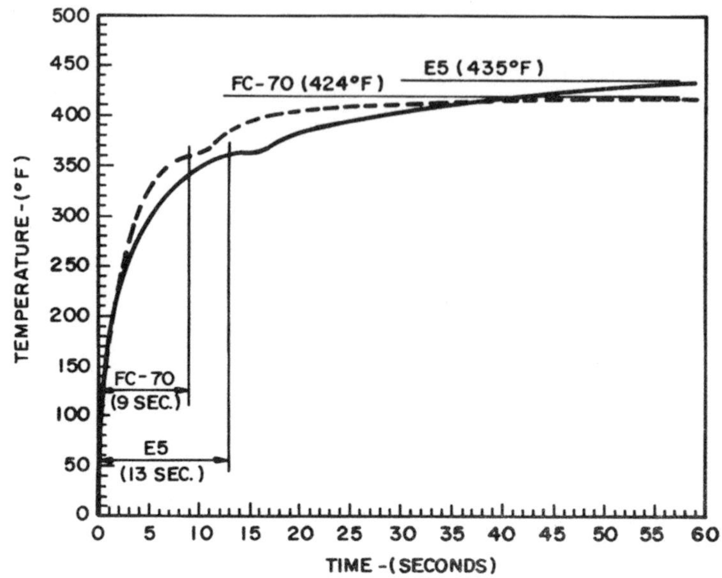

Fig. 9.16 Transient temperature response of a solder preform in the condensation soldering process for two different heating fluids

Different fluids may be used for different solders that have different melting points. Figure 9.16 shows the transient temperature response of a solder preform in the condensation soldering process for two different fluids. The constant temperature at the melting point is clearly seen, and the time taken to reach this point is indicated. By the use of different fluids and heating methods, a wide range of the heat transfer coefficient h can be obtained. For instance, h for natural convection in air is of order 5–10 W/m^2 K, forced convection in air 50–100, forced convection in common liquids 500–2500, radiation and condensation up to around 10^4, and fluidized bed up to around 5000 W/m^2 K. Thus, all the preceding aspects are taken into account for the design of an appropriate soldering system (Jaluria 2008). The process described here is used in different configurations for a variety of bonding applications, as well as for other processes where uniform, rapid, heating is important.

We have discussed heating, cooling, melting, and solidification for various materials processing techniques earlier, and similar considerations apply in soldering, brazing, and welding as well. However, the flow of molten filler material is often dominated by surface tension effects, as mentioned earlier. The equilibrium shape of the solder or welding joint is largely determined by surface tension and gravity forces. Figure 2.5b showed a weld fillet and Fig. 9.13b a solder joint. The shape is determined by the wetting characteristics of the fluid for the surfaces involved, with the surface tension force balancing the weight of the filler material (Chu 1975; Kou 1996). For instance, the height H of a column of liquid in a vertical circular tube of radius r is given by the force balance:

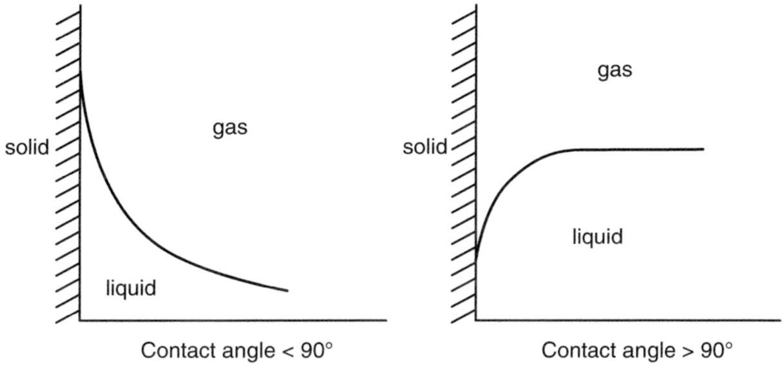

Fig. 9.17 Effect of contact angle on the shape of the meniscus

$$2\pi r\sigma \cos \phi = \rho g\pi r^2 H \tag{9.8a}$$

Thus, the capillary rise or depression is given by

$$H = \frac{2\sigma \cos (\phi)}{\rho g r} \tag{9.8b}$$

where σ is the surface tension and ϕ is the contact angle, which is the angle made by the liquid at the junction of the liquid-vapor interface and the solid surface, as shown in Fig. 9.17. It indicates the wettability of a solid surface by a liquid, an angle of less than 90° arising for wetting cases and more than that for non-wetting ones.

The shape of the meniscus may be obtained from the equation (Szekely 1979)

$$h(x) = -\left(\frac{\sigma}{\rho g}\right)^{1/2} \cosh^{-1} 2\left(\frac{\sigma}{2gx^2}\right)^{1/2} + \left(\frac{2\sigma}{\rho g}\right)^{1/2}\left(2 - \frac{\rho g x^2}{2\sigma}\right)^{1/2} + \text{constant} \tag{9.9}$$

where $h(x)$ is the height at a given coordinate distance x and the constant is determined by the value of the contact angle. The flux is used in soldering to clean the surfaces in order to improve the wettability of the surfaces by the molten solder. The time taken for the flux to act depends on its physical and chemical characteristics and on the materials involved. The flux also has to drain so that the molten solder can reach the surface. The time for it to drain depends on its viscosity μ and surface roughness h_r. A lower viscosity helps in draining and reduces the soldering time. An increase in temperature reduces the viscosity and decreases the draining time τ, which can be estimated from the following expression obtained from a flow analysis based on lubrication approximation and considering a piece of molten solder above a flat surface at an initial height of h_i

$$\tau = \frac{3\mu\pi R^4}{4mg}\left[\frac{1}{h_r^2} - \frac{1}{h_i^2}\right] \quad (9.10)$$

where R is the characteristic dimension of the solder and m its mass. The initial distance h_i of the solder from the surface may be neglected in many cases. Then a time of around 25 s is obtained in typical solder joint formations for the flux to drain. The final equilibrium shape results from the various forces acting on the free surface and the minimization of the free energy, as outlined in the preceding.

The flow of solder is driven by gravitational, viscous, and surface tension forces. The important dimensionless parameters that arise are

$$\text{Weber number, } We = \frac{\rho V^2 L}{\sigma} \quad (9.11)$$

$$\text{Froude number, } Fr = \frac{V^2}{gL} \quad (9.12)$$

$$\text{Bond number, } Bo = \frac{We}{Fr} = \frac{\rho g L^2}{\sigma} \quad (9.13)$$

$$\text{Capillary number, } Ca = \frac{\mu V}{\sigma} \quad (9.14)$$

where V and L are characteristic velocity and dimension, respectively. For different soldering, brazing, or welding problems, these dimensionless numbers may be calculated to determine the dominant mechanisms. For instance, We represents the importance of inertia forces compared to surface tension forces. Similarly, Fr represents inertia versus gravity, Bo represents gravity versus surface tension, and Ca represents viscous forces versus surface tension forces. The characteristic value of L typically ranges from 0.2 to 3 mm for a solder joint, less than 0.15 cm for a capillary, and around 30 cm for wave soldering. The velocity V is of order 1.2 m/min for a capillary and 20 m/min for wave soldering. The board typically moves at a speed between 1 and 5 m/min. Thus, the various dimensionless parameters may be estimated.

Another soldering process, which is extensively used for solder plating and for the formation of solder joints on electronic circuit boards, is wave soldering, which was mentioned earlier and shown in Fig. 2.5a. A schematic of the flow in wave soldering with and without a circuit board is shown in Fig. 9.18. As shown, the circuit board is passed over molten solder with a pump producing an upwelling of solder that looks like a wave. By bringing the board in contact with the solder wave, the components become soldered to the board. Solder coating for electronic connections is also achieved. The thickness h_o of the film obtained adjacent to the surface in this process may be estimated by considering the withdrawal of a solid body from a liquid. The expressions given by Szekely (1979) for small and large values of the capillary number Ca for a vertical surface moving at speed U are, respectively

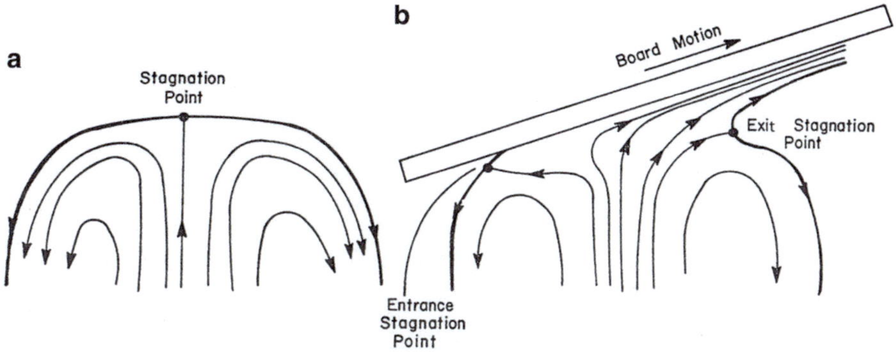

Fig. 9.18 A schematic of the flow in a wave soldering process. (**a**) Without a circuit board and (**b**) with a circuit board

$$h_o = 0.95\left[\frac{\mu^2 U}{\sqrt{\sigma}\rho g^{3/2}}\right]^{1/3} \text{ and } h_o = \frac{\mu U^{1/2}}{\rho g} \tag{9.15}$$

Good agreement with experimental data was obtained. Further details on this and similar flow problems may be obtained from Levich (1962), Gutfinger and Tallmadge (1965), and Szekely (1979). Wave soldering is of particular interest in through-hole printed circuit boards, shown in Fig. 1.2c, though it is also used for surface-mounted circuitry.

Figure 9.19 shows the hydrodynamic domain in which processes such as through-plate soldering, backwash flow, and wave soldering lie. Thus, laminar flow arises in capillaries and is surface tension dominated. The flow in wave soldering is gravity dominated. The governing equations would then focus on the relevant mechanisms. For instance, the capillary penetration of solder into circular channels may be simulated. Figure 9.20 shows the results for different tube diameters. As expected, the capillary action is stronger for smaller diameter tubes and is reflected in faster penetration. The effect of gravity is seen to be small for tubes of smaller diameters since the upward and downward penetration results are quite close, indicating the dominant effect of surface tension.

9.2.2 Welding

Similar considerations arise in brazing and welding (Easterling 2013; Kou 2003; Lancaster 1986; Lippold 2015). As mentioned earlier, the filler material in brazing has a lower melting point than the work pieces being joined. The filler material is similar or the same as the materials being joined in welding, which also involves melting the base materials to form a bond with the filler material. Because of the importance of welding in manufacturing, considerable information exists on different welding methods, the effect on the base materials being joined, strength and

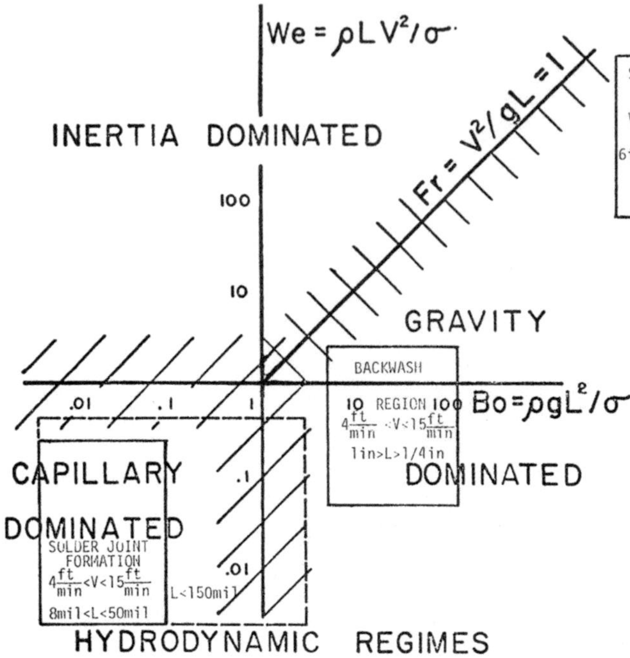

Fig. 9.19 Hydrodynamic regimes of solder flow

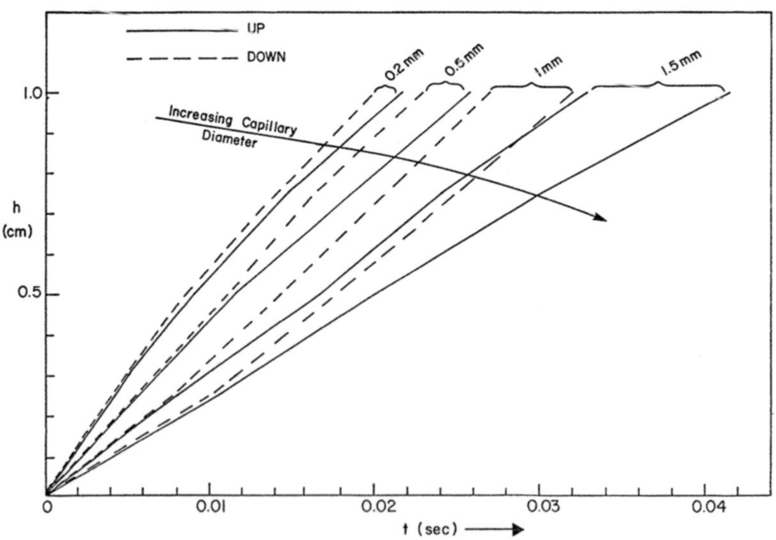

Fig. 9.20 Capillary penetration of molten solder into circular tubes

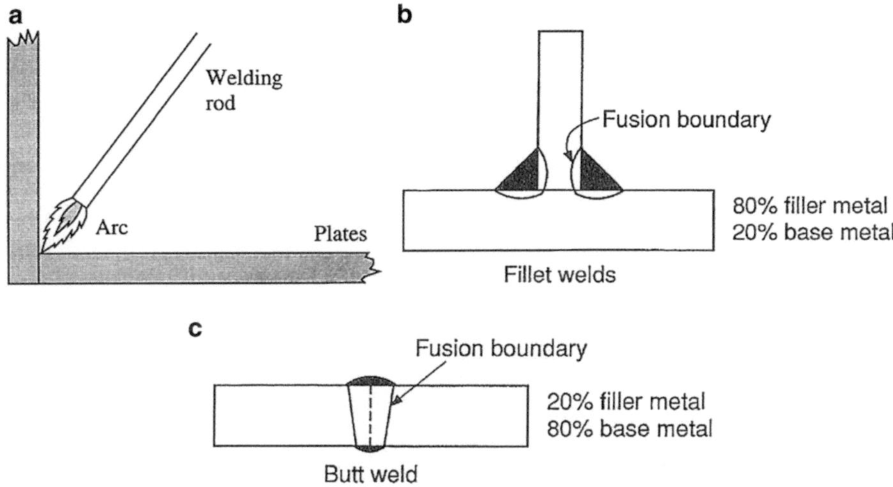

Fig. 9.21 (**a**) Sketch of the arc welding process; (**b**) Sketch of a fillet weld, showing the fusion boundary; (**c**) Sketch of a butt weld, showing the fusion boundary

characteristics of the joints formed, temperature and flux needed, the equilibrium shape of the welded joint, and systems needed for achieving the desired quality of the welded joint.

Figure 9.21a shows a sketch of the commonly used process of arc welding in which an electric arc is employed to melt the filler material, along with the flux, and the work pieces close to the location where the bond is to be formed. A fusion bond is formed by the molten materials, as shown for two common cases of fillet and butt-welded joints in Fig. 9.21b, c, respectively, the former one being shown earlier in Chap. 2. The fusion boundary of the molten materials is also shown. The joint configuration is determined by the forces acting on the fluid, particularly gravitational and surface tension forces, as discussed earlier for soldering. The welding process is completed by the solidification of the bond. The welded joint may be stronger than the base materials. and thermal stresses and resulting distortion are of concern due to the high temperatures involved. Figure 9.22 shows two different systems for arc welding, with a filler rod in one case and a consumable electrode in the other, the former system requiring lower power density. A shielded environment is employed in both cases to avoid oxidation of the molten materials. Obviously, the energy needed for melting may be supplied by a variety of sources, such as laser, gas flame, electric arc, electron beam, and resistance heating, leading to different welding systems for different materials and applications. The energy supplied can be calculated by considering the chemical reactions in a gas flame, the power supplied in arc welding, and so on. A heat transfer analysis of the parts to be joined, along with the filler material, is to be carried out to determine the temperatures and thermal stresses that arise. Numerical methods, particularly finite element methods, are particularly suitable for the modeling and simulation.

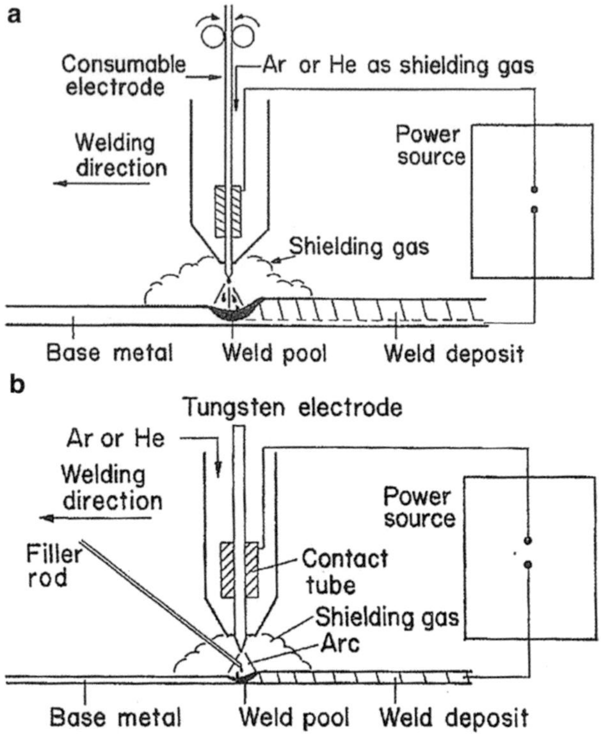

Fig. 9.22 (**a**) Gas tungsten arc welding; (**b**) Gas metal arc welding (Kou 2003)

A weld pool is created, which forms the bond after establishing an equilibrium profile and solidification. Various forces arise and generate flows in the weld pool. Figure 9.23 shows sketches of flows generated by buoyancy, electric current, and surface tension. The last effect is similar to the one discussed earlier and sketched in Fig. 2.5c for Marangoni convection. A flow visualization of the Marangoni flow in a weld pool from Kou (2012), who has done extensive fundamental work in this area, is shown in Fig. 9.24. These flows affect the solidification process and thus the characteristics of the bond formed. Attempts are usually made to reduce these effects by minimizing temperature differences, using magnetic fields, varying the dimensions, etc. Another major concern is the presence of voids that arise due to density change on solidification. This aspect was considered earlier with respect to casting processes, and the basic considerations and analysis are similar to those brought out earlier. The strength of welded joints is compromised due to the presence of voids, and if the voids appear at the surface, the joint is rejected. Radiographic and ultrasonic testing are often employed to test for voids and other discontinuities in the weld.

In many cases, a moving heat source is involved as the welding proceeds to join two components over an overlapping boundary or over a contact line or surface. This is similar to the circumstances shown in Fig. 9.12. For a heat source moving at speed

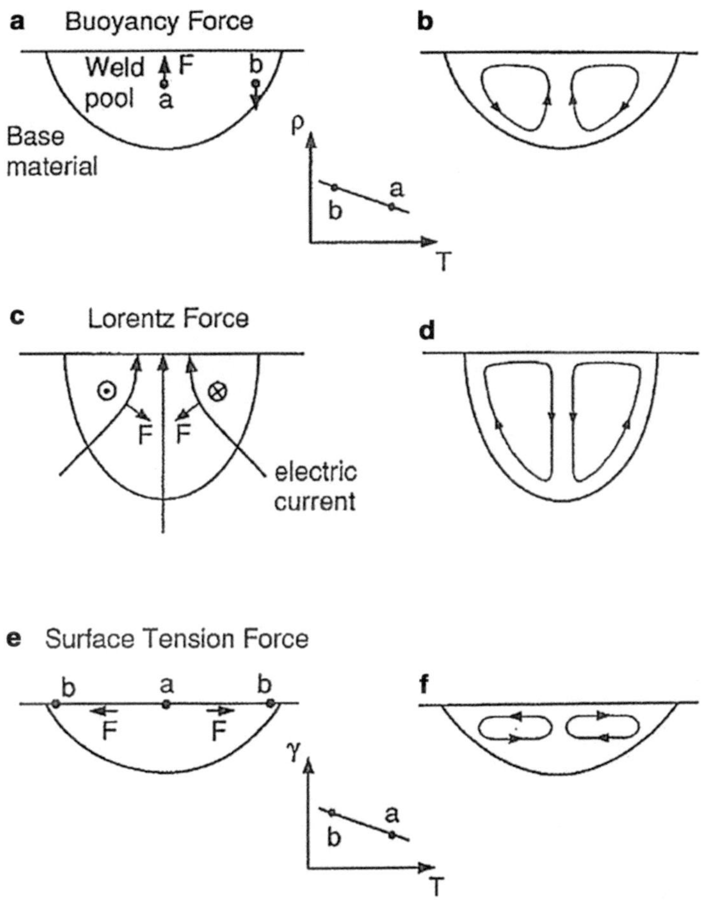

Fig. 9.23 (**a–f**) show forces on the *left* and the resulting flows on the *right* for the corresponding circumstances (Kuo 1996)

u, a coordinate transformation $\xi = x - ut$ is used to convert the transient problem into a steady-state one, governed by the equation

$$\frac{\partial^2 T}{\partial \xi^2} + \frac{\partial^2 T}{\partial y^2} + \frac{\partial^2 T}{\partial z^2} + \frac{\dot{Q}}{k} = -\frac{u}{\alpha}\frac{\partial T}{\partial \xi} \tag{9.16}$$

where $\dot{Q}$ is the heat source per unit volume. It may be replaced by a delta function $\delta(x)$ for a localized point, line, or planar source. This equation can be solved analytically for semi-infinite bodies and numerically for finite bodies. For instance, if g_s is a continuous plane surface source, the temperature decay away from the source is given by

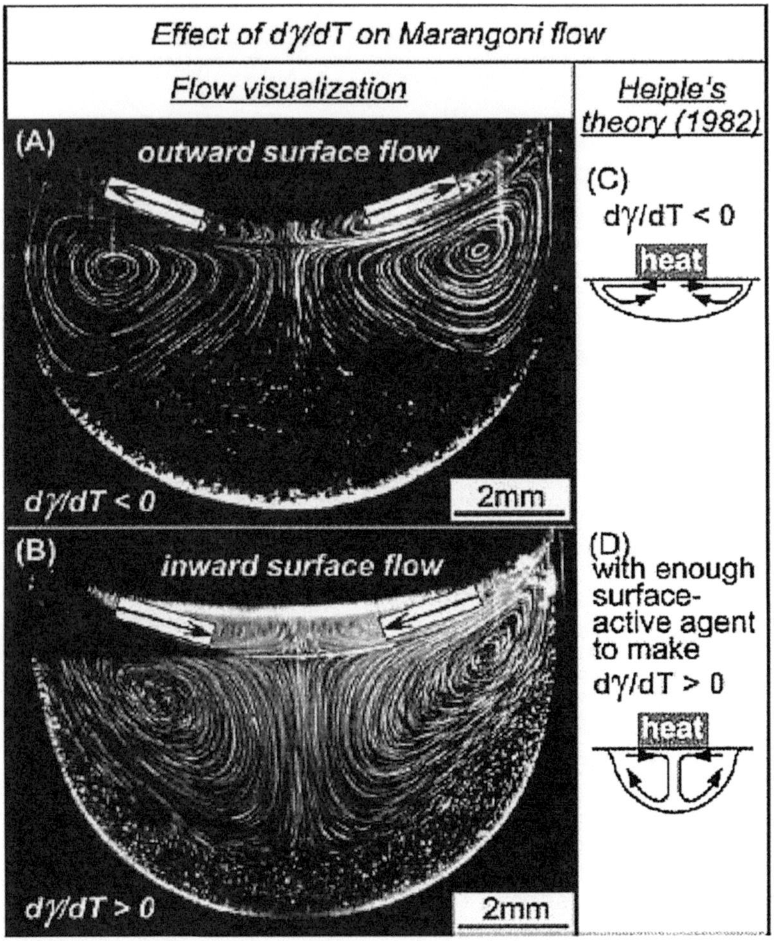

Fig. 9.24 Flow visualization of Marangoni convection in a simulated weld pool of $NaNO_3$ (Kou 2012). (**A**) and (**B**) show the flows for two different variations of surface tension γ with temperature and (**C**) and (**D**) indicate the corresponding theoretical behavior

$$T - T_i = \frac{\alpha}{uk} g_s \exp\left(-\frac{u}{\alpha}\xi\right) \tag{9.17}$$

where T_i is the uniform initial temperature and no energy losses are taken into account. Similarly, point and line sources may be considered, a line source g_L yielding the result

$$T(r, \varsigma) = \frac{1}{2\pi k} g_L K_o\left(\frac{4r}{2\alpha}\right) \exp\left(-\frac{u}{2\alpha}\xi\right) \tag{9.18}$$

where K_o is the modified Bessel function of second kind of order zero. It varies as $\sqrt{\frac{\pi}{2z}}\exp(-z)$ at large z. For further details, see Hahn and Ozisik (2012). For an instantaneous heat input q, the temperatures outside the fusion zone are given for thick (3D) and thin (2D approximation) plates, respectively, as

$$T - T_{\mathrm{i}} = \frac{q}{2\pi kut}\exp\left(-\frac{r^2}{4\alpha t}\right) \tag{9.19}$$

$$T - T_i = \frac{q}{d(4\pi k\rho Ct)^{1/2}}\exp\left(-\frac{r^2}{4\alpha t}\right) \tag{9.20}$$

where d is the plate thickness and r the distance from the point of application of heat. These expressions, which apply for idealized semi-infinite and infinite regions, are given to indicate the basic trends. Numerical methods are needed for practical problems with finite regions.

There are several other welding processes that do not need a filler material. High pressure is applied to form the bond or material is softened and bonded. Friction welding, forge welding, and diffusion welding heat the materials below the melting point and press them together to form a bond. This approach is also used for joining plastics, rubber, and various other materials. For example, insulated wires may be bonded by heating the surfaces in the region where a bond is to be formed and pressing the two wires together. Similarly, explosive and ultrasonic welding use high pressure and heating to form the bond. These processes are often called solid-state welding processes. Surface modification through micro- or nanostructures may be used to obtain better contact and stronger bonds. Chemical bonding is also used for various materials. In many cases, the chemical paste is applied and then baked or cured to form the bond. This process involves combined heat and mass transfer, and methods similar to those discussed earlier for polymers may be used to model the process.

9.3 Fabrication of Microchannels and Microdevices

A manufacturing process, which has grown significantly in recent years, is the fabrication of micro- and nanochannels and devices. Interest has also increased in micro- and nanostructured surfaces to enhance thermal transport and to provide greater stability and control to processes such as boiling. Many manufacturing techniques are available for these applications. The use of CVD to obtain micro- or nanostructures was outlined earlier. Similarly, coating and deposition processes, for instance, by means of combustion, may be used.

There are several materials that can be used for the fabrication of microchannels, whose diameter or width is typically in the range 50–200 μm. These include silicon, diamond, iron, steel, stainless steel, and aluminum. Single-crystal silicon is often employed in the fabrication of microdevices, as well as for heat sinks for the cooling of electronic circuitry, because of its well-established electronic properties for semi-conductors. Many microfabrication technologies have been developed using single-crystal silicon for its high precision, high strength, and high reliability (Petersen 1982).

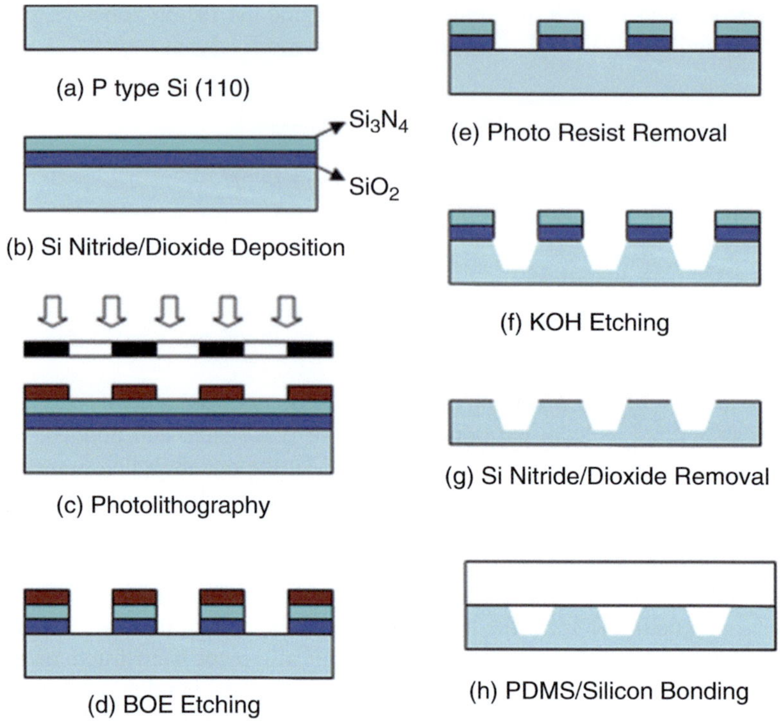

Fig. 9.25 Fabrication and packaging process of microchannel heat sinks. The sequence of operations is shown in parts (**a**) through (**h**)

Two different methods used to fabricate microchannels are wet etching and dry etching. The advantages and disadvantages of both methods are well established. The most important ones for micromachining are as follows: wet etching is usually isotropic, which can have a selectivity that depends on crystallographic direction and can be selective in masking surfaces and underlying layers. Plasma etching (dry etching) can be vertically anisotropic, allowing the patterning of narrow lines. Hence many high aspect ratio microelectromechanical (MEM) devices are made by dry etching. Wet etching is relatively economical and the chemical mixtures are easier to prepare. For instance, 30% potassium hydroxide solution (KOH) may be used as an etchant to etch silicon wafers. The wafer is masked with silicon nitride to create the desired configuration of microchannels. KOH etching is orientation dependent, and this anisotropic etching scheme allows tailoring of sidewall profiles. The fabrication and packaging process is illustrated in Fig. 9.25 as a process flow diagram. A SEM micrograph of the microchannels after KOH etching was shown earlier in Fig. 3.13a.

As shown in Fig. 9.25, bare silicon is first cleaned to remove the organic contamination, followed by a HF dip immediately to remove native oxide on the silicon in order to reduce undercutting of the nitride mask. A plasma-enhanced CVD system is used to deposit a 2000 Å layer of silicon dioxide and a 2500 Å layer of

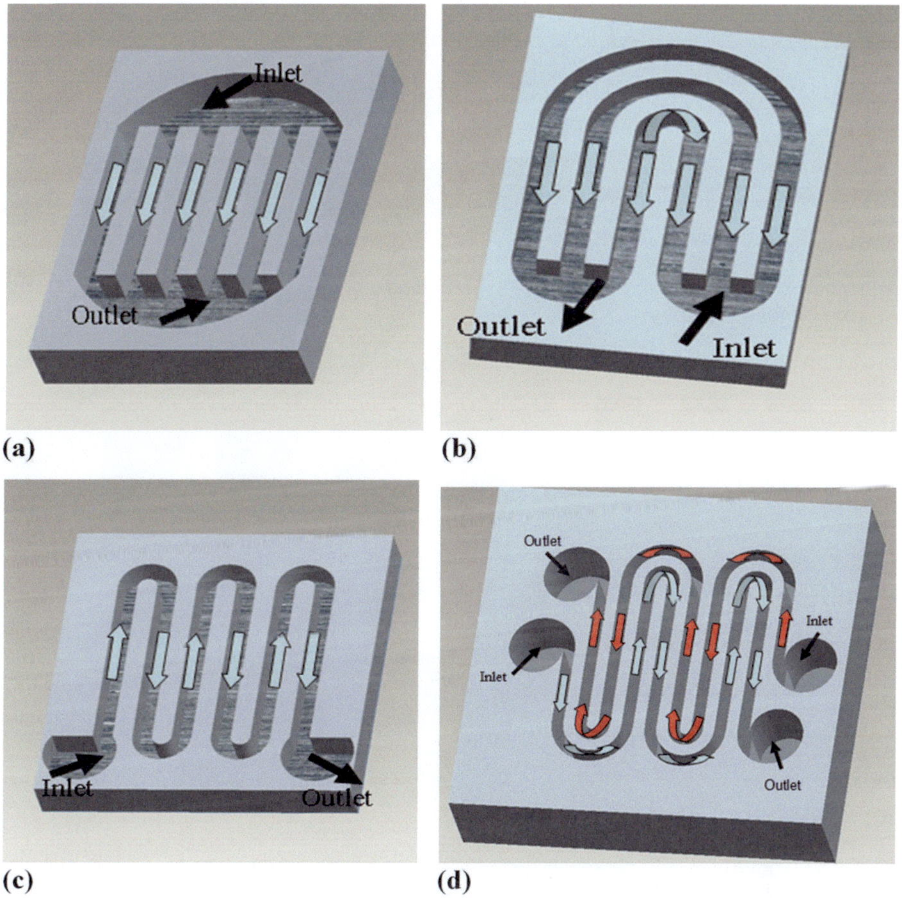

Fig. 9.26 (**a–d**) Schematic of different microchannel heat sink configurations

silicon nitride. A buffered oxide etch is used to open window of silicon nitride after conventional ultraviolet (UV) photolithography defines the microchannel pattern on the photo resist. The silicon wafer is then dipped into an appropriate solution to remove the rest of the photo resist before it is dissolved in the KOH solution. A magnetic stirrer is used to agitate the KOH solution to prevent the etch rate variation from the top to the bottom. 10%~15% isopropanol is added to KOH solution to improve the etch uniformity. In order to simulate a heat sink, a PDMS layer is bonded on top of the silicon microchannels after treating with oxygen plasma at 200 W for 15 s at room temperature. Openings are then punched in the PDMS for fluid connections. Figure 9.26 shows a few configurations of microchannel systems that simulate heat sinks for the cooling of electronic equipment.

Wet etching provides a relatively economical method to fabricate microchannel heat sinks when plasma etching equipment is not available. The downside of the wet

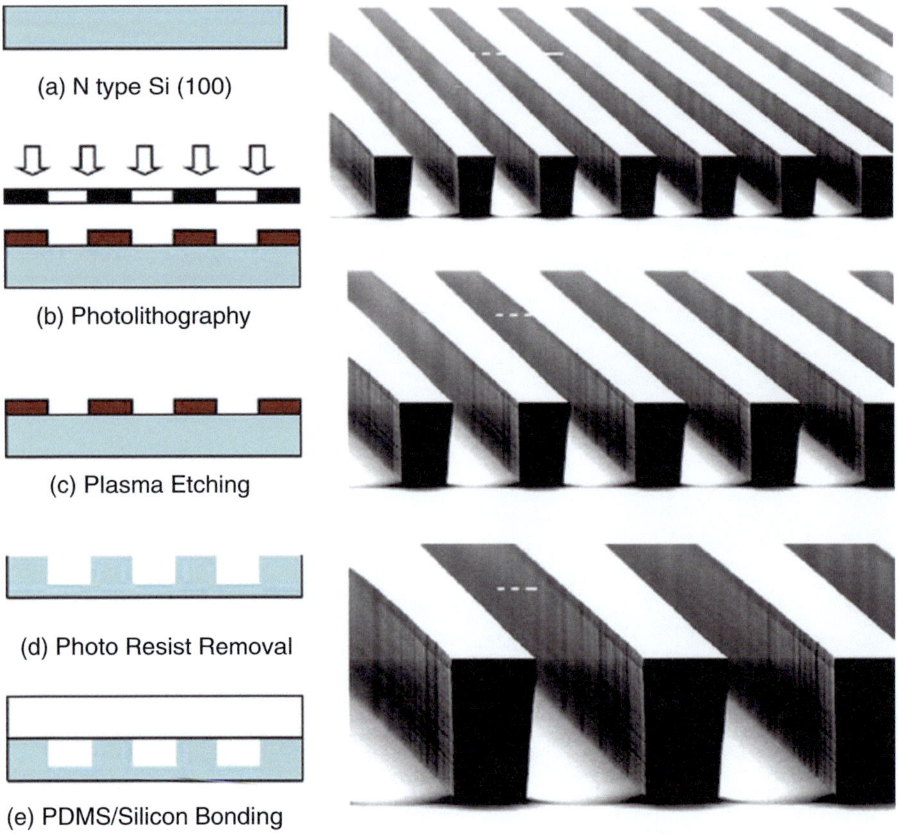

Fig. 9.27 Fabrication and packaging process of microchannel heat sinks by plasma etching. Parts (**a**) through (**e**) give the sequence of operations in this manufacturing process

etching process is the challenge in controlling the undercutting (Fig. 3.13a) and sidewall profile, which tend to be more controllable with dry etching. Moreover, plasma etching is also more efficient. Plasma etching is useful in the fabrication of complicated structures. The wafer preparation for dry etching is easier since it includes fewer steps compared with wet etching. The cleaning process is the same, then followed by photolithography process, as shown in Fig. 9.27. A mask is used to protect fins from etching. The figure also shows a SEM of the resulting microchannels, which are better defined than those obtained with wet etching. Figure 9.28 shows two different microchannel configurations fabricated by this method. Thus, both of these methods, as well as others mentioned earlier, may be used to obtain micro- and nanochannels, microdevices, and micro-/nanostructured surfaces.

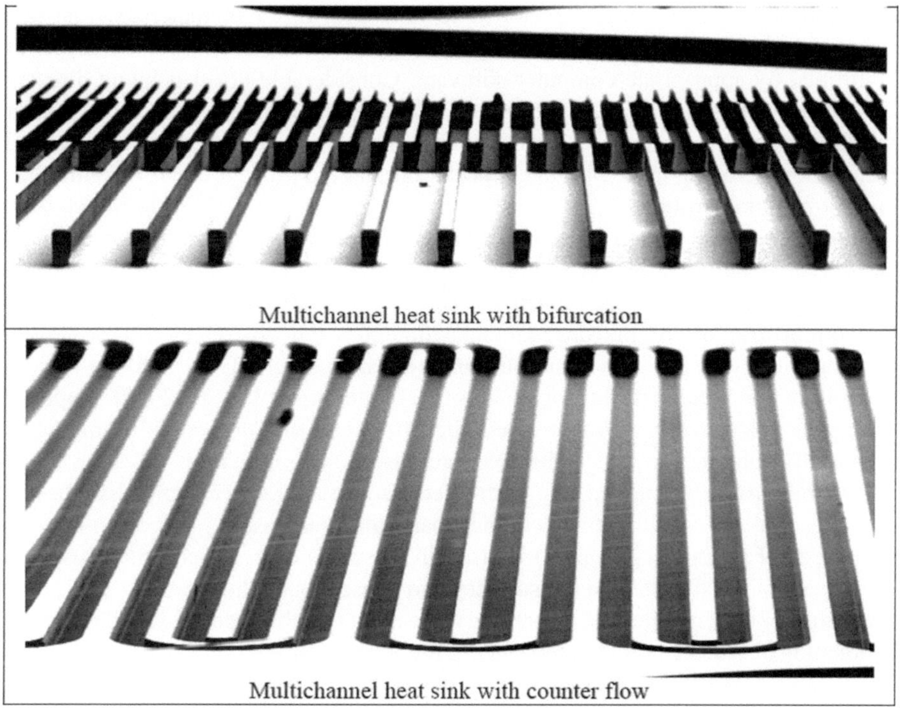

Fig. 9.28 SEM micrograph of microchannel heat sinks obtained with plasma etching

9.4 Additional Manufacturing Processes

There are many more manufacturing processes that are important and that are widely used to obtain a wide variety of components and products. Though beyond the scope of this book, many of these processes have been extensively studied and applied in different fields. Some are traditional processes, and many others are new and emerging techniques that have generated a lot of interest in recent years. Among these are additive manufacturing, including stereolithography and 3D printing, thermal sprays, laser processing, cutting and machining, powder metallurgy, and microgravity materials processing.

An important emerging manufacturing process is *additive manufacturing*, where the part is fabricated layer by layer by adding material to obtain the exact, complex, shape and dimensions (Chua and Leong 2017; Gibson et al. 2015). This is in contrast to removal processes like drilling, cutting, and grinding where the material is removed to obtain the desired product. Often known as 3D printing, the process can be used for the fabrication of parts for a wide range of applications. Many of the steps involved in the fabrication are governed by the same basic mechanisms as discussed earlier such as heating, melting, cooling, and solidification. Different techniques such as laser sintering, stereolithography, extrusion, and electronic beam fabrication have been developed for use with different materials. The raw material is melted and deposited to fabricate the desired item layer by layer, as controlled by appropriate computer software. Considerable effort is being directed by industry at this process due to its potential to fabricate complex geometries as near net shape so that further processing is reduced or eliminated. Different materials and techniques are being used with the expectation that the process could revolutionize current manufacturing.

Use of *thermal sprays* in manufacturing is another area which has received considerable attention as a viable process for coatings and manufacturing near net shape structured materials (Fauchais et al. 2014; Powlowski 2008). Sprays containing droplets of the desired deposition material are directed at a chosen substrate. The droplets spread out on impingement and solidify to yield the desired deposition for manufacturing or coating (Pasandideh-Fard et al. 2002). Superior properties associated with fine-grained microstructures and non-equilibrium phases are usually obtained. A wide variety of materials, such as aluminum, tin, and various alloys, have been employed in the droplet-based manufacturing process. Plasma spraying is used for fabricating ceramics, particularly nanostructured ceramics, and various other materials. Much of the effort in these and other papers has focused on the impact of the droplets on the surfaces and subsequent solidification because the properties of the final product are strongly dependent on these processes. The overall problem is obviously complicated since it involves complex flows with free boundaries, moving surfaces, phase change, and rapid changes with time. However, because of the advantages it offers in material fabrication, in terms of speed and quality, substantial research effort is currently directed at the thermal spray process.

Machining or material cutting and removal processes have been around for ages, though many new techniques, such as electron beam machining, laser machining, and fluid jet cutting, have been developed in recent years. Processes such as turning, milling, and grinding are traditional processes which are extensively used to convert a given part of raw material into desired final shape and size by a material removal process, using various machine tools. Though largely used for metals and alloys, machining is also used for materials such as plastics, ceramics, composites, and wood. The literature is replete with information on machine tools, tool wear, forces acting during material removal, heat generated in the process, and the effects of various parameters like feed, rotational speed, and depth of cut. Most books on manufacturing have details on these processes and may be consulted for further information (DeGarmo et al. 2007; Ghosh and Mallik 1986; Kalpakjian and Schmid 2016).

Laser, plasma, electron beam, electric discharge, and gas machining are used for material cutting and removal processes for drilling holes, etching, engraving, and shaping. The material is removed by melting and vaporization and each technique has its advantages and disadvantages. The energy input may be continuous or pulsed, and the total energy supplied goes into raising the temperature of a localized region to its melting point and then melting it. The molten material may be removed or blown away, or the region may be heated further to the boiling point and vaporizing it. Depending on the material and the size of the engraving, cavity, or hole, the source of energy and the diameter of the beam or flow stream supplying the energy may be chosen (Grigoropoulos et al. 1996; Steen and Mazumder 2010). The process follows much of the discussions and analysis given earlier for phase change processes. Ultrasonic machining is used for ceramics and hard materials. Fluid jet cutting was initially used for relatively softer materials like plastics but is now used for a much wider range of materials. It is used for cutting, scribing, stripping, and etching. Basically, the force exerted by the jet as it impinges on a given surface is used for material removal. It has the advantage of minimal thermal deformation and flushing away of the removed material. Abrasive particles may also be added to the jet to reduce the pressure, obtain a cleaner cut, and apply to harder materials. Ghosh and Mallik (1986) have given an excellent treatment of unconventional machining processes.

Powder metallurgy is another important manufacturing process, which starts with the powder of the chosen material and uses compaction and forming to obtain the desired product. Heat and pressure are applied without melting the material in a process known as sintering. The process may be used for metals, plastics, ceramics, and other materials. Powder production is obviously critical since it affects the properties and characteristics of the final product. Atomization, crushing, electrolysis, chemical reactions, etc. are used to obtain the desired particle shape, structure, and size distribution. Atomization involves melting the material, discharge of a stream of molten material in a gas, formation of droplets, and solidification. Compaction is achieved by applying high pressure, and sintering is obtained by heating the material to high temperature in a protective environment and then applying the pressure to achieve the desired compact density. Uniformity is an important concern, and mixing and blending of the powder are employed to homogenize it. Several pharmaceutical manufacturing processes deal with powders that are compacted to obtain tablets. The process may be carried out at room temperature or higher, depending on the medicines and materials involved. Further deformation processing of sintered powders may be undertaken to increase the density and obtain the desired final shape (Kuhn and Lawley 1978).

Microgravity materials processing has been of considerable recent interest in materials processing. Such an environment would be obtained, for instance, in laboratories orbiting in space, such as a space station, and would allow the processing of materials to be carried out with reduced effects of the terrestrial gravitational field. This would also reduce buoyancy-induced flow and affect local transport rates and the characteristics of the liquid-solid interface. This could lead to defect-free crystals of high purity and other products of enhanced quality due to bubble suppression, stable interfaces, etc. Both experimental and numerical studies have been carried out (Naumann 2005; Wilcox and Regel 2001).

9.5 Summary

This chapter discusses several manufacturing processes, besides those discussed in detail in earlier chapters. These are employed in industry for the fabrication of a wide range of parts, using different materials. Most of these processes have strong similarities with the processes considered earlier, and therefore, the methods for analysis, simulation, and experimentation outlined earlier may be employed for these as well. Also, several of the results presented earlier indicate the trends, concerns, solutions, and approaches that may be adopted for these. The chapter presents heat treatment, material bonding, and fabrication of microchannels, in some detail, whereas many other additional processes are briefly outlined. Most of the processes mentioned in this chapter, as well other similar ones that have not been mentioned, are important materials processing techniques in their own right, and extensive work has been done on understanding the basic mechanisms involved, the effects of various parameters, the desired characteristics of the product, and methods to improve product quality and production rate. Many of the references given here may be consulted for further information on these processes, since a more detailed discussion is beyond the scope of this book. However, the basic considerations and detailed discussions of several manufacturing processes given in earlier chapters will provide the basis for further work on these additional processes.

References

T.L. Bergman, A.S. Lavine, F.P. Incropera, D.P. Dewitt, *Introduction to Heat Transfer*, 6th edn. (Wiley, New York, 2011)

J.T. Black, R.A. Kohser, *DeGarmo's Materials and Processes in Manufacturing*, 12th edn. (Wiley, New York, 2017)

T.Y. Chu, A hydrostatic model of solder fillets. West. Elect. Eng. **19**, 31–42 (1975)

C.K. Chua, K.F. Leong, *3D Printing and Additive Manufacturing: Principles and Applications*, 5th edn. (World Scientific Pub. Co., Hackensack, 2017)

J.W. Dally, *Packaging of Electronic Systems: A Mechanical Engineering Approach* (McGraw-Hill, New York, 1990)

E.P. DeGarmo, J.T. Black, R.A. Kohser, *Materials and Processes in Manufacturing*, 10th edn. (Wiley, New York, 2007)

K. Easterling, *Introduction to the Physical Metallurgy of Welding*, 2nd edn. (Butterworth-Heinemann, London, 2013)

E.R.G. Eckert, R.M. Drake, *Analysis of Heat and Mass Transfer* (McGraw-Hill, New York, 1972)

P.L. Fauchais, J.V.R. Heberlein, M. Boulos, *Thermal Spray Fundamentals: From Powder to Part* (Springer, New York, 2014)

A. Ghosh, A.K. Mallik, *Manufacturing Science* (Ellis Horwood, Chichester, 1986)

I. Gibson, D. Rosen, B. Stucker, *Additive Manufacturing Technologies: 3D Printing, Rapid Prototyping, and Direct Digital Manufacturing*, 2nd edn. (Springer, New York, 2015)

C.P. Grigoropoulos, T.D. Bennett, J.-R. Ho, X. Xu, X. Zhang, Heat and mass transfer in pulsed-lased-induced phase transformations. Adv. Heat Transf. **28**, 75–144 (1996)

C. Gutfinger, J.A. Tallmadge, Films of non-Newtonian fluids adhering to flat plates. AIChE J **11**, 403–413 (1965)

D.W. Hahn, M.N. Ozisik, *Heat Conduction*, 3rd edn. (Wiley, New York, 2012)
G.F. Harvey, Mathematical simulation of tight coil annealing. J. Aust. Inst. Metals **22**, 28–37 (1977)
Y. Jaluria, An experimental study of the interface of a condensing vapor region. Int. J. Heat Mass Transf. **22**, 865–873 (1979)
Y. Jaluria, Numerical study of the thermal processes in a furnace. Numer. Heat Transf. **7**, 211–224 (1984)
Y. Jaluria, Numerical simulation of the transport processes in a heat treatment furnace. Int. J. Num. Meth. Eng. **25**, 387–399 (1988)
Y. Jaluria, *Design and Optimization of Thermal Systems*, 2nd edn. (CRC Press, Boca Raton, 2008)
S. Kalpakjian, S. Schmid, *Manufacturing Processes for Engineering Materials*, 6th edn. (Pearson, Boston, 2016)
S. Kou, *Transport Phenomena and Materials Processing* (Wiley, New York, 1996)
S. Kou, *Welding Metallurgy*, 2nd edn. (Wiley-Interscience, New York, 2003)
S. Kou, Fluid flow and solidification in welding: Three decades of fundamental research at the University of Wisconsin. Weld. J. **91**, 287–302 (2012)
H.A. Kuhn, A. Lawley, *Powder Metallurgy Processing: New Techniques and Analyses* (Academic Press, New York, 1978)
J. F. Lancaster (ed.), *The Physics of Welding*, 2nd edn. (Pergamon Press, London, 1986)
V. Levich, *Physicochemical Hydrodynamics* (Prentice-Hall, Englewood Cliffs, 1962)
J.C. Lippold, *Welding Metallurgy and Weldability* (Wiley, New York, 2015)
R. Naumann, Microgravity processing of materials, in *Encyclopedia of Chemical Processing*, ed. by S. Lee (Ed), (CRC Press, Boca Raton, 2005), pp. 1633–1642
H.R.B. Orlande, Inverse problems in heat transfer: New trends on solution methodologies and applications. J. Heat Transf. **134**, 031011-1-13 (2012)
H. R. B. Orlande, O, Fudym, D. Maillet, R. M. Cotta (eds.), *Thermal Measurements and Inverse Techniques (Heat Transfer)* (CRC Press, Boca Raton, 2011)
M.N. Ozisik, *Inverse Heat Transfer: Fundamentals and Applications* (Taylor & Francis, Philadelphia, 2000)
M. Pasandideh-Fard, S. Chandra, J. Motaghimi, A three-dimensional model of droplet impact and solidification. Int. J. Heat Mass Transf. **45**, 2229–2242 (2002)
K.E. Petersen, Silicon as a mechanical material. Proc. IEEE **70**, 420–457 (1982)
L. Powlowski, *The Science and Engineering of Thermal Spray Coatings*, 2nd edn. (Wiley, New York, 2008)
W.M. Steen, J. Mazumder, *Laser Materials Processing*, 4th edn. (Springer, New York, 2010)
U.O. Stikker, Numerical simulation of the coil annealing process. Math. Models Metall. Proc. Dev. Iron Steel Inst. Spec. Rep. **123**, 104–113 (1970)
J. Szekely, *Fluid Flow Phenomena in Metals Processing* (Academic Press, New York, 1979)
W.R. Wilcox, L.L. Regel, Microgravity effects on materials processing: A review, in *Proceedings of the EUROMAT 2001*, Rimini, Italy, Metall. Assoc. Italy, Milan, Italy, 2001

Part III
System Considerations

The preceding chapters have discussed the basic considerations in manufacturing; modeling of manufacturing processes; analytical, numerical, and experimental solutions and results; product characteristics; and process feasibility. In order to achieve the desired process, we need an appropriate system whose operating conditions may be varied to obtain a given product. Thus, the design, operation, control, and optimization of the manufacturing system are critical to obtaining a high-quality product at the least cost. However, despite extensive modeling and simulation, the results obtained have, in most cases, not been used to design and optimize the process. This section reviews the basic concepts of design of manufacturing systems, based on simulation as well as experimentation, and discusses strategies that may be employed to design and optimize the system. Many complexities that arise in practical processes, such as strong property variations, complicated domains, conjugate mechanisms, chemical reactions, and intricate boundary conditions, have been presented earlier for several important processes, along with the basic approaches that may be used to accurately simulate them. The link between the process and the resulting output is critical. Thus, it is important to couple the modeling and experimental data with the performance and design of the system. Optimization in terms of the operating conditions, as well as of the system hardware, is needed to minimize costs and enhance product quality and system performance. Different optimization strategies that are currently available and that may be used are outlined. Several practical processes that were discussed earlier are again considered to illustrate these approaches, as well as to present typical results. Feasibility of the process and the choice of operating conditions from inverse solutions were discussed earlier and some additional issues are brought out here. Knowledge-based design, sensitivity analysis, uncertainty, and other important aspects are also presented. Most systems have more than one objective of interest, leading to multi-objective optimization, which is briefly presented.

Chapter 10
Simulation, Design, and Optimization of Manufacturing Systems

It has been seen that many applications, such as those involved in the fabrication of electronic and optical components, demand a high level of product quality and purity. Therefore, it is important to design the relevant systems and processes in order to achieve the desired output characteristics. The designs need to be optimized in order to reduce the energy consumption, enhance product quality, and reduce the environmental impact. However, as mentioned in the preceding chapters, detailed results from simulation and experimentation on the effects of operating conditions and governing parameters have often not been extended to system design and optimization (Stoecker 1989; Bejan et al. 1996; Jaluria 2008). This chapter focuses on the design of thermal systems, considering numerical and experimental simulation, acceptable design, and finally optimal design.

The main steps in system design may be specified as:

1. Formulation of the design problem
2. Conceptual design
3. Modeling and simulation
4. Choice of governing parameters and operating conditions
5. Feasible or acceptable design
6. Design optimization

The design process starts with the formulation of the problem in terms of requirements and constraints. Fixed or given quantities must also be specified. The variables include the system hardware, such as dimensions, geometry, components, and materials, as well as the operating conditions, which include quantities like flow rate, heat input, inlet temperature, speed, and pressure. In manufacturing, the desired characteristics of the product are specified as requirements. The limitations on space available, pressure, temperature, weight, cost, etc, must also be given. Constraints due to environmental impact of the system must also be included in many cases. Considerations also arise due to conservation principles and other basic laws, such as the second law of thermodynamics. The governing equations generally take such constraints into account.

Y. Jaluria, *Advanced Materials Processing and Manufacturing*, Mechanical Engineering Series, https://doi.org/10.1007/978-3-319-76983-7_10

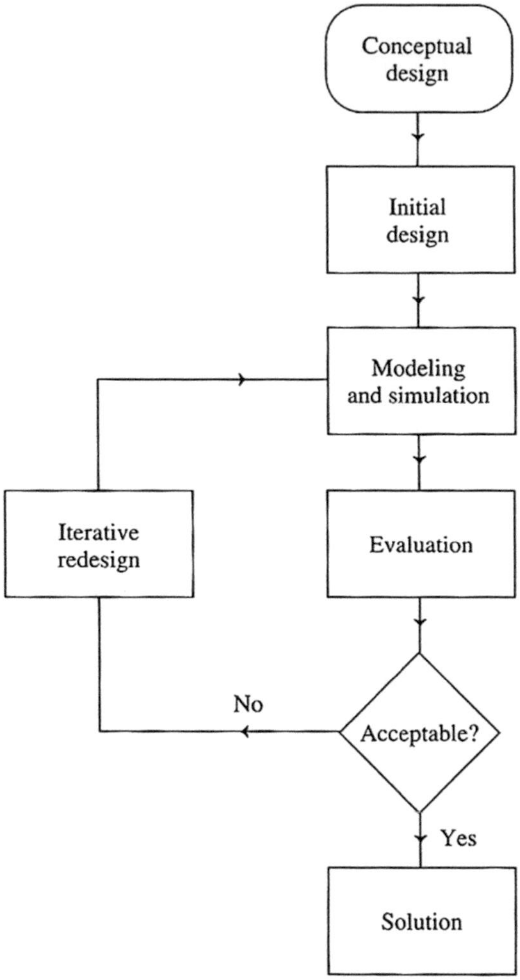

Fig. 10.1 Iterative process to obtain an acceptable or feasible design

The next step is conceptual design, which is often the most difficult part of the design process. The basic concept may be based on existing or similar systems, or it may be a new or an innovative design to meet the given problem. Usually, several concepts are considered, and finally one or more designs that have the strongest chance of success are chosen. Common approaches and examples are given by Jaluria (2008). The conceptual design is then subjected to modeling and simulation to study its behavior in detail and evaluate if it would meet the design problem statement. The initial design is varied, and iterative redesigns are developed till an acceptable design that satisfies the problem is obtained. Figure 10.1 shows a schematic of this process. Of course, there are cases where an acceptable design is not obtained and other conceptual designs may have to be considered.

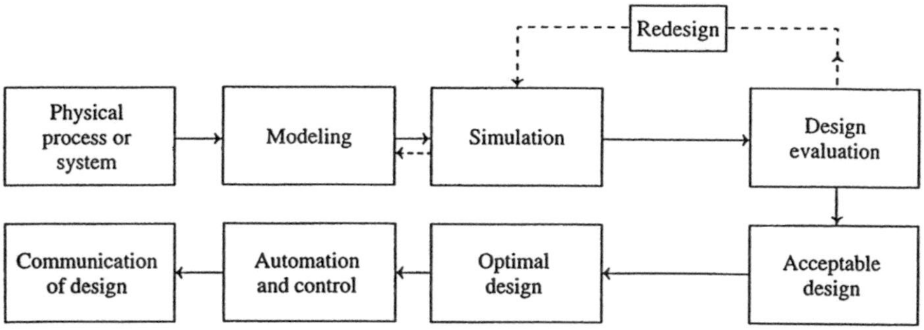

Fig. 10.2 Various steps involved in the design and optimization of a manufacturing system and in the implementation of the design

Generally, a domain of acceptable designs is obtained from which readily available components, materials, and dimensions may be selected for an appropriate acceptable design. However, an acceptable design would not usually be the optimal design. Thus, optimization of the design may be undertaken to maximize a chosen quantity, such as output, product quality, and efficiency, or minimize an item, such as cost, environmental effect, and energy consumption. The quantity chosen for maximization or minimization is known as the *objective function*. In several cases, the optimization focuses on a single criterion, such as product quality or efficiency. However, in most manufacturing systems, several objective functions are of interest. In some cases, the different objective functions may be combined into a single objective function, which is then optimized. In other cases, optimization is carried out with different objective functions, yielding a *Pareto front*, which gives the non-dominant solutions and which may then be used, with trade-offs, to achieve the desired optimal design (Deb 2001). Uncertainties that arise in material properties and in boundary conditions are also important in design and are considered. Different strategies and results are presented here. Optimization is considered in terms of both the operating conditions and the design, or hardware, parameters. Figure 10.2 shows the main steps involved in the design of a manufacturing system, going from the basic process to the communication of the final design.

10.1 Modeling and Simulation

Modeling of manufacturing processes and systems involves obtaining a simplified analytical, numerical, or experimental representation of a system so that a study of the model may be used to predict the response and behavior of the given system. A systematic approach may be used, employing characteristic dimensions, scale analysis, and simplifying assumptions, to develop a satisfactory model, as discussed in detail in earlier chapters. Though analysis may be used for a few components or for idealized circumstances, numerical modeling is essential for most practical systems.

However, mathematical models form the basis for the numerical simulation and also help with the interpretation and consolidation of experimental data. Various simplifications, idealizations, and assumptions are used to make the model amenable to analytical or numerical methods. Experimental data obtained from physical models, prototypes and from existing or similar systems are also often cast in terms of equations through curve fitting and are used along with numerical and analytical results for the design process. Most of these aspects have been discussed earlier and are mentioned here only for completeness and to tie in design and optimization with simulation results.

Simulation refers to the process of obtaining information on the behavior and characteristics of an actual system by subjecting the mathematical, numerical, or experimental model to changes in the governing parameters and operating conditions. A one-to-one correspondence is established between the model and the real system to obtain inputs that may be used for design and optimization. Simulation is used for:

Evaluation of different designs for selecting an acceptable design
Study of system behavior under off-design conditions and establishment of safety constraints
Determination of effects of various design parameters on system performance
Improvement and modification of existing systems
Study of sensitivity of the system to different operating conditions

In manufacturing, most of the simulation approaches are dynamic, or time dependent, and deterministic, though uncertainties need to be considered for realistic results. The process itself may be a continuous one, as in optical fiber drawing, or a discrete one, as in mold casting. The application of modeling and simulation are schematically shown in Fig. 10.3 for three different strategies to obtain an acceptable design.

In many practical systems, different length and time scales are involved, and a multiscale simulation is necessary. As mentioned earlier, the changes in the structure, properties, and characteristics of the material being processed generally occur at the micro- or nanometer scale in the material. Several examples of this aspect were seen, for instance, in processes that occur at the solid-liquid interface in crystal growing, at the heated susceptor surface in a CVD reactor, or in local chemical conversion of reactive materials. However, the dimensions of the final product are generally at the commercial or engineering scale, involving dimensions on the order of centimeters or meters. The overall manufacturing systems are also at engineering or macroscales, requiring the design parameters and operating conditions to be at these length scales as well. Therefore, it is imperative to investigate processes at different length and corresponding time scales, often by different methodologies, and to couple these to obtain the overall behavior of the manufacturing system.

There are several approaches that may be adopted to solve such multiscale problems (He and Tao 2012; Joshi 2012; Jaluria 2014). In many cases, separate experimental, numerical or analytical methods may be used to obtain the results at the microscale level and then use these as inputs to the macroscale simulation or experimentation. An example of this approach was demonstrated in Chap. 8 for

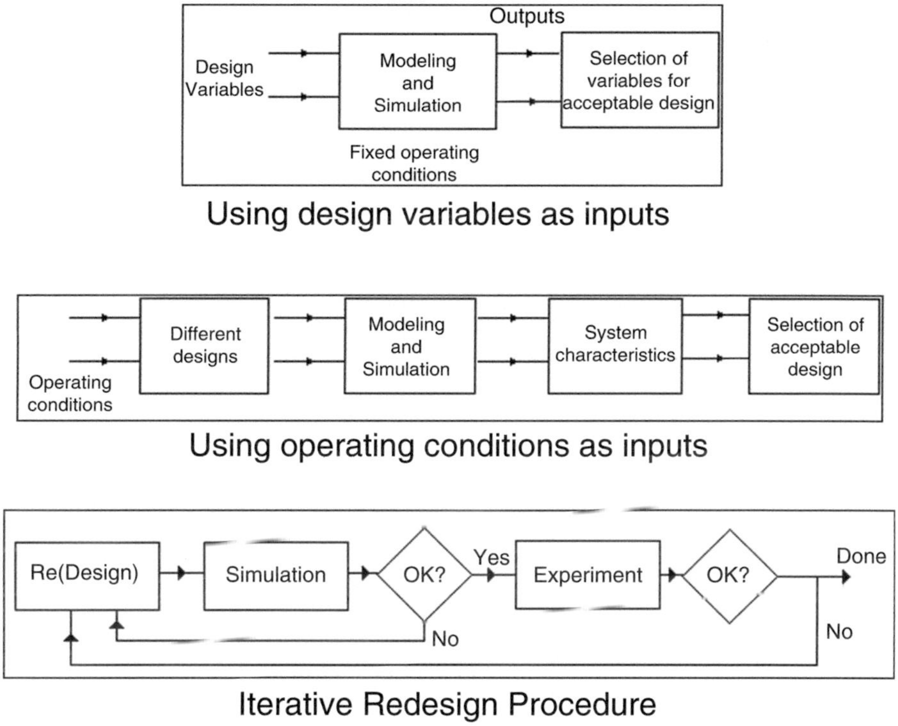

Fig. 10.3 Different strategies to obtain an acceptable design from modeling and simulation

optical fiber coating, where the meniscus was taken as known from experiments. The meniscus, which is at much smaller length scales, being around 100 μm in height, than the coating applicator, which is about 2 cm in height, is thus obtained separately and provided as an input to the relatively simpler simulation of the transport processes in the coating fluid; see Fig. 8.33. Similarly, in CVD, the chemistry model can be applied at the micro- or nanoscale level to obtain the reaction rates and the deposition, while the boundary conditions, in terms of inlet velocity, susceptor temperature, inlet concentration, rotational speed, etc., are imposed at the system level. In the numerical scheme, the grid near the surface is much finer than that in the main flow, resulting in control volumes that would allow the capture of deposition which is in microns. The chemistry is applied to each control volume, and the results linked with the transport processes at the larger scales to obtain the final deposition results. Some characteristic results were shown in Chap. 7. In optical fiber drawing, thermally induced E' defects are generated at the nano and molecular levels and linked with the transport processes in the silica preform/fibers. Dopants added to the preform cause changes in the local properties, as mentioned earlier, and need to be treated similar to the point defects, since both are at micro-/nanoscale.

Similar multiscale considerations arise in other materials processing applications. The chemical conversion of a reactive polymer like food occurs at the molecular or nanoscale level, whereas the boundary conditions are imposed at the macroscale. Some of the results on the flow and temperature were shown earlier for polymer extrusion. In food extrusion, both shear and temperature affect the conversion; see Fig. 6.19. Thus, the conversion is studied at the microscale to obtain the dependence on shear and temperature. These results are then provided to the modeling of the entire extrusion process. The results from the microscale modeling are linked with the macroscale modeling and finally to the characteristics of the product. Further work is needed on linking the micro-/nanoscale processes with the macro- or engineering scale in order to accurately and realistically model the processes undergone by the material and to predict the resulting properties and characteristics.

10.2 Inverse Problem and Process Feasibility

Both of these aspects were considered in earlier chapters in the discussion of a few manufacturing processes. However, these are particularly relevant to most manufacturing systems and are thus briefly given here to stress their importance and to link these with the modeling and simulation of the process.

10.2.1 Inverse Problem

An important aspect in manufacturing is the choice of boundary conditions and governing parameters in order to achieve the desired thermal process. This is an inverse heat transfer problem, which involves finding the conditions that would lead to a given result. Generally, several conditions may result in the same outcome. Thus, the solution of inverse problems is not unique, though the domain of uncertainty may be reduced through optimization and other techniques (Ozisik 2000; Orlande 2012). Generally, the direct solution, with specified conditions, is solved, and the results are used to develop a scheme to converge to a solution to the inverse problem in a fairly narrow domain.

An example of an inverse problem in manufacturing was discussed earlier and shown in Fig. 9.6, which gives a sketch of a batch annealing furnace for steel sheets rolled up into cylindrical coils. The material has to be heated beyond its recrystallization temperature of around 723 °C and then cooled very slowly to achieve the desired annealing, as sketched in Figs. 1.2b and 9.7. Thus, the desired temperature cycle is given, along with an envelope of acceptable variation. The boundary conditions, particularly the time-dependent flow of flue gases from the blast furnace, need to be determined to obtain the desired temperature variation with time in the coils. This problem is numerically modeled, and the dependence of the temperature distributions on the boundary conditions is established. This information is then used to solve the inverse problem to determine the temporal variation of a temperature sensor located at the base of the furnace that controls the flow rate of the hot gases in order to achieve the desired annealing process (Jaluria 1984). Another example was provided by the modeling of the optical fiber draw furnace, as shown in Figs. 8.20 and 8.21. The temperature distribution at the axis of a graphite rod mounted at the center of the furnace was used in the solution of the inverse problem to obtain the wall temperature distribution. Similar approaches have been used for other materials processing systems.

10.2.2 Feasible Domain

Another major issue in manufacturing is the feasibility of the process, which is determined by the physical constraints, such as temperature, pressure, and tension limitations, attainment of the desired product characteristics, and availability and cost of resources and materials needed. The boundary conditions and design parameters may be varied to determine if the process is physically feasible. An example of this aspect is given in Fig. 8.22, which showed the drawing process for optical fibers. The main limitation is due to the viscous rupture of the fiber due to excessive strain, resulting from low temperature or high draw speed. For hollow fibers, another limitation arises from the collapse of the central core, which results in a solid fiber rather than the desired hollow one. A high draw temperature and a slow draw speed are conducive to the collapse of the hole. Thus, a feasibility domain can be obtained, as shown in the figure.

Another example is shown in Fig. 6.22, which gives the results for the extrusion of starch in a twin screw extruder. At a given rotational speed, a very small throughput, or flow rate, implies a small die opening, resulting in recirculation, high temperatures, and damage to the material. A very high flow rate implies a forced flow, with greater pressure at the inlet than at the outlet. This is not an acceptable solution for an extruder, which should build up pressure downstream.

Clearly, the solution of the inverse problem to determine the appropriate operating condition and the feasibility of the process are important aspects in materials processing, though only limited work has been done on these for many practical processes. The examples given here were discussed in earlier chapters, and similar considerations can be included for other processes as well, such as coating, soldering, and deposition, where various constraints may make the process infeasible and where inverse solutions are needed to choose the relevant conditions.

10.3 Design Optimization

As presented in the preceding, modeling and experimentation are employed to gain a better understanding of a given manufacturing process and quantitatively determine the effect of operating conditions on the transport process undergone by the material. The effects of the physical system parameters such as dimensions, materials, and geometry are also obtained. The resulting changes in the material can then be obtained by linking the manufacturing process with material characteristics and transformations. The next step is to use this information to design an appropriate manufacturing system. It is necessary to first develop acceptable or feasible designs that meet the given requirements without violating constraints, as discussed in the preceding section. Thus, a feasible design domain is generated, and any point within this domain would yield a satisfactory, though not optimal, process. This is followed by optimization of the process, in terms of the operating conditions, and of the manufacturing system, in terms of the hardware design parameters. An objective function is chosen and this is minimized or maximized for an optimum. Interest obviously lies in obtaining high product quality, low cost, high efficiency, and reduced environmental effect.

To summarize, the main engineering considerations involved in obtaining the desired manufacturing process are:

Modeling and simulation of the process
Parametric study and determination of feasibility
Design of the thermal system to achieve an acceptable design
Optimization of the process with respect to operating conditions
Optimization of the system with respect to physical parameters and hardware
Sensitivity analysis
Effect of uncertainties

The first three items have already been discussed earlier and the focus now shifts to optimization.

As mentioned above, a feasible design would generally not be the best or optimal design. In recent years, it has become essential to optimize the process and the system in order to address growing global competition. A particularly important aspect in optimization is the objective function U, which is to be minimized or maximized. This function considers important aspects of a given process such as quality, cost, efficiency, etc. Because of the complexity of most manufacturing processes and the nonlinearities associated with thermal processes, search methods, which seek the best solution over the feasible domain, are among the most common optimization strategies used for materials processing. A number of feasible or acceptable designs are generated, and the best among these is selected by searching in the design domain. Search methods such as exhaustive, Fibonacci, and golden section use trial runs to narrow the domain to reach the optimum. The steepest ascent/descent method and other gradient-based methods, along with genetic algorithms, are widely used for thermal systems (Bejan et al. 1996; Rao 2009; Arora 2011; Jaluria 2017). Figure 10.4 shows sketches of the steepest ascent method, in which search is carried out in the direction of maximum rate of change, and of the univariate search method, in which the optimum is sought by varying one design variable at a time, keeping the other constant.

The objective function U is thus a critical aspect in the optimization of materials processing systems. It is not an easy parameter to define or quantify. The optimal design depends on the chosen objective for optimization. In many cases, there are several objectives that are of interest, and a multi-objective optimization is needed.

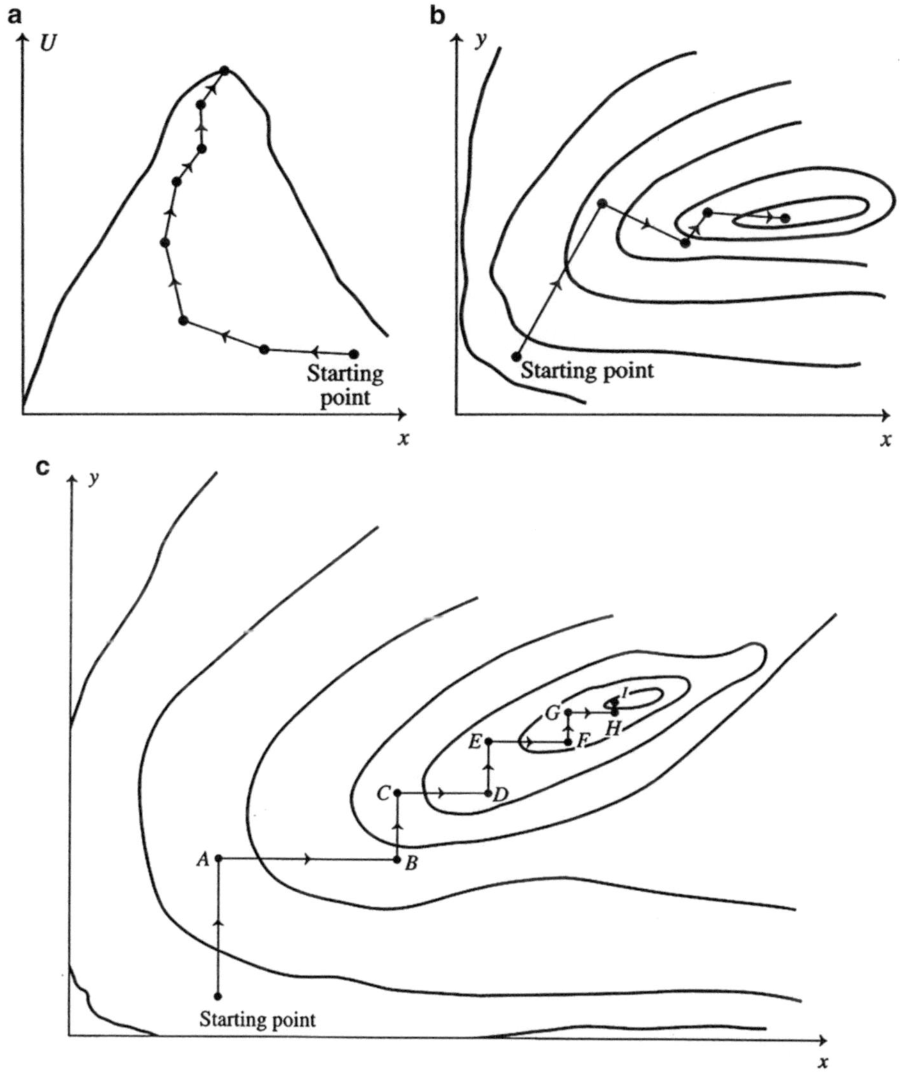

Fig. 10.4 Sketches of the optimization processes based on search methods, showing steepest ascent in (**a**) and (**b**) and univariate search in (**c**)

One approach is to carry out optimization for different criteria separately, leading to an optimization curve. A combination of the solutions known as a *Pareto front* is generated. This curve, for two objectives, or surface, for three objectives, represents a set of non-dominated solutions being chosen as optimal, if no objective can be improved without sacrificing at least one other objective. Then, the optimization is based on trading off between different criteria on the basis of available information and desired results for a given system (Zhao et al. 2007; Zhang et al. 2014).

In materials processing, interest lies mainly in the product quality and the cost per unit of the product. In the optical fiber drawing process, for instance, the objective function could be taken as the fiber quality, which depends on the tension, defect concentration, and velocity difference across the fiber, all these being scaled to obtain similar ranges of variation. The objective function U may then be taken as the square root of the sum of the squares of these three quantities and minimized for the highest fiber quality (Cheng and Jaluria 2005). Several search methods, such as golden section for single variable and univariate search for multivariable cases, can be employed. Figure 10.5 shows typical results from golden section search for the optimal draw temperature and draw speed. The golden section search is convenient because it retains the function values from previous points and the trial runs are taken at locations based on the golden ratio 0.618. The results from the first search are used in the second search, following the univariate search strategy, to obtain optimal design in terms of these two variables. Several other results were obtained on this complicated problem. In this case, only the E' defects were considered. Other defects and imperfections may similarly be considered. For instance, both E' and non-bridging oxygen hole center defects (NBOHC) may be considered, along with tension and velocity lag. The corresponding results for hollow fiber drawing are shown in Fig. 10.6 (Yang and Jaluria 2009). Similar approaches may be used for different processes and fibers to obtain the optimal conditions, as well as optimal design parameters.

For a more detailed study, let us consider a CVD system. The main aspects of interest are product quality, often characterized by the uniformity of the film, and the production rate, characterized by the deposition rate (Chiu et al. 2002; Lin et al. 2009). These may, in turn, be represented by the percentage working area (PWA), which gives the percentage acceptable area from film uniformity, and the mean deposition rate (MDR). Similarly, the root mean square (RMS) and kurtosis of the deposition distribution over the surface may be used to represent uniformity; see Fig. 7.9. These quantities may be combined to yield a single objective function, which is optimized with the given constraints, or each objective function may be considered separately to obtain the Pareto set, as mentioned earlier (Miettinen 1999; Deb 2001). In the first case, the composite objective function may be formulated in different ways. The optimal design will generally depend on the function chosen. Some examples of such an objective function are

$$U = \frac{(\text{Product Quality}) \times (\text{Operating Cost})}{(\text{Production Rate})} \qquad U = (\text{PWA}) \times (\text{MDR}) \tag{10.1}$$

We can also choose one of the quantities for maximization, while the other is specified by a given constraint. For example, we could

$$\text{Maximize } U = \text{PWA for given constraint on MDR} \tag{10.2}$$

$$\text{Maximize } U = \text{MDR for given constraint on PWA} \tag{10.3}$$

Thus, we could specify that the deposition rate should not be lower than a given value while we maximize the percentage working area and vice versa. This approach

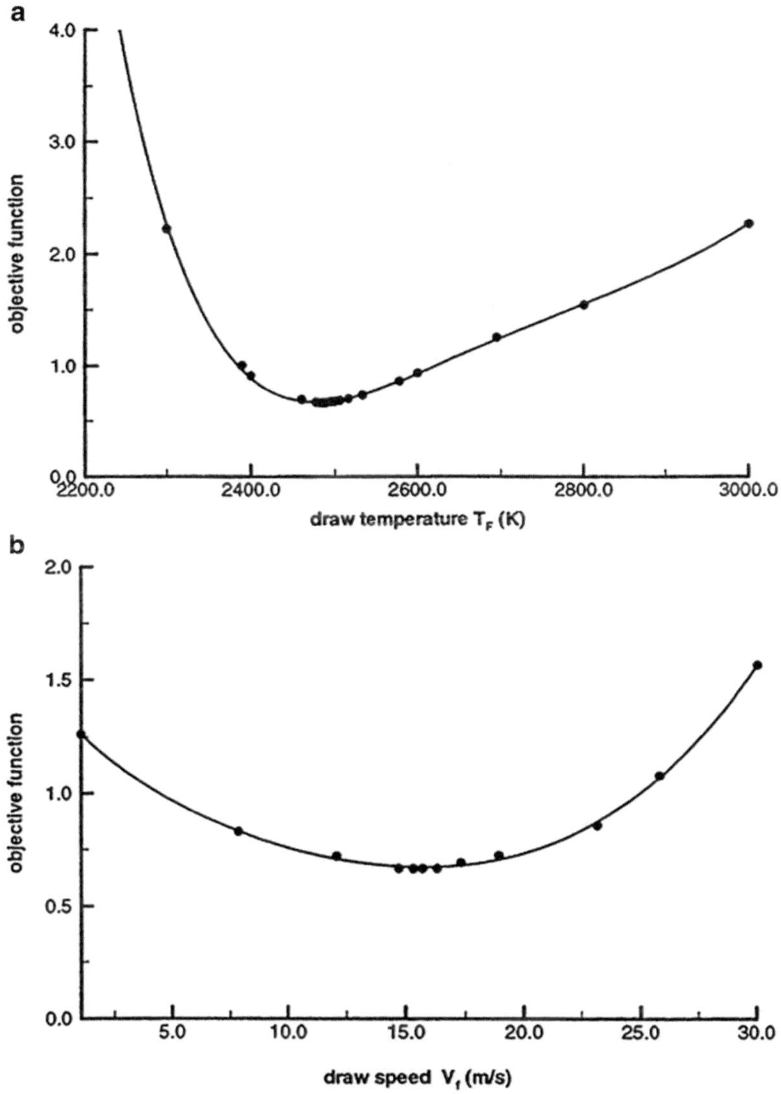

Fig. 10.5 Determination of optimal draw temperature for optical fiber drawing at a draw speed of 15 m/s and the optimal draw speed at a draw temperature of 2489.78 K, obtained in the first part (**a**), by using the golden section search method

was found to be quite definitive and to yield useful and consistent results (Lin et al. 2009).

On the basis of simulation and experimental results, response surfaces may be generated to represent the overall behavior of the process or the system. Response surfaces of different orders can be generated, though lower order surfaces, such as

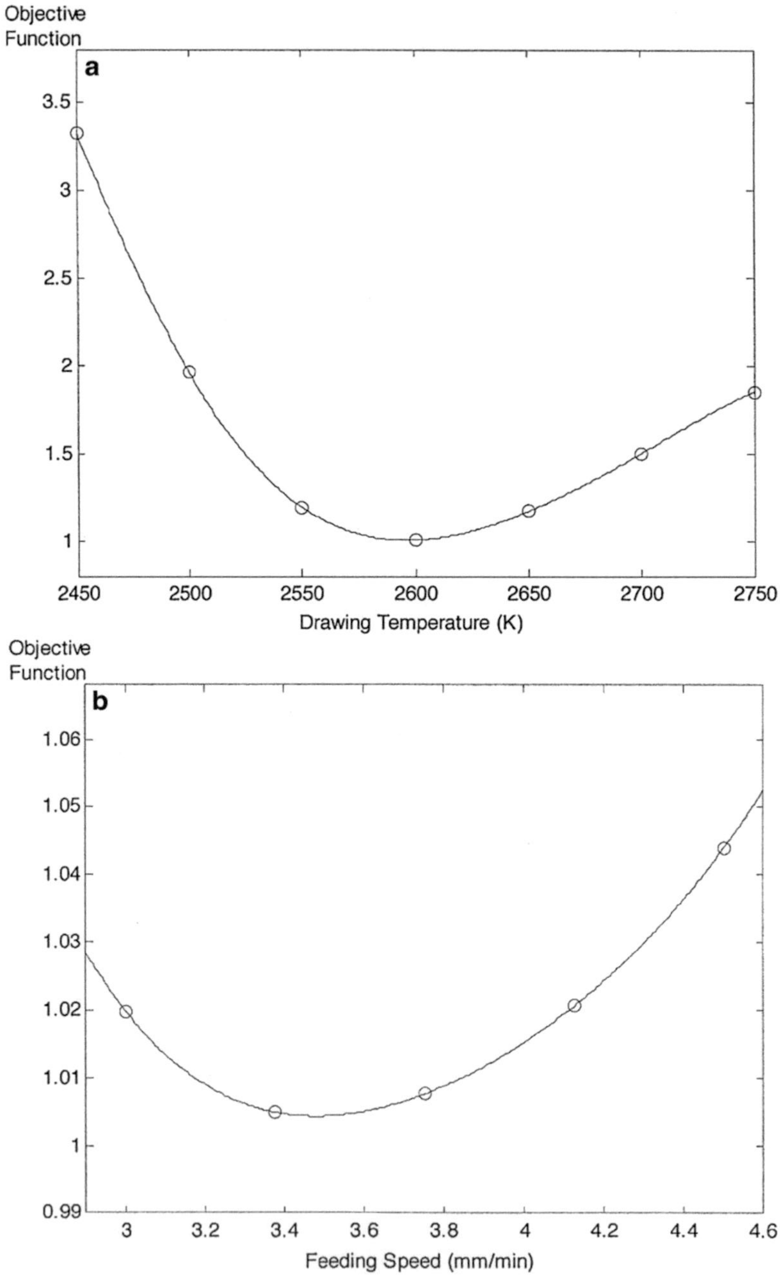

Fig. 10.6 Optimization of hollow fiber drawing. (**a**) Variation of the objective function with the drawing temperature using a polynomial curve fit, (**b**) variation of the objective function with the feeding speed at a fixed drawing temperature of 2599.1 K, the optimal point from (**a**), and the corresponding polynomial curve fit

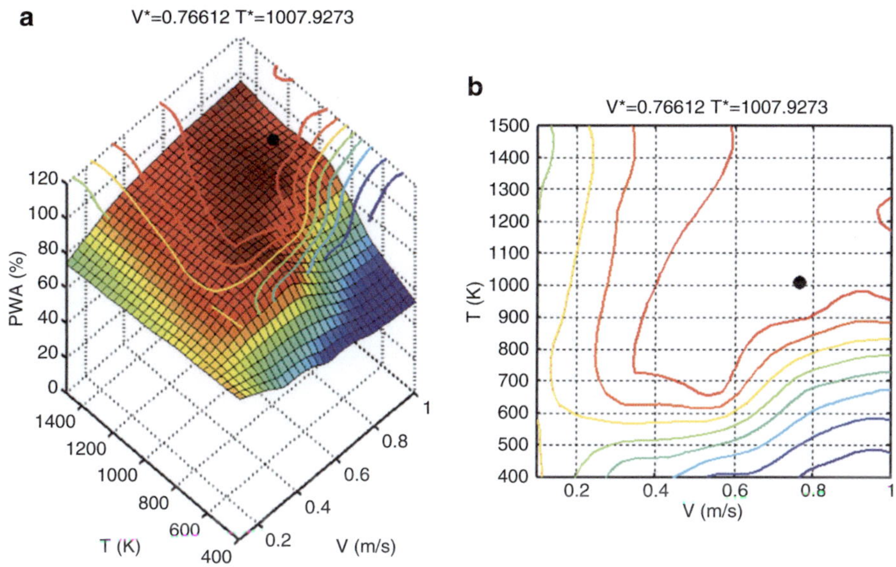

Fig. 10.7 Response surface for the percentage working area and the optimal conditions in terms of susceptor temperature and average inlet flow velocity for an impingement chemical vapor deposition reactor for the deposition of silicon

second and third order, are desirable for ease in optimization (Myers and Montgomery 2002; George et al. 2015). Figure 10.7a shows a typical response surface obtained from the numerical simulation of the CVD process for the deposition of silicon. It is desirable to use a small number of simulation runs since each is expensive in such complicated systems. This figure is based on only 25 runs, selected to cover the feasible domain. The figure shows the dependence of the objective function on the operating conditions. The existence of an optimal circumstance is also indicated, as also shown on a two-dimensional plot in Fig. 10.7b. Many such cases were investigated, and the operating conditions for optimal deposition were determined. Figure 10.8 shows the response surfaces for the average deposition rate and uniformity for the deposition of gallium nitride. Such curves may then be used to obtain the optimal conditions.

Figure 10.9 shows the results for silicon deposition from optimization based on Eqs. (10.2) and (10.3). Several deterministic constraints are given in order to define the infeasible domains, leaving the white region as the feasible domain. Then the optimum is obtained as maximum uniformity, given by PWA, or maximum deposition rate, given MDR (Lin et al. 2010). The resulting optimal points are shown in the figure.

Similar approaches can be used for other materials processing methods. However, because of the complexity of the processes, particularly in chemically reactive systems, few results are available in the literature on optimal conditions. Recent studies have employed advances like genetic algorithms, concurrent experimental and numerical methods, and expert systems to make the optimization process more efficient.

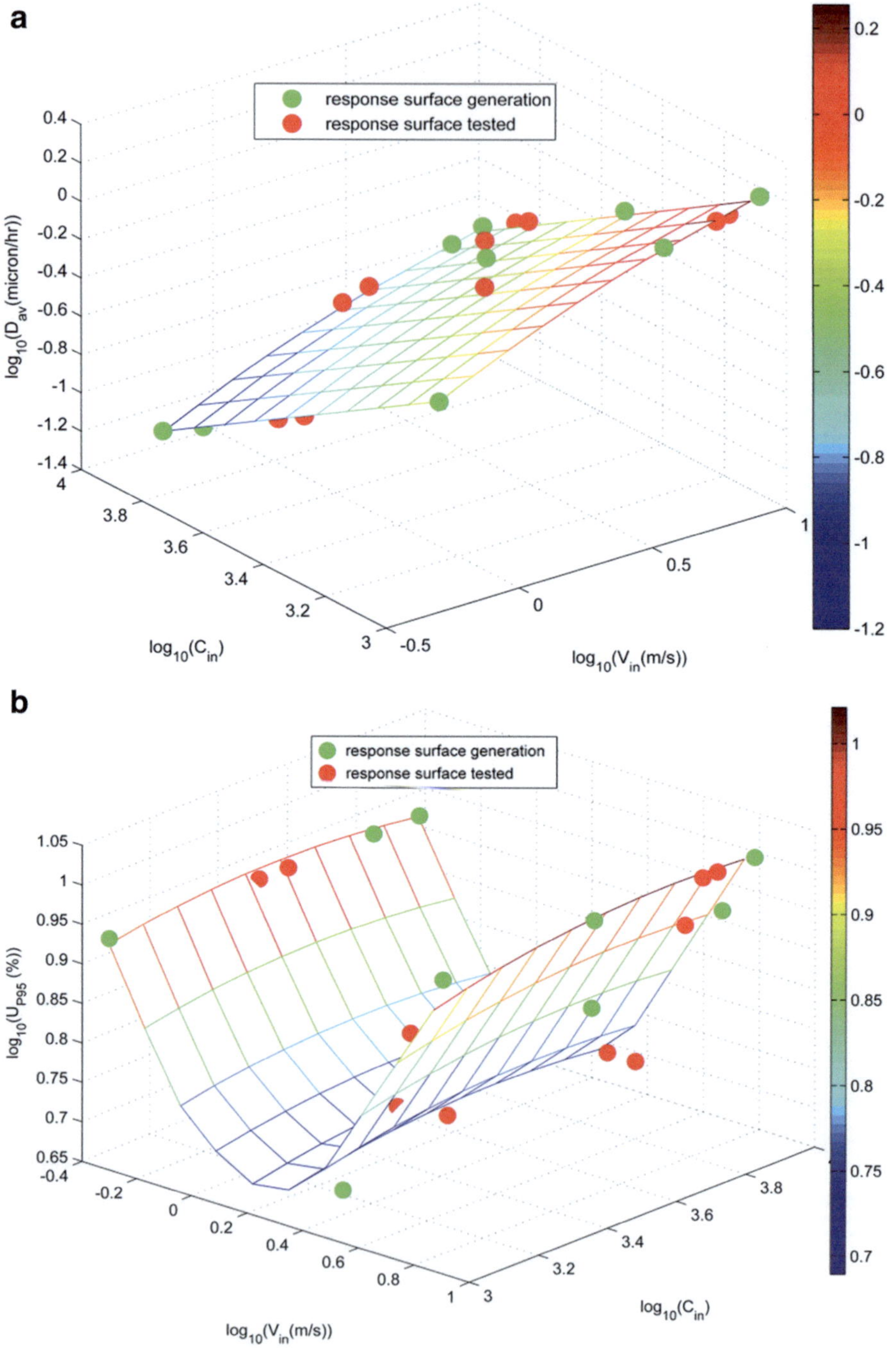

Fig. 10.8 Response surfaces for (**a**) average deposition rate and (**b**) uniformity parameter, for the deposition of gallium nitride in a vertical impingement CVD reactor

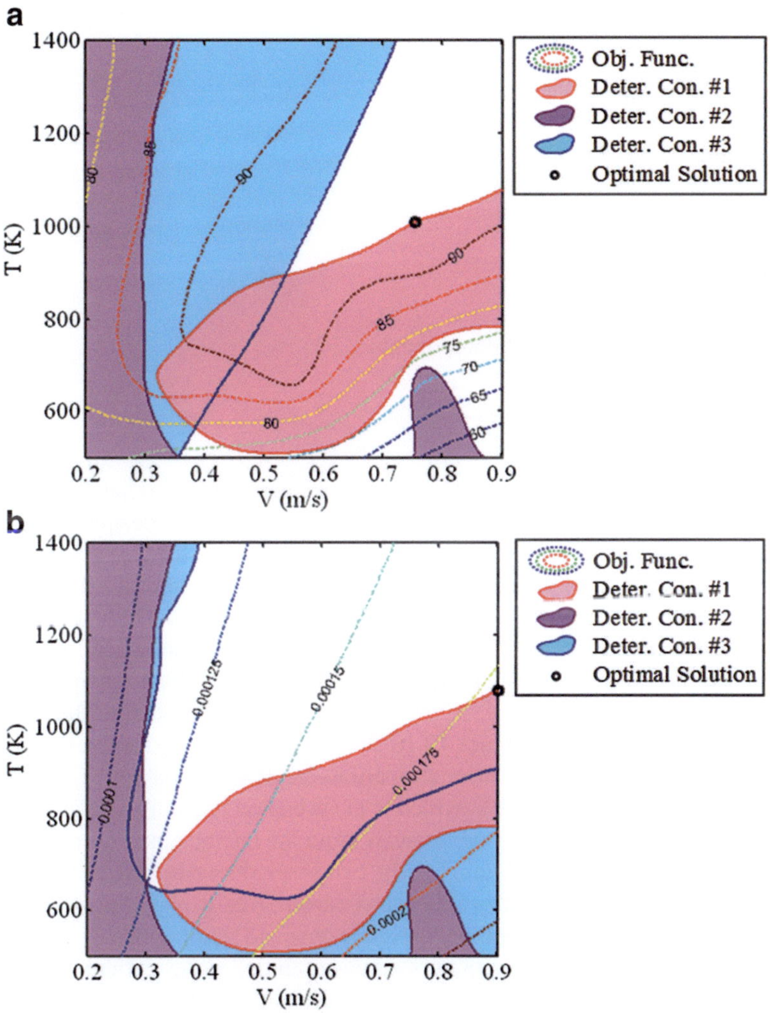

Fig. 10.9 (**a**) Optimization of percentage working area (PWA) at a chosen value of the mean deposition rate, specified as constant; (**b**) Optimization of the mean deposition rate (MDR) with a specified constraint on PWA

10.4 Uncertainties

In actual practice, all the design parameters and operating conditions are subject to uncertainties due to unforeseen variations. These must be considered for a practical solution because the failure of a given thermal system can be dangerous and expensive. The review paper by Lin et al. (2010) presents a systematic strategy for the modeling and optimization of a thermal system including the effects of uncertainties. An impingement type CVD reactor, on which several results have been presented earlier, is taken as an example. Some of the major uncertainties that arise in this process are:

Uncertainties in operating conditions:
Inlet flow rate or velocity
Temperature of susceptor or heat flux input
Mass fractions of reactive gases at inlet
Initial conditions
Environmental conditions

Uncertainties in system design:
Dimensions of the system and components
Location of inlet and susceptor
Geometry, configuration, and symmetry
Material and gas properties

The design and optimization with uncertainties are based on reliability and, thus, on the percentage failure rate of the design. Reliability-based design optimization (RBDO) algorithms are generally employed, with chosen distribution of uncertainty in the parameters (Tu et al. 1999; Youn and Choi 2004; Lin et al. 2011). Again, objectives such as percentage working area and mean deposition rate may be considered for optimization. Probabilistic constraints are established, with respect to either normally or non-normally distributed random variables. Then the optimal solutions are obtained, using techniques discussed above, subject to the allowable level of failure probability. Figure 10.10 shows a sampling of results for different constraints of Fig. 10.9, with normal distribution of the variables. The failure is brought down to less than 0.13%, which is the accepted level in RBDO. Due to the uncertainties, the optimal point moves away from the one obtained for deterministic conditions, to satisfy this condition. For the two cases shown in Fig. 10.9, the failure rate was over 40% if uncertainties are not considered. But by including uncertainties, a more realistic and practical optimal design is obtained. Further details are given by Lin et al. (2010, 2011). The importance of uncertainties in design and optimization has only recently been considered for thermal systems.

Uncertainties are particularly critical in the manufacture of microscale and nanoscale devices. We have discussed fabrication of optical fibers, solid as well as hollow, and of thin films. These are largely in the micrometer scale, and variations in the operating conditions or in the system hardware can have a significant effect on the product. Non-concentricity is, for example, a problem in optical fiber manufacture and also in several CVD systems. Similarly, uncertainties in draw speed for fiber fabrication or in rotational speed for thin film fabrication can affect the results significantly. Another important example that was considered earlier is the fabrication of microchannels for use in microfluidics, in heat removal, or in other applications (Zhang et al. 2013). The microchannels are typically in the range 20–100 μm in diameter, width, or height. Then uncertainties in the fabrication process can substantially affect the final shape and dimensions obtained. Uncertainties in the various steps of the process, such as the dimensions of the mask and of plasma beam and the

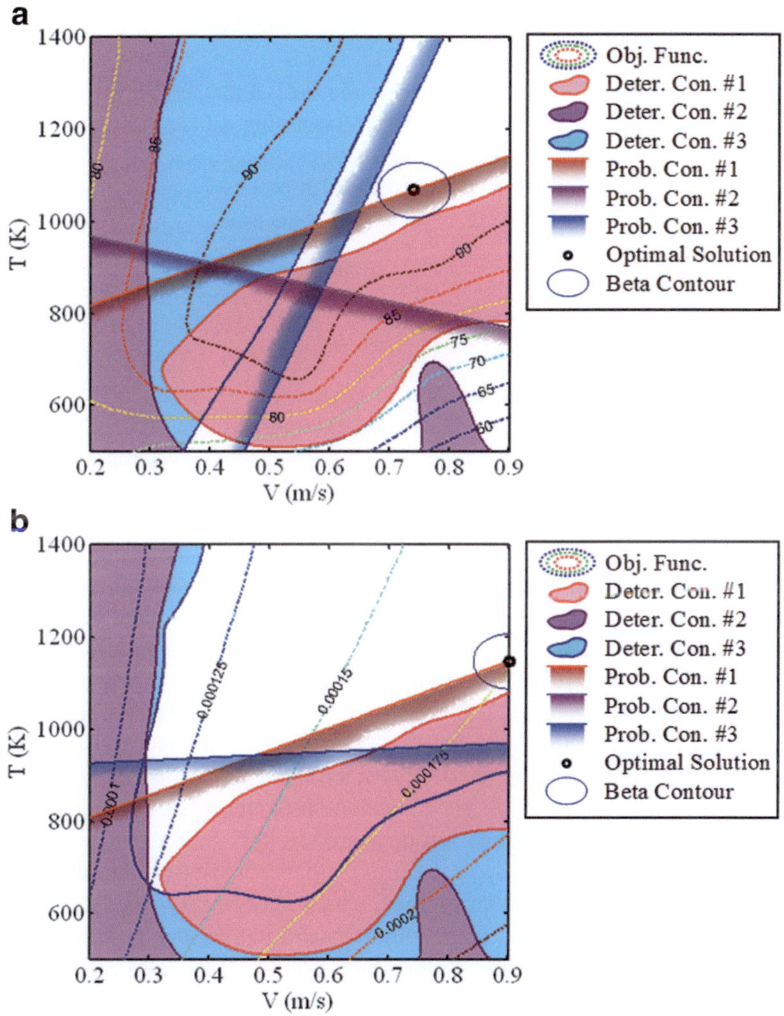

Fig. 10.10 Optimization with deterministic and probabilistic constraints for normally distributed variables, indicating the optimization with variables with given uncertainties to achieve a specified level of reliability, for the circumstance of Fig. 10.8

etching speed, must be considered in the fabrication process and corresponding system design.

Sensitivity analysis is also an important tool that is generally used to study the importance of various parameters and focus on the dominant ones for design and optimization. In the absence of the information from a sensitivity study, the work would consider essentially all the parameters, making it a fairly prohibitive task. Clearly, much work is needed for the materials processing systems to obtain optimal designs that are more reliable and realistic.

10.5 Additional Design Aspects and Strategies

Several new aspects have been brought into the design process in the recent years. These help in developing an initial design, in obtaining a faster convergence to an acceptable design, and also in making the final design realistic and practical. These strategies can also be extended to the optimization process.

10.5.1 Knowledge-Based Design

An important consideration in the design of thermal systems is the use of the available knowledge base to guide the design and optimization process. This knowledge base typically includes relevant information on existing systems and processes, current practice, knowledge of an expert in the particular area, material property data, and empirical data on equipment and transport, such as pump characteristics and heat transfer correlations. It is advantageous to use any such available information in developing initial, acceptable, and optimal designs, as well as in selecting operating conditions. Though available knowledge on the system is generally used in design anyway, some focused effort has been directed in recent years at streamlining the design process, improving the design methodology, automating the use of existing information, and developing strategies for rapid convergence to the final design (Jaluria 2008, 2017).

Consider, for instance, the initial design of a polymer extrusion die. Die design is an important consideration in plastic extrusion since the operation of the extruder system and the quality of the final product are strongly influenced by the die, as discussed in detail in Chap. 6. Typical extrusion dies consist of different parts, such as entrance, flow channel, and exit regions, which may be fabricated as one unit or may be attached to each other through couplings and screws and the heating/cooling arrangement to maintain a desired temperature. Figure 10.11 shows the sketch of a typical, generic, extrusion die that may be considered for design and optimization with respect to the geometry, dimensions, and materials. Two strategies may be used for generating an initial design. The first is based on a library of designs from earlier design efforts and from existing systems. The design closest to the given problem may be selected by comparing the designs in the library with the desired specifications. The second approach is based on expert rules for die design. Employing knowledge and experience used by an expert, rules may be set down to generate a design for the given requirements and constraints (Jamalabad et al. 1994). Preliminary evaluations and estimates are used to develop a possible design that is used as initial design. Of course, the user can always enter a personal selection for initial design if the output from the library or the expert rules is not satisfactory. The expert knowledge may also be used in redesign in case the initial design is not acceptable. Figure 10.12 shows these basic ideas schematically. Figure 10.13 shows the convergence of the redesign process if selective rules, based on the knowledge base, are

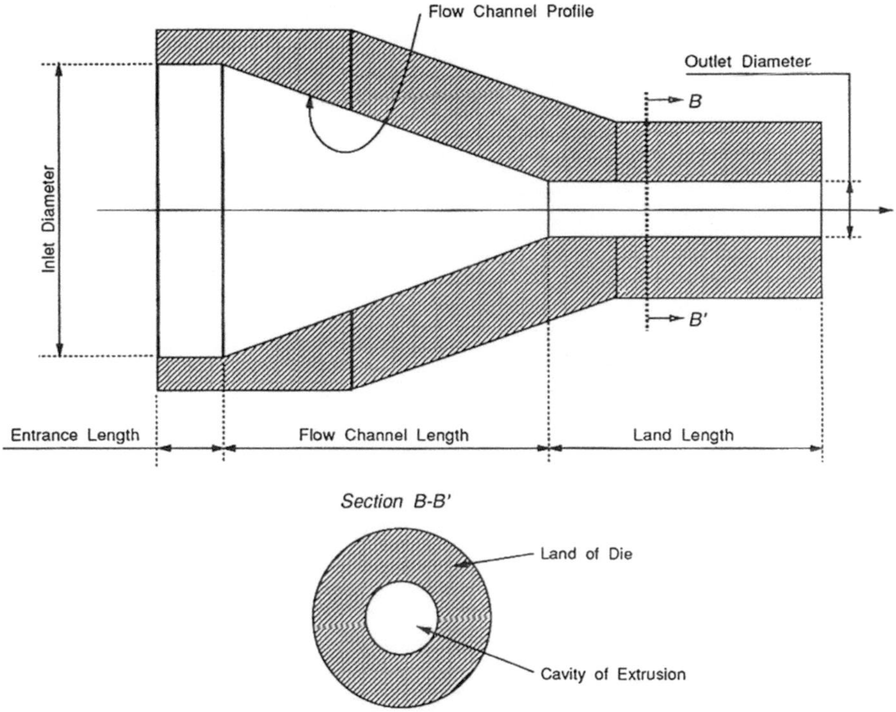

Fig. 10.11 A generic extrusion die with a circular exit

used. An objective function *Z*, based on the pressure, flow, and uniformity, is used, yielding the optimal design at its minimum value. Clearly, the convergence is accelerated by using a knowledge-based approach.

Another example is the casting of a material by pouring the molten material into an enclosed region and allowing it to solidify. A large number of design parameters arise in this problem, such as materials, initial melt pour temperature, cooling fluid and its flow rate, and dimensions, as discussed in Chap. 4. Viswanath and Jaluria (1991) considered this problem, using knowledge-based design methodology. Several models, with different simplifications, are available in the literature for the study of solidification (Ghosh and Mullick 1986). As presented in detail earlier, these models include one with steady conduction heat transfer to an isothermal mold, Chvorinov model, which takes the entire thermal resistance in the mold, lumped mold model, one-dimensional conduction model, two- and three-dimensional conduction models, models with buoyancy effects, and more complicated models for alloys, defects, and complicated geometries. Each model has its own level of accuracy and range of validity. Different models may be chosen, depending on the application and materials involved.

An initial design may be developed with, say, a desired solidification time for a given casting as the requirement and constraints from allowable thermal stresses and defects in the casting. The process can start with relatively simple models and move

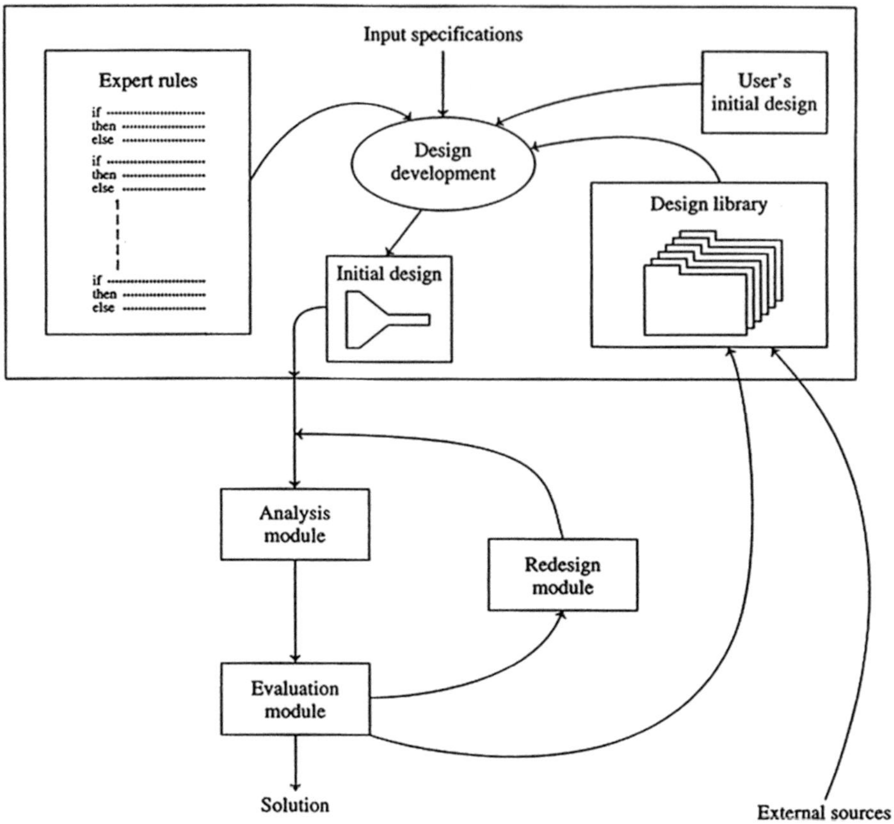

Fig. 10.12 Schematic of the development of an initial design, followed by evaluation and redesign

to more complicated models if the desired information on solidification time, stresses, etc. is not obtained. It may also start with the simplest model and keep on moving to models with greater complexity till the results remain essentially unchanged from one model to the next. Thus, models may be automatically selected using decision-making based on accuracy considerations. If the desired solidification time is not achieved by varying the cooling arrangement, the pour temperature of the melt may be varied. If even this does not satisfy the requirement, the thickness of the mold wall may be changed. The material of the wall may also be varied, if needed. Thus, by first varying the operating conditions and then the dimensions and materials, the solidification time may be brought to a desired value, thereby yielding an acceptable design. Figure 10.14 shows a typical run for the design of the given system, indicating model change as the design proceeds. Each successful design may be stored for help in future designs. This implies improving the design process through learning from past experiences.

Thus, it is seen that the use of the available knowledge base on the system is important in the design, as well as in choosing the operating conditions for a desired product. In actual practice, this knowledge, which may simply be called experience,

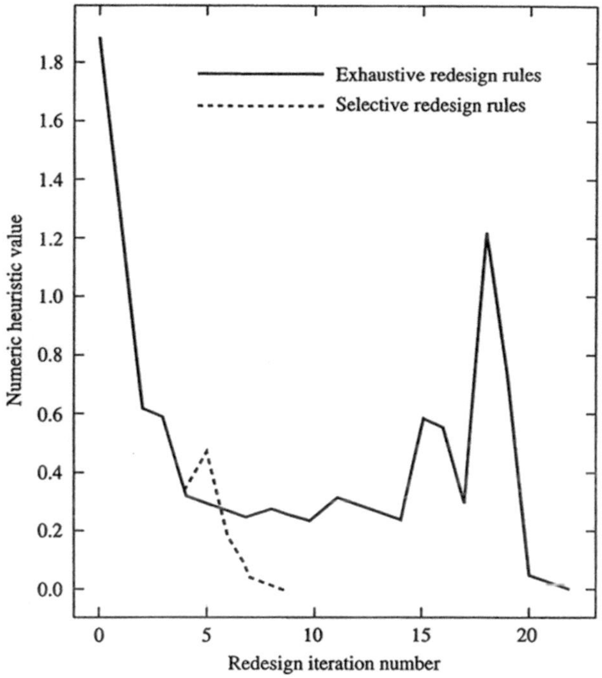

Fig. 10.13 Iterative convergence of the design-redesign process, monitored in terms of an objective function *Z*, for the design of a 1 cm diameter circular die for polymer extrusion (Jamalabad et al. 1994)

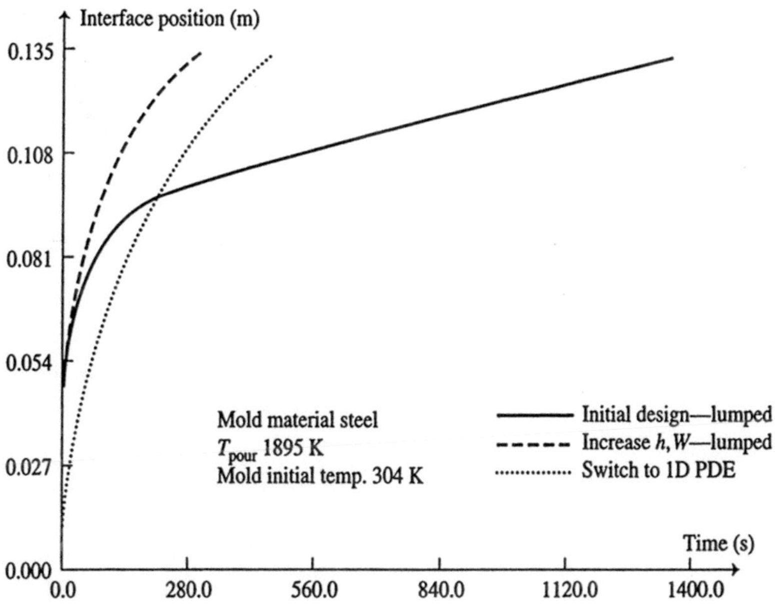

Fig. 10.14 Design of a casting system, using change of models during the design and optimization process

is commonly used by engineers involved with thermal systems and processes. Decisions are often made on the basis of what is known about existing processes, past trials, and other similar systems. Therefore, it is desirable to build this knowledge base into the design process.

10.5.2 Concurrent Experimentation and Simulation

As presented in the preceding sections, most approaches for design of thermal systems are based on the use of numerical simulation, which is validated by experiment. Thus, a sequential numerical and experimental approach is commonly adopted. This strategy fails to use certain advantages of using both tools concurrently where the outputs from each approach drives the other to achieve better engineering design in a more efficient way. The hardware is often difficult to vary in practice, and numerical simulation can be used advantageously. However, for given hardware, operating conditions can be easily changed in an experimental setup, and the corresponding inputs may be obtained experimentally. Also, numerical simulation is accurate and dependable under certain conditions, such as low flow rates, whereas in the transitional and turbulent regimes at higher flow rates, experimental data may be more reliable. Switching from one approach to the other can then be based on flow transition to turbulence and on the design parameter being varied (Knight et al. 2002; Icoz and Jaluria 2004).

The results from simulation and experiment are combined to build a surrogate model, and the concurrent inputs are used to achieve an acceptable or optimal design, using a system such as the one sketched in Fig. 10.15. By making the two approaches concurrent, the models are automatically validated, and the experiments can guide the selection of conditions for the next simulation in an iterative process. Similarly, simulation helps in the selection of conditions at which experimental data should be obtained, leading to faster convergence and a more realistic design. Results have been obtained on a few thermal systems, such as those for cooling of

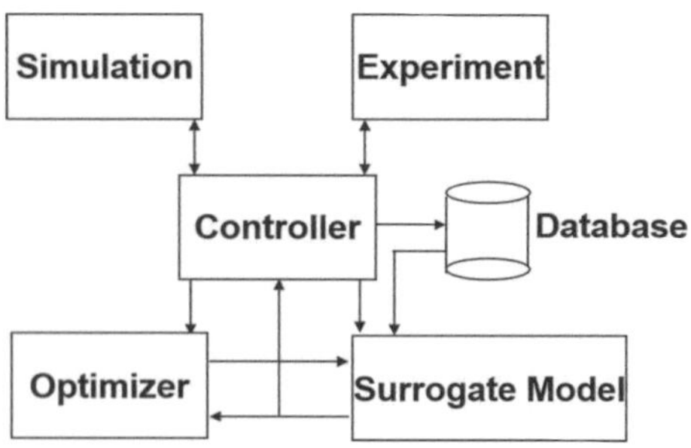

Fig. 10.15 Schematic representation of a design methodology based on concurrent simulation and experimentation

electronic equipment, and the concept has been shown to be a useful and efficient technique for design and optimization of practical problems.

10.6 Summary

This chapter focuses on the system considerations related to manufacturing processes. As discussed in detail earlier, material transformations generally occur at the micro-/nanometer length scales, whereas the boundary and initial conditions are imposed at the engineering or macroscale. The manufacturing systems are also at engineering scales, even though the devices like thin films and optical fibers are microns in thickness or diameter. Thus, a multiscale problem arises in modeling and simulation, linking the processes at different length and time scales. To complement the simulation, experimentation is needed for validation of the models, for physical insight, and for providing inputs to simplify the analysis, if the model is unavailable, difficult, or inaccurate. An important aspect in manufacturing is the design and optimization of the thermal system needed to achieve the desired process. The inverse problem that arises in order to choose the appropriate boundary conditions is briefly discussed. The resulting feasible design domain and optimization of the process are also discussed. In many cases, multiple objectives are of interest, and multi-objective optimization with trade-offs is needed. Uncertainties in design parameters and operating conditions may arise and must be considered to obtain realistic and practical results. These aspects are discussed in detail.

It is necessary to carry out further detailed work on feasibility, control, design, and optimization if the existing processes are to be improved and new techniques are to be developed. A consideration of appropriate objective functions for different manufacturing processes, along with multi-objective optimization using available and new methods, is needed. Obviously, future work must particularly address the needs for new materials and devices, particularly for new and emerging applications. Since uncertainties inevitably arise in practical systems, it is important to consider this aspect, along with sensitivity to the governing parameters, to obtain realistic designs.

References

J.S. Arora, *Introduction to Optimum Design*, 3rd edn. (Academic Press, New York, 2011)

A. Bejan, G. Tsatsaronis, M. Moran, *Thermal Design and Optimization* (Wiley, New York, 1996)

X. Cheng, Y. Jaluria, Optimization of a thermal manufacturing process: Drawing of optical fiber. Int. J. Heat Mass Transf. **48**, 3560–3573 (2005)

W.K.S. Chiu, Y. Jaluria, N.G. Glumac, Control of thin film growth in chemical vapor deposition manufacturing systems. J. Mfg. Sci. Engg. **124**, 715–724 (2002)

K. Deb, *Multi-objective Optimization Using Evolutionary Algorithms* (Wiley, New York, 2001)

P. George, J. Meng, Y. Jaluria, Optimization of gallium nitride metalorganic chemical vapor deposition process. J. Heat Transf. **137**, 061007-1–061007-8 (2015)

A. Ghosh, A.K. Mallik, *Manufacturing Science* (Ellis Horwood, Chichester, 1986)

Y.L. He, W.Q. Tao, Multiscale simulations of heat transfer and fluid flow problems. J. Heat Transf. **134**, 031018-1–031018-13 (2012)

T. Icoz, Y. Jaluria, Design of cooling systems for electronic equipment using both experimental and numerical inputs. J. Elect. Pkg. **126**, 465–471 (2004)

Y. Jaluria, Numerical study of the thermal processes in a furnace. Numer. Heat Transf. **7**, 211–224 (1984)

Y. Jaluria (ed.), *Design and Optimization of Thermal Systems*, 2nd edn. (CRC Press, Boca Raton, 2008)

Y. Jaluria, Numerical modeling of multiple length scales in thermal transport processes. Int. J. Num. Meth. Heat Fluid Flow. **24**, 781–796 (2014)

Y. Jaluria, *Design of Thermal Systems, in Handbook Thermal Science Engineering*, F.A. Kulacki. (Springer, New York, 2017)

V.R. Jamalabad, N.A. Langrana, Y. Jaluria, Rule-based design of a materials processing component. Eng. Comput. **10**, 81–94 (1994)

Y. Joshi, Reduced order thermal models of multiscale microsystems. J. Heat Transf. **134**, 031008-1–031008-11 (2012)

D. Knight, G. Elliott, Y. Jaluria, N. Langrana, K. Rasheed. Automated optimal design using concurrent integrated experiment and simulation, AIAA Paper No. 2002–5636, 2002

P.T. Lin, H.C. Gea, Y. Jaluria, Parametric modeling and optimization of chemical vapor deposition process. J. Manuf. Sci. Eng. **131**, 011011-1-7 (2009)

P.T. Lin, H.C. Gea, Y. Jaluria, Systematic strategy for modeling and optimization of thermal systems with design uncertainties. Front. Heat Mass Transf. **1**, 013003-1-20 (2010)

P.T. Lin, H.C. Gea, Y. Jaluria, A modified reliability index approach for reliability-based design optimization. J. Mech. Design. **133**, 044501-1–044501-7 (2011)

K.M. Miettinen, *Nonlinear Multi-objective Optimization* (Kluwer Academic Press, Boston, 1999)

R.H. Myers, D.C. Montgomery, *Response Surface Methodology, Process and Product Optimization Using Designed Experiments*, 2nd edn. (Wiley, New York, 2002)

H.R.B. Orlande, Inverse problems in heat transfer: New trends on solution methodologies and applications. J. Heat Transf. **134**, 031011-1-13 (2012)

M.N. Ozisik, *Inverse Heat Transfer: Fundamentals and Applications* (Taylor & Francis, Philadelphia, 2000)

S.S. Rao, *Engineering Optimization: Theory and Practice*, 4th edn. (Wiley, New York, 2009)

W.F. Stoecker, *Design of Thermal Systems*, 3rd edn. (McGraw-Hill, New York, 1989)

J. Tu, J. Cho, Y.H. Park, A new study on reliability based design optimization. J. Mech. Des. **121**, 557–564 (1999)

R. Viswanath, Y. Jaluria, Knowledge-based system for the computer aided design of ingot casting processes. Eng. Comput. **7**, 109–120 (1991)

J. Yang, Y. Jaluria, Feasibility and optimization of the hollow optical fiber drawing process. Int. J. Heat Mass Transf. **52**, 4108–4116 (2009)

B.D. Youn, K.K. Choi, An investigation of nonlinearity of reliability-based design optimization approaches. J. Mech. Des. **126**, 403–411 (2004)

J. Zhang, Y. Jaluria, T. Zhang, L. Jia, Combined experimental and numerical study for multiple microchannel heat transfer system. Numer. Heat Transf. **64**, 293–305 (2013)

J. Zhang, P.T. Lin, Y. Jaluria, Design and optimization of multiple microchannel heat transfer systems. J. Thermal Sci. Eng. Appl. **6**, 011004-1-10 (2014)

H. Zhao, T. Icoz, Y. Jaluria, D. Knight, Application of data driven design optimization methodology to a multi-objective design optimization problem. J. Eng. Design. **18**, 343–359 (2007)

Index

A

Adaptive grid, 54
Additive manufacturing, 69, 103, 317
Adsorption phase, 211
Advanced materials processing methods, 13–16
Adverse pressure gradient, 42, 153, 155, 165
Alloys, 4, 6
  binary phase diagram, 95
  combined heat and mass transfer processes, 94
  cooling curve, 93
  double-diffusive convection, 94
  equiaxed and columnar dendritic crystals, 98
  equilibrium partition coefficient, 95
  flow through porous media, 95
  lead-tin alloy, 95
  macrosegregation, 99
  microsegregation, 98
  phase diagram, 93
  solidification, 93
  tin-lead alloy, 96
  transient variation of the maximum velocities, 97
Alternating direction implicit (ADI) method, 54
Amioca/pure starch, 176–179
Analytical solution, 41–44
Annealing, 10–11
Appropriate boundary conditions, 45–47
Atomic force micrometer (AFM), 61, 232
Atomization, 319

B

Band absorption model, 255
Band radiation model, 257
Batch annealing furnace, 292–297
Bingham fluid, 147
Biot number, 21, 90, 91, 93, 111, 112, 114, 117, 191
Bond number, 306
Boundary-layer formulation
  downstream variation of the centerline temperature, 120–122
  flat plate, 120
  parabolic boundary-layer formulation, 120
  Reynolds number, 122
  temperature gradient, 123
  temperature uniformity, 123
Boussinesq approximations, 28, 30, 126
Buoyancy-driven flows, 5, 28, 29, 83, 84, 103
Buoyancy effects
  and forced flow effects
    Boussinesq approximations, 126
    computed flow and thermal field, 130–131
    computed velocity profiles, 130, 132
    cooling of glass fibers, 129
    dimensionless temperature, 130, 132
    flow geometry, 126, 128
    Grashof number, 127
    mixed convection parameter, 128
    moving axisymmetric solid, 130
    opposing forced flow streams, 131
    Reynolds number, 127
    velocity profiles, 128–129
  mathematical modeling, challenges in, 27–30
Buoyancy-induced flow, 27, 48, 82

Y. Jaluria, *Advanced Materials Processing and Manufacturing*, Mechanical Engineering Series, https://doi.org/10.1007/978-3-319-76983-7

C

Cam-driven thermocouple system, 55
Carreau model, 147
Carrier gas, 211
Case hardening, 300
Casting
  continuous, 11–12, 107–108
    Biot number, 90–91
    energy balance, 88
    one-dimensional, three-zone model, 89–90
    Peclet number, 90–91
    temperature distributions, 91–92
  ingot, 99
  mold, 10–11, 76
  system design, 344–345
Ceramics, 4, 7, 318, 319
Chemical conversion, 4, 53
  in reactive materials, 175–180
Chemically reactive materials, 34–35
Chemical vapor deposition (CVD), 13–15
  applications and materials, 198
  chemical kinetics and properties, 203
  chemical reactions, 203
  reactor
    basic considerations, 201
    configurations, 199, 200
    flow and heat transfer problem, 202
    transport process, governing equations for, 201
Chvorinov constant, 77
Chvorinov model, 75–77, 343
Cladding, 240
Cold rolling, 13, 107, 140
Combined mechanisms and complex transport phenomena, 49–52
Combined transport mechanisms, 37
Complex boundary conditions, 37–38
Complicated geometry, 36
Complicated regions, 47–49
Composite materials/composites, 4, 8
Computational fluid dynamics (CFD), 54
Conduction models
  Chvorinov constant, 77
  Chvorinov model, 75–77
  convective heat transfer coefficient, 77
  energy balance, 77
  energy equation, 76
  heat flux, 77
  knowledge-based design methodology, 78
  lumped mold model, 75
  mold casting, 76
  mold temperature, 77–78
  one-dimensional conduction model, 75
  quality of the casting, 74
  solidification time, 77
  sophisticated models, 75
  steady conduction in solid model, 75
  two-and three-dimensional models, 75
Conjugate transport, 85–87
  boundary-layer formulation
    downstream variation of the centerline temperature, 120–122
    flat plate, 120
    parabolic boundary-layer formulation, 120
    Reynolds number, 122
    temperature gradient, 123
    temperature uniformity, 123
  elliptic formulation
    adiabatic surface boundary condition, 125–126
    back-conduction effects, 126
    computational domain, 124
    for the fluid, 123
    heated aluminum plate, 124–125
    isothermal surface boundary condition, 125–126
    for the moving solid material, 124
    upstream boundary conditions, 124–125
    variation of centerline temperature, 125, 127
Conservation laws, 21
Contact angle, 305
Continuous casting, 11–12, 107–108
  Biot number, 90–91
  energy balance, 88
  one-dimensional, three-zone model, 89–90
  Peclet number, 90–91
  temperature distributions, 91–92
Continuous materials processing
  basic process
    continuously moving material, 108
    convection, 108
    extrusion process, 108–109
    heat transfer coefficients, 110
    operating conditions, 110
    radiation, 108
    transient process, 108–109
    velocity and temperature profiles, 108–109
  buoyancy and forced flow effects
    Boussinesq approximations, 126
    computed flow and thermal field, 130–131
    computed velocity profiles, 130, 132

cooling of glass fibers, 129
dimensionless temperature, 130, 132
flow geometry, 126, 128
Grashof number, 127
mixed convection parameter, 128
moving axisymmetric solid, 130
opposing forced flow streams, 131
Reynolds number, 127
velocity profiles, 128–129
conjugate transport
boundary-layer formulation, 120–123
elliptic formulation, 123–127
continuous casting, 107–108
continuous chemical vapor deposition, 137, 139
convective transport due to a moving surface
boundary-layer thickness, 118
convection coefficient, 120
heat transfer coefficient, 117
no-slip conditions, 119
Prandtl number, 119
2D laminar boundary-layer flow, 118
velocity and temperature boundary layers, 118
drawing, 137, 139
experimental work, 136–138
extrusion, 137, 139
heat transfer within moving material, 110
analysis, 111–113
analytical and numerical results, 113–117
hot rolling, 107–108, 140
roll bending, 137, 139
rolling, 137, 139
transient effects, 133–135
wire drawing, 107–108
Continuum enthalpy, 73
Convection, 50
double-diffusive, 94
Marangoni, 31, 310, 312
thermocapillary, 32
Convective heating, 44
Convective heat transfer coefficient, 43
Convective transport due to a moving surface
boundary-layer thickness, 118
convection coefficient, 120
heat transfer coefficient, 117
no-slip conditions, 119
Prandtl number, 119
2D laminar boundary-layer flow, 118
velocity and temperature boundary layers, 118
Coordinate transformation, 49–50
Co-rotating twin screw extruders, 169
Couette flow, 28, 42
Counter-rotating twin screw extruders, 169
Crystal growing process, 99–101
Czochralski crystal growing process, 13–14, 70–71, 99–101

D

Design optimization, 331–339
Desorption phase, 211
Dies, in extrusion system
coupling of extruder with, 180–181
finite element method, 181, 183
heat transfer and flow, 180
Newton-Raphson correction scheme, 180
numerical modeling, 181, 184
transport in complicated dies, 181–185
variation of pressure, 181, 182
Dilatant fluid, 147
Dimethyl silicone fluid, 147
viscosity measurements for, 148
Dopant gas, 211
Dopants, 241
Drag flow, 42
Drawing, optical fibers
basic equations, 246–248
characteristics, 248–252
furnace wall, 264–265
hollow fibres, 259–264
multilayer fibers, 253–259
neck-down region, 245
Dry etching, 314, 316
Dufour effect, 50, 201, 202
Dynamic contact angle, 275

E

Electric furnace, 56–57
Elliptic formulation
adiabatic surface boundary condition, 125–126
back-conduction effects, 126
computational domain, 124
for the fluid, 123
heated aluminum plate, 124–125
isothermal surface boundary condition, 125–126
for the moving solid material, 124
upstream boundary conditions, 124–125
variation of centerline temperature, 125, 127

Energy transport, 10
Engineering materials and products, 6–8
Enthalpy formulation, 83–85
Enthalpy method, 73–74
Enthalpy model, 33
Enthalpy-porosity model, 45
Experimentation
  extrusion of polymers, 165–168
  numerical modeling, 59–62
Explosion, 262
Extrusion, 137, 139
  and hot rolling processes, 11–12
  of polymers
    chaotic mixing effect, 165
    combined thermal and mass transport mechanisms, 172–180
    experimentation, 165–168
    extrusion dies, 180–185
    feasibility, 177, 179
    mixing inside extruder channel, 165, 166
    powder flow and compaction, 185, 186
    residence time distribution, 163–165
    schematic illustration, 185, 186
    single screw, 149–162
    twin screw, 169–172

F
Favorable pressure gradient, 42, 153, 155, 177
Fiber coating, 274–282
Fiber cooling, 272–274
Finite difference methods, 54, 79, 119, 153, 170, 298
Finite element grid, 49–50, 159
Finite element methods, 36, 49, 54, 157, 170, 181, 309
Finite volume, 49, 54, 79, 170, 298
Flame material synthesis, 235
Floating-zone method, 99–100
Flow behavior, Newtonian *vs.* non-Newtonian fluids, 146
Fluid flow, 5
Food materials, extrusion of
  chemical reactions, 175–180
  moisture transport, 173–175
Froude number, 306
Full equations, *see* Elliptic formulation
Furnace wall drawing, 264–265

G
Gallium arsenide (GaAs), 209, 210
Gallium nitride (GaN) thin films
  chemical kinetics, 212
  comprehensive chemistry model, 213, 214
  gas phase reactions, 212–213
  MOCVD
    buoyancy effects, 224
    computed thermal field, 219
    conjugate heat transfer problem, 222
    contours and thickness contours, 233, 234
    flow and temperature, 220–222
    flow guide configurations, 219
    impinging two-dimensional reactor, 217
    inlet velocity, 224, 225
    isothermal surface for deposition, 222
    mass fraction profiles, of trimethylgallium, 230, 231
    vs. MBE deposition techniques, 212
    reactor chamber with rotating susceptor and upper flow guide, 218, 219
    reactor pressure effect, 224, 225
    steady-state conditions, 224
    stiff equations, 214
    susceptor temperature, 224–227
    temperature and velocity distributions, 217, 218
    time-dependent temperatures, 228, 229
    transient deposition rate, 231, 232
    wafer loading, 218, 219
  numerical model validation and verification, 214–216
  uniformity measurement, 233, 234
Gas metal arc welding, 309–310
Gas tungsten arc welding, 309–310
Generalized Newtonian fluid, 31
Generic extrusion die, 342–343
Genetic algorithm, 332, 337
GeO_2 concentrations, isotherms for, 258
Glass, 4, 5, 7, 9, 10
  fluoride and phosphate, 240
  silica, 24, 50, 240, 256, 265, 272
Graded index, 240

H
Heat sinks, 314–317
Heat transfer, 5, 10
Heat transfer within the moving material, 110
  analysis
    Biot number, 112
    heat transfer coefficient, 112
    heat treatment, 113
    hot rolling of steel sheets, 111

metal forming, 113
Peclet number, 112
steady-state 1D conditions, 113
thermal diffusivity, 111
time-dependent temperature, 111
two-dimensional transport, 111
analytical and numerical results
Biot number, 114, 117
isotherms, 114, 116
optical fiber drawing, 115
Peclet number, 114, 117
steady-state one-dimensional problems, 113
steady-state temperature distributions, 114–115
Heat treatment, 63–64
batch annealing furnace, 292–297
case hardening, 300
quenching, 288, 300
surface heat treatment, 297–299
wave soldering, 306, 307
welding, 307–313
High-pressure liquid-encapsulated Czochralski crystal growing system, 100–101
Hollow fibres, 259–264
Hollow optical fiber drawing, 52
Horizontal CVD process, 198, 199
Horizontal CVD reactor
flow entrainment, 205
flow pattern in converging channel, 204, 205
heated susceptor, 199, 204, 207
reaction-controlled deposition mode, 208, 209
silicon deposition rate, 206, 208
streamlines for ceramic susceptor, 203, 204
temperature distributions, 205, 206
TiN deposition, results for, 205, 207
Hot rolling, 11–12, 107–108, 140

I
Idealizations, 21
Impinging type vertical CVD reactor systems, 209
Ingot casting, 99
Injection molding process, 186–188
Interface tracking, 70–72
Intermeshing twin screw extruder, 169
Inverse problem, 330–331

J
Joining and bonding processes, 300–301
soldering, 301–307
welding, 307–313

K
Knowledge-based design, 342–346

L
Landau's transformation, 36, 49, 252, 254
Lumped mass approximation, 21, 43
Lumped mold model, 75

M
Machining, 318
Macrosegregation, 99
Mass transfer, 5
Material cutting and removal process, 318
Mathematical modeling
additional transport mechanisms and considerations
chemically reactive materials, 34–35
combined mechanisms, 37
complex boundary conditions, 37–38
complicated geometry, 36
non-Newtonian fluids, 30–31
particulates and powdery materials, 32–33
phase change, 33–34
surface tension effects, 31–32
challenges, 38
buoyancy effects, 27–30
variable properties, 23–26
viscous dissipation, 26–27
computational models, 19
conservation laws, 21
governing equations, 22–23
idealizations, 21
important considerations in transport phenomena, 19–20
lumped mass approximation, 21
material properties, 21
simplification of boundary conditions, 21
simplified physical arrangement, 20
spatial dimensions, 21
steady *vs.* time-dependent, 21
Melting
of pure gallium, 84, 85
of pure tin, 87, 88
of solid preform, 301, 302

Metal-organic chemical vapor deposition (MOCVD), 209, 210
basic process, 211
gallium nitride semiconductors (*see* Gallium nitride (GaN) thin films)
Metals, 6, 33, 63, 85
Microchannels, 60–61
fabrication
dry etching, 314, 315
wet etching, 314–316
heat sinks, 314–317
Microelectromechanical (MEM) devices, 314
Microgravity materials processing, 319
Micro/nanoscale fabrication, of thin films, *see* Chemical vapor deposition (CVD)
Microsegregation, 98
Microstructured polymer optical fiber (MPOF), 263
Mixing characteristics, in single screw extruder channel, 165, 166
Modified chemical vapor deposition (MCVD), 243–244
Mold, 10, 70
casting, 10–11, 75, 76, 81, 103
Moldflow, 55
Molecular beam epitaxy (MBE) techniques, GaN film deposition, 212
Multigrid methods, 54
Multilayer fibers and dopants, 252–259
Multiscale modeling, 52–53
Multizone models, 47
Mushy zone, 93

N

NBOHC defects, 271, 272
Neck-down region, 245, 251
Newtonian fluids, 31, 42, 146
Non-bridging oxygen hole center (NBOHC) defects, 272, 334
Non-intermeshing twin screw extruder, 169
Non-Newtonian fluids, 30–31, 146
fully developed flow results, 153–155
vs. Newtonian fluid, flow behavior of, 146
Numerical modeling
accurate imposition of appropriate boundary conditions, 45–47
analytical solution, 41–44
combined mechanisms and complex transport phenomena, 49–52
complicated regions, 47–49
experimentation, 59–62
manufacturing systems, 56–57
multiscale modeling, 52–53
numerical simulation, 57–59
numerical solution, 53–55
transport processes *vs.* product characteristics, 63–64
verification and validation of the model, 55–56

O

Objective function, 334, 336, 337, 343, 345
One-dimensional conduction model, 75
One-dimensional solidification, 70, 72, 79–82
Optical fiber manufacturing
band radiation model, 257
bubble entrapment, 275
cladding, 240
coating, 274–282
collapse ratio, 260, 261
cooling, 272–274
dopants, 241, 253–259
drawing, 5, 13–14
basic equations, 246–248
characteristics, 248–252
furnace wall, 264–265
hollow fibres, 259–264
multilayer fibers, 253–259
neck-down region, 245
dynamic contact angle, 275
E′ defects, 272
exit die, 277, 280, 281
exit meniscus, 276, 278
fabrication process, 243–244
feasibility process, 265–268
graded index, 240
holey fiber, 259
inlet die, 277, 278
inlet meniscus, 276
meniscus breakdown, 272
multiscale modeling, 282–283
NBOHC defects, 271, 272
refractive index, 257, 259, 272
saw-tooth instability, 275
silica glass, 265, 272
thermally induced defects, 268–272
unclad fiber, 240
viscous rupture, 265–267
Optical fiber systems, 239–242
Optical materials, 4, 7, 13
Outside vapor deposition (OVD), 243–245, 259

P
Pareto frontier, 333, 334
Particle motion, 32
Particulates and powdery materials, 32–33
Phase change, 33–34
  alloys
    binary phase diagram, 95
    combined heat and mass transfer processes, 94
    cooling curve, 93
    double-diffusive convection, 94
    equiaxed and columnar dendritic crystals, 98
    equilibrium partition coefficient, 95
    flow through porous media, 95
    lead-tin alloy, 95
    macrosegregation, 99
    microsegregation, 98
    phase diagram, 93
    solidification, 93
    tin-lead alloy, 96
    transient variation of the maximum velocities, 97
  continuous casting
    Biot number, 90–91
    energy balance, 88
    one-dimensional, three-zone model, 89–90
    Peclet number, 90–91
    temperature distributions, 91–92
  crystal growing, 99–101
  enthalpy method, 73–74
  interface tracking, 70–72
  multidimensional problems
    conjugate transport, 85–87
    convective heat transfer coefficients, 90
    enthalpy formulation, 83–85
    experimental studies, 87–89
    transformed grid method, 82–83
  one-dimensional solidification, 79–82
  processes and materials, 101–102
  simple conduction models, 74–78
Phase enthalpies, 73
Plasma etching, 314–316
Plastics, 4, 145
  extrusion (*see* Extrusion, of polymers)
  thermoplastics, manufacturing of (*see* Thermoforming process)
Poiseuille flow, 42
Polymers, 4, 6–7
  extrusion process (*see* Extrusion, of polymers)
  processing of
    injection molding process, 186–188
    thermoforming process, 189–192
  solidification, 101–102
  thermoset, 145
Polymethyl methacrylate (PMMA), 241
Porosity, 263
Porous media, 95, 96
Powder metallurgy, 319
Power-law index, 31, 43, 147, 180
Practical issues, in manufacturing, 16
Prandtl number, 27, 81, 82, 86, 119
Precursor gas, 211
Preform fabrication process, 243–244
Production rate, 4, 16, 22, 57, 69, 107, 110, 141, 169, 185, 192, 198, 202, 235, 246, 320, 334
Product quality, 3–5, 16, 17, 69, 70, 133, 149, 325, 327, 334
Purely viscous fluid, 30, 146
Purely viscous fluids, 30
Pure starch (Amioca), 176–179

Q
Quenching, 288, 300

R
Radiation, 21, 24, 37, 50, 110, 112, 217, 244, 246, 247, 252, 254, 258, 284, 290, 292, 302–304
Rayleigh number, 86, 87
Reaction phase, 211
Reactive gas, 211
Reactive thermal processing, 53
Reflow soldering, 301
Refractive index, 257, 259, 272
Reliability-based design optimization (RBDO) algorithms, 340
Residence time distribution (RTD), 163–165
Riser, in mold casting system, 70
Roll bending, 137, 139
Rolling, 137, 139
  cold rolling, 13, 107, 140
  hot rolling, 11–12, 107–108, 140

S
Saw-tooth instability, 275
Scanning electron microscope (SEM), 60, 232, 314, 316, 317

Self-wiping co-rotating twin screw extruder, 169
Semiconductor materials, 4, 8
  See also Gallium arsenide (GaAs); Gallium nitride (GaN) thin films
Semi-infinite approximation, 76–77
Sensitivity analysis, 24, 52, 341
Shear thickening fluid, 147
Shear thinning materials, 147
Silica (silicon dioxide SiO_2) glass, 239, 265, 272
Silicon carbide (SiC) deposition, 233
Simple mold casting system, 70–71
SIMPLER algorithm, 53
Simplification of boundary conditions, 21
Single and twin screw polymer extruders, 55
Single-screw extruder, 34–35
Single screw extrusion
  axial formulation, 160
  mixing characteristics, 165, 166
  non-Newtonian fluids, fully developed flow results, 153–155
  single screw extruder, schematic illustration of, 149, 150
  solid conveying zone/region, 151, 185
  stages, 149
  tapered screw, 160–162
  temperature profiles, 157
  three-dimensional transport process, 157–159
  two-dimensional model, 150–153
    curvature effects, 150
    dimensionless parameters, 153
    fixed coordinate system, 150
    governing equations and boundary conditions, 152
    homogeneous fluid, two-dimensional flow, 151
    iterative Newton-Raphson method, 153
    viscous dissipation effects, 151
  velocity and temperature fields, 155, 156
Softening point, 50
Soldering process, 11
Solidification, 45, 71, 186–188
  one-dimensional, 70, 72, 79–82
  polymers, 101–102
  two-dimensional, 72
Solid-liquid interface, 83–84
Soret effects, 50, 201, 202
Starch, 173, 175
  isotherms and constant conversion contours, 176, 177
  shear-and temperature-driven conversion, 176
Steel and plastic thermoforming, 63
Steepest ascent method, 332–333
Stefan number, 86, 87
Step index, 240
Surface heat treatment, 297–299
Surface tension driven flow, 309, 310, 312
Surface tension effects, 31–32
System design
  acceptable design, 326–328, 332
  conceptual design, 326
  CVD process, 337
  design optimization, 331–339
  feasible domain, 331
  genetic algorithm, 332, 337
  inverse problem, 330–331
  iterative process, 326
  knowledge-based design, 342–346
  modeling and simulation, 327–330
  objective function, 334, 336, 337, 343, 345
  Pareto frontier, 333, 334
  RBDO algorithms, 340
  search methods, 332–334
  simulation and experiment, 346–347
  steepest ascent/descent method, 332
  surrogate model, 346
  univariate search method, 332–334
System simulation, 58

T

Temperature coefficient, 31
Temperature-dependent coefficient, 43, 180
Thermal conductivity, 21, 24–26, 72, 73, 111, 151, 187, 192, 214, 288, 293
Thermally induced defects, 268–272
Thermal sprays, 318
Thermoforming process, 63–64, 189–192
Thin film deposition
  characteristics of, 198, 232–234
  quality of, 198
  See also Chemical vapor deposition (CVD)
Thin film fabrication, *see* Chemical vapor deposition (CVD)
Thin film thickness, uniformity of, 232
3D printing, 317
Traditional manufacturing processes, 10–13
Transformed grid method, 82–83
Transient effects, 133–135
Transmission electron microscope (TEM), 232
Transport phenomena, 8–9
Transport processes *vs.* product characteristics, 63–64

Twin screw extruders
finite element method, 170
flow domain, 169
isothermal heavy corn syrup, fluid velocity field for, 172
mesh discretization for mixing region, 170, 171
types, 169
Twin-screw extrusion, 13–15
Two-dimensional solidification, 72
Two-phase interface tracking, 73–74

U
Ultrasonic machining, 319
Unclad fiber, 240
Univariate search method, 332–334

V
Vapor-phase axial deposition (VAD), 243–244
Variable properties, 23–26
Vertical CVD reactor systems
impinging type, 209
with rotating susceptor, 209
Viscoelastic fluid, 146
Viscoinelastic fluid, 146
Viscoinelastic fluids, 30
Viscosity coefficient, 23
Viscous dissipation, 26–27
Viscous rupture, 265

W
Wave soldering, 306, 307
Weber number, 306
Welding, 307–313
Wet etching, 314–316
Wire drawing, 107–108

X
X-ray diffraction (XRD), 232

MIX
Papier aus verantwortungsvollen Quellen
Paper from responsible sources
FSC® C105338

If you have any concerns about our products,
you can contact us on
ProductSafety@springernature.com

In case Publisher is established outside the EU,
the EU authorized representative is:
Springer Nature Customer Service Center GmbH
Europaplatz 3, 69115 Heidelberg, Germany

Printed by Libri Plureos GmbH
in Hamburg, Germany